MANY SIDES: A PROTAGOREAN APPROACH TO THE THEORY, PRACTICE AND PEDAGOGY OF ARGUMENT

Argumentation Library

Volume 5

MANY SIDES:
A PROTAGOREAN APPROACH TO THE THEORY, PRACTICE AND PEDAGOGY OF ARGUMENT

by

MICHAEL MENDELSON
Iowa State University,
U.S.A.

KLUWER ACADEMIC PUBLISHERS
DORDRECHT / BOSTON / LONDON

A C.I.P. Catalogue record for this book is available from the Library of Congress.

ISBN 1-4020-0402-8

Published by Kluwer Academic Publishers,
P.O. Box 17, 3300 AA Dordrecht, The Netherlands.

Sold and distributed in North, Central and South America
by Kluwer Academic Publishers,
101 Philip Drive, Norwell, MA 02061, U.S.A.

In all other countries, sold and distributed
by Kluwer Academic Publishers,
P.O. Box 322, 3300 AH Dordrecht, The Netherlands.

Printed on acid-free paper

Printed in the Netherlands

for Susan, uxori carissimae

CONTENTS

Part II Roman Developments in Practice and Pedagogy

ACKNOWLEDGEMENTS

Protagorean theory would argue that books, like ideas, need opposition to flourish. Of course, it doesn't hurt to receive a little encouragement along the way, and this book has been the recipient of much of the latter. In particular, *Many Sides* has been supported over several years by the research program of Iowa State University, first in the form of a research leave and then in steady research support from the Department of English. Among my friends and colleagues at Iowa State, Rich and Julie Freed have been steadfast in their interest in me and my work, David Russell has prodded me gently, and Scott Consigny has helped me to think about Protagoras through his own fine work on Gorgias. I have also been lucky in my students, many of whom have understood antilogic by instinct and shown me what it can do when practiced well.

In many ways, this book is an effort to develop a "thick pedagogy," a concept that relies on gifted teachers able to realize the special *kairos* of the classroom. I am privileged to have seen three such teachers at work—John Ehrstine, Uli Knoepflmacher, and Michael Leff. Their professional decorum has been an inspiration to this book and my career.

From the people at Kluwer, I have received not only professional advice but also the kind of goodwill that has made publication a pleasure. In particular, Susan Jones, Jacqueline Bergsma, and Frans van Eemeren have been supportive at every step. Additional thanks to my anonymous reader, whose good judgment improved more than my title.

In preparing my manuscript, I have had the benefit of some very professional friends. Donna Kain gave this book more than its graceful design; her good sense, hard work, and patience are on every page. David

Roberts was a scrupulous copy editor who, between naps, was always ready to clip my wings. Irene Faas, my proofreader, used red pencils and rhetorical expertise in the constant battle against "hateful error." And James McGlew has done all he could to nudge me towards correctness in my ancient language.

There is also my friend Jane Smiley, who works very fast and who encouraged me when I was very slow. There are my sons, Aaron and Noah, the A and N of the text, who have never had any difficulty taking the other side. And, most of all, there is my wife, Professor Susan Carlson, my closest dialogical partner, who has discussed every new development in this book's history and then read every new draft that followed. I believe that almost everything can and should be argued. But there is simply no argument about the importance of her role in the development of *Many Sides* or about the sincerity of its dedication.

Introduction

> *"For every human presupposition and every enunciation has as much authority as another, unless reason shows the difference between them. Thus they must all be put in the scales, and first of all the general ones, which tyrannize over us."*
>
> Montaigne, "Apology for Raymond Sebond"

John Dewey, in *Experience and Nature*, writes that "language is specifically a mode of interaction of at least two beings, a speaker and a hearer; it presupposes an organized group to which these creatures belong, and from which they have acquired their habits of speech. It is, therefore, a relationship, not a particularity" (1925, 153). The same may be said more specifically of argument: it is always a relationship, an interaction between at least two participants who possess or assume more than one view on a matter of controversy; it cannot adequately be conceived of as a "particularity," the isolated ruminations of the individual grappling with uncertainty. Put another way, the locus of argument is in the exchange of ideas, the bi-play of opinions that invariably attach themselves to controversy. Which is not to say that a single individual cannot engage in argument, cannot ponder a problem in the sanctity of her own study; when she does so, however, she will always mimic the presence of another interlocutor in some form of internal *prosopopoeia*, or impersonation.[1] The presence of a simulated "other" in such a case is merely the acknowledgment of the relational dynamics that are central to the argumentative process.

Perhaps all this seems obvious in a scholarly arena that includes Bakhtinian dialogism, social construction, Burkean dramatism, philosophical hermeneutics, paralogic rhetoric, pragma-dialectics, deconstruction, and a myriad of collaborative theories. And admittedly, the relational basis of argument is hardly a new idea. In 1958, Chaim Perelman and Lucie Olbrechts-Tyteca introduced their "new rhetoric" with the assertions that an argument develops "in terms of an audience" and that "intellectual contact" among participants in argument is therefore crucial (5, 14). Thirteen years later, Richard Johannesen inaugurated the discussion of "communication as dialogue" by arguing that the process of interaction between interlocutors takes precedence over the formal elements of argument (the structure of claims or standards for rationality) and that a theory of dialogical communication would foreground the contact among interlocutors (1971, 375-80). Perelman's emphasis on the rhetor/audience relationship is by now almost universally embraced in rhetorical studies, and the concept of discourse as dialogical has become wide-spread in rhetorical theory and composition studies (see Ward). And yet, despite this theoretical interest, I remain unconvinced that our general conception of argument and—of particular importance to this study—the brand of argumentative discourse presented in contemporary classrooms is truly "relational" in the sense that Dewey promotes.

Consider, for example, Deanna Kuhn's comprehensive study of the actual argumentation skills of a group of Americans (160+) drawn from diverse age and ethnic groups and spanning the educational spectrum from high school students to college professors (1985, 18-20). The study revealed that only a small minority of the test group (9-22%) could mount a complete argument, meaning that they could generate a thesis, provide supporting evidence, compare one thesis with other positions, form a reasoned judgment, and meet objections (265). Especially interesting with regard to the relational dynamics of argument is the fact that Kuhn's subjects were generally unable to imagine arguments opposed to their own (139-44). If we conceive of argument as the interaction of alternative views in controversy rather than simply a self-generated claim supported by some accompanying justification, then the reasoning that follows from the independent, monological process will invariably be myopic and incomplete, the result of what Kuhn refers to as the "my-side bias" (282). Kuhn's study indicates that our capacity to evaluate our own positions is "at best limited" when we operate in isolation, without the benefit of counter claims that inevitably arise in the process of dialogue (265).

Consider also the general conception of argument held by the public at large. Lakoff and Johnson's famous indictment of "argument as war" as one of the "metaphors we live by" is now two decades old. But as Deborah Tannen indicates (in 1998), the popular conception of argument continues to be characterized by "unrelenting contention" and a "lust for opposition" (3). In other words, the relationship that distinguishes argument in the public mind is thoroughly agonistic and certainly without pretense of understanding, much less accommodating the other side (25). Tannen documents the prevalence of this degraded conception of argument in the press, on television and radio, in political discourse, in modern litigation practices and developing trends in electronic communication, in conventions of gender, and in educational traditions. She notes, however, that "smashing heads does not open minds" and argues that more constructive ways of expressing opposition and managing difference can be achieved by replacing the debate model with a model of argument as dialogue (26, 288-90).[2]

Or consider finally the way we present argument in college composition courses. I take as representative examples two American textbooks that advertise themselves as the most popular of their kind and that introduce tens of thousands of college students annually to the process of argumentation. Ramage and Bean's *Writing Arguments* (4th ed.) is an unusually sophisticated text, particularly alert to contemporary trends in rhetorical theory. Nonetheless, this textbook defines argument as the "justification of claims," it emphasizes argument as a formal product rather than a dynamic process, and it posits "truth seeking" and persuasion as the twin aims of argument (3). All of these elements serve to subordinate the role of dialogical exchange: "justification" implies that a claim has been established by the writer and is to be defended against modification by others; the precedence given to a formal product consigns interaction with others to something we do after the claim itself has been crafted; and the goal of "truth seeking" (despite authorial scruples) gives the false impression that most arguments can be resolved with certainty. As for "persuasion," the authors note that their goal is to help the student become "a more powerful arguer" (3); but power over whom and for what purpose? If power is the aim of argument, we remain in the eristic mode that Deborah Tannen claims runs counter to the interests of dialogue.

Our second textbook, Annette Rottenberg's *Elements of Argument* (5th ed.), is designed, says the author, to teach students how to "defend" their claims "as directly and efficiently as possible" (v). Student writers are asked to assume that "there is a reader who may not agree with you" and, in response, to "persuade the unconvinced, to acquaint them with good reasons for changing their minds" (13, 4). Given such a model, dialogue becomes a method for engineering conciliation on the part of a passive listener; it is certainly not a means by which rhetors acquaint themselves with opposing positions in order to enhance mutual understanding or transcend limited preconceptions. Indeed, the goals of "defense" and "justification" both invoke the metaphor of confrontation. In 1979, Daniel O'Keefe, in his well-known classification of argument as "something one person makes" and, alternatively, "something two or more people have," noted that "the emphasis of textbooks and coursework in argumentation" is on teaching students *to make* effective artifacts rather than *to have* productive argumentative encounters (121, 126). My own survey (with Neil Lindeman) of contemporary textbooks in written argument corroborates the notion that we have yet to pioneer pedagogical procedures that successfully shift the balance of attention from a single person making a claim to a group of people engaged in the give-and-take of argumentation (2000).

This brief sketch of general practice, popular attitudes, and contemporary pedagogy implies that, despite recent efforts to develop productive theories of dialogical exchange, much contemporary thinking about the process of argumentation continues to be founded on the eristic, monological model. Such resistance is hardly surprising, however, because the image of argument as "something one person makes" is deeply entrenched in the tradition of Western Rationalism. This tradition has its theoretical base in the Cartesian, Enlightenment conception of the independent, objective subject who, by following the methods of formal logic, is able to achieve an impartial assessment of the "truth", an assessment independent of the persons and circumstances involved in any particular argumentative situation (see Bernstein 1985, Bineham 1990, Toulmin 1995). That is, we begin the thinking process alone, and having once codified our own propositions in response to a matter of controversy, we then (and only then) engage in argumentative interaction with others in an effort to uphold the primacy of our own position and debunk the alternatives.

According to this "traditional" conception, the function of argument is what Douglas Ehninger has called "correction" (1970, 101); i.e., having taken possession of the truth through the formal rigors of logical reasoning, we are in a position to correct the mistaken opinions of others whose thinking deviates from the universal, objective standard which governs carefully reasoned judgment. This monological, debate-oriented model can only be called "relational" in the most tangential sense: we are related to the social matrix out of which the controversy grows (Dewey's "organized group"), and we have an adversarial relationship with anyone who would maintain a different position from our own. When it comes to interacting with others in the process of investigating a controversy and generating a proposition in response, there is no real relationship at all. Like Athene, daughter of Zeus, argumentative propositions appear from the heads of their makers full-grown and armed for combat. The eventual conflict between opposing positions may unavoidably entail contact; but the interactive process begins only after one's own argument has been invented, refined, and readied for debate.

This sequential conception of the argumentative process (independent reasoning followed by the rhetorical adjustment to circumstances) has become a defining feature of the modern rhetorical tradition, a tradition with very deep roots. Indeed, Cicero ascribes the process of thinking first/responding later to Socrates and the dialectical effort to arrive at abstract truth as a prelude to rhetorical exchange (*De Oratore* 3.160-61). Throughout antiquity, of course, the dialectical model shares the stage with other, less formal approaches to argumentation, approaches that feature practical reasoning and contextual relations, the rhetoric of *pros to kairon* or reasoning according to the situation. At the dawn of the early modern era, however, a more purely logical model begins to dominate educational practice, most notably perhaps in the influential curriculum of Peter Ramus (1515-72), who separates logic from rhetoric, confers on the former the responsibility for critical thinking, and confines the latter to matters of style and delivery. Shortly after, with Descartes (1596-1650) and the Port Royal educators, logic takes on new dimensions as a quasi-mathematical practice whose procedures are guaranteed by their assumed congruence with the underlying structure of the world itself. In the process, the methods of formal logic assume pride of place as the accepted standard for serious thinking. To paraphrase Stephen Toulmin's terse description of this watershed moment, with the advent of Descartes and the Rationalist tradition, logic is in and dialogue is out (1988, 139).

In the classroom, the legacy of formal reasoning is confirmed in the influential rhetorics of the English Enlightenment, most notably Thomas Campbell's *The Philosophy of Rhetoric* (1776) and Richard Whately's *Elements of Rhetoric* (1828). For Campbell, the process of critical reasoning remains the province of logic, while the work of rhetoric is to find the means by which discourse achieves persuasive effect. Whately reinforces this categorical division not only by considering the "elements of Logic" and those of "Rhetoric" in separate volumes (1826 and 1828, resp.) but also by insisting that rhetoric is "in truth an offshoot of Logic" and that the "Rhetorician must labour under great disadvantages who is not only ill-acquainted with that system, but also utterly unconscious of his deficiency" (*Elements of Rhetoric* 284). Whately's distinctive contribution to argument and its pedagogy is to provide what James Berlin calls "explicit principles for demonstrating propositions which have been established outside of the rhetorical process," i.e., developed without the benefit of contact among the people involved in controversy (1980, 13-14).

This circumscribed conception of rhetoric was advanced in the United States in the late 19C by academic reformers in the growing field of composition and rhetoric. Scholar-teachers such as Adams Sherman Hill, Barrett Wendell, and John Genung not only perpetuate Enlightenment epistemology they also promote a significant transfer of pedagogical attention from argument to exposition. With the consolidation of English Departments in American colleges between 1885 and 1910, and with the enthusiastic support of a burgeoning textbook industry, an emerging paradigm characterized by simple taxonomies (the four modes, the three levels) and formalist abstractions (unity, coherence, emphasis) came to dominate academic instruction in rhetoric (Connors 1997, 1-22). In one form or another, this now-traditional approach to argument remained in place through the expansion of American higher education that continued into the 1960s. In the last half-century, of course, this reigning paradigm has been challenged by a multitude of theories endeavoring to reclaim rhetoric's role as an epistemic discipline and to reconceive knowledge in terms that are historical, contingent, and culturally diverse rather than permanent, absolute, and universal. However, despite the ambition and sophistication of these theoretical efforts, the preeminent model for the practice and pedagogy of argumentation continues to reflect, in one guise or another, the formal, abstract, non-relational procedures of logical rather than dialogical reasoning, procedures that ask the rhetor to construct a formally valid proposition as the necessary prelude to actual contact with the dialogical "other." Given the persistent dominance this formal approach to critical reason, it is little wonder that the concept and practice of dialogical argument remain obscure and undeveloped, unnoticed by the public and almost unacknowledged in the classroom. The hegemony of formal argument represents a formidable tradition, coeval with the birth of the Modern, ascendant as the method of science, and entrenched in the lore of instruction.

And yet, there are, in fact, alternative possibilities for argumentative practice, alternatives as old as rhetoric itself, traditions with their own deep roots in Western humanism and pedagogy. In this book, I explore what I take to be the primary alternative to the Rationalist model for argumentative rhetoric, a tradition originally known in ancient Greece as **antilogic** and in Rome as ***controversia***.[3] This antilogical alternative might well take as its motto Blake's notion that "opposition is true friendship"; i.e., antilogic assumes that there will always be contrasting perspectives on any topic open to social discourse and that the most comprehensive, enlightened approach to such controversy is inevitably founded on the orderly, even amicable interaction of opposing positions. Antilogic, therefore, serves to highlight difference not as a drawback but as a fact of social life; and, in response to this fact, it posits argument as the species of dialogue designed for the successful accommodation of unavoidable differences. The origins of this unique rhetorical perspective lie in the Sophistic tradition of 5C BCE Athens and, in particular, in the philosophy of Protagoras, the first and perhaps the most enduring of the major Sophists. A few words on Protagoras and his rhetoric are in order here by way of introduction to the main ideas to be developed in *Many Sides*.

According to the Greek record, Protagoras was the first to say that "on every issue there are two arguments opposed to each other" (*DK* 80 A12).[4] Out of this concept, Protagoras not only crafts the first set of instructions on the art of debate, he also initiates a process of reasoning by which contrary positions (anti-*logoi* or opposing reasons) are purposefully juxtaposed. As a result, resolution in any controversy conducted along antilogical lines is sought through a concerted effort to examine each stand (or *logos*) **in relation to** its opposite number(s). This Sophistic approach to argumentation is itself founded on another ground-breaking Protagorean dictum, the famous "human-measure" concept that "humanity is the measure of all things" (*DK* 80 B1). Often seen as the *locus classicus* of a pernicious relativism, the human-measure doctrine is more fairly approached as the starting point of an anti-foundational philosophy that concentrates on the relation of individual perception to the nature of knowledge. The doctrine does indeed imply that all knowledge should be seen as "relative to" the source or perspective from which it is derived and that different perspectives will yield different (re)constructions of any shared experience. But variability in individual perception does not entail the naive belief that all views are equal; rather, as Protagoras put it, goodness is "diverse and multiform" and while one set of thoughts may not be "truer" than another by any absolute standard, one position (or *logos*) may be "stronger" in the sense of more advantageous or useful (Plato's *Protagoras* 334b).

Translated into the realm of rhetoric and practical argumentation, the human-measure doctrine yields a rhetorical practice based on the assumption that there are not just two sides to every question, but that on any significant point of controversy there are likely to be multiple, competing positions, multiple "measurements" resulting from the different perspectives which individuals bring to a particular issue. As a result of this multiplicity, antilogic would have us first acknowledge the "partiality" of any and all opening positions in controversy (cf. Ehninger 1970, 104). Once we recognize that all relevant *logoi* (reasons) are both *part of* what can be said and *partial to* certain aspects of the case, we can begin to work toward understanding by appeal to antilogic's primary strategies: the dialogical consideration of opposing positions in relation to one another, the suspension of judgment while alternatives are tested, and the crafting of a resolution based on *pros to kairon*, what is appropriate to this particular situation. The antilogical method, therefore, runs counter to the Platonic drive towards universal principles, to the Cartesian emphasis on "clear and distinct" premises, and to the more recent insistence on formal rigor as the preeminent method of reasoning one's way to the "truth." Instead, antilogic would embrace the ambiguity, diversity, and temporality that condition practical arguments and, in response, would encourage tolerance and dialogical relationships among partners in controversy as necessary features in the operation of practical reasoning (*phronesis*). In other words, antilogical practice comes with its own ethics of dialogical exchange.

In Rome, disparate rhetorical practices were transformed by Cicero into a concrete program of argumentation, a program commonly referred to as *controversia* and filled out in considerable detail in the *De Oratore*. In this remarkable dialogue, we find Cicero experimenting with a host of practical strategies that promote the productive interaction of oppositional voices in dispute (*multiplex ratio disputandi*). These Ciceronian innovations become, in turn, the model for Quintilian's pedagogy and, in particular, for an approach to instruction based on the *controversia*-related theory of *in utramque partem*; i.e., preparing for argument by assessing all sides in controversy. Only when Roman students were well practiced in contemplating all sides of an argument were they deemed ready to think critically and respond prudently to the issues that dominated public discourse. In Protagoras, Cicero, and Quintilian, then, the antilogical/controversial tradition develops a comprehensive approach to the rhetorical experience by exploring, in turn, the theoretical, practical, and pedagogical extensions of the subject. And yet, despite the distinction of its major figures, antilogic as an alternative rhetorical tradition has been all but ignored in the standard accounts of rhetorical history, marginalized by the ascendancy of analytical reasoning as the primary guide to validity in argument, and (most importantly) left out of the general reappraisal of argumentation that has been underway at least since Perelman and Olbrechts-Tyteca's *The New Rhetoric*.

Given this historical situation, I might well claim that my intention in this book is to introduce students, teachers, and scholars to a method of argumentation, inquiry, and pedagogy that was initiated and developed by some of the rhetorical tradition's most distinguished figures. And, certainly, I will argue in behalf of an expanded presence for antilogic as an alternative to the "traditional" practice of formal and informal argument, an alternative that seeks to accommodate diversity and difference as a source of strength rather than as obstacles to clarity and resolution, an approach as old as rhetoric itself. But the rehabilitation of antilogic is only part of this book's agenda; for despite my own fascination with the historical record, the overarching motive behind this reclamation of the past is to make some contribution to the present practice of argumentation. In consequence, this book, like antilogic itself, is Janus-like. Janus, you may recall, was the Latin god of doors and gates, the god who looked both forward and back, who was alert to the transformation of "what was" into "what is to come." It was common for Romans to place the emblem of Janus at the main door of the house where he could look in both directions. Janus, therefore, is a fitting deity for a form of rhetorical practice which, as Thomas Sloane points out, habitually looks both ways (1985, 57-63; see n.5). I follow Professor Sloane by invoking Janus at the portal of this book and at the gateway to a new century as an emblem of how the past can cross the boundary into the present and beyond. And indeed, the antilogical tradition is already claiming some contemporary intellectual space as an ancient corollary of significant trends in current philosophical and rhetorical thought.

Among those areas that resonate with antilogical implications and sympathies are Bakhtinian dialogics, Deweyan pragmatism, Buber's dialogue, Gadamer's philosophical hermeneutics, and the relational dynamics of paralogic rhetoric. As a result of these and other affinities, the revival of antilogic becomes more than just the remembrance of things past; rather, in the context of these analogous intellectual currents, antilogic contributes to a movement beyond the limits of rationalism, beyond the dominance of objectivity, beyond the demand for certainty, and towards what has been called the "new sophistic" (Corbett 1995, 9).

In these pages, then, I will argue that antilogic is not only an alternative tradition of argumentation, a tradition obscured by the dominance of the Platonic, Cartesian, Rationalist mode, it is also a method of rhetorical practice especially well suited to the pluralism of our own age and compatible with contemporary efforts to move beyond the foundational assumptions that have grounded what Dewey called "the Quest for Certainty." In addition, the study of antilogic in its original Greek and Roman forms is inseparable from the study of rhetorical pedagogy (see *Protagoras* 317b). Consequently, *Many Sides* will also spend considerable time exploring this distinctive pedagogical tradition, a tradition that places argument at the apogee of advanced education and sees rhetoric itself as the key to personal development and civic order. Indeed, for Protagoras, Cicero, and Quintilian, the ability to address the practical and theoretical controversies that routinely condition both private and public life is the true mark of *humanitas*. For those teachers who are disenchanted with the logico-formalistic emphasis of available texts and who would adopt a more dialogic, collaborative approach to the conduct of argument, the ancient practice of antilogic offers a pragmatic counter-tradition for addressing concrete human differences by attending first of all to the people involved. To return to Dewey again, the antilogical alternative is "relational" to the core.

My intention is to offer an introduction to the practice of antilogical/controversial reasoning, an introduction that both appraises the distinguished past of this neglected tradition and revises its methods in accordance with an evolving rhetorical epoch. I make no effort to disguise my opinion that argument is the prototype of all serious discourse and that any approach to the rhetorical process which ignores argument's appropriate role runs the risk of a Ramistic split in the discipline—a split between the "brain" and the "tongue," thinking and speaking, ideas and expression—a split that abandons rhetoric's epistemic potential and leaves it with little more to work with than the "flowers of ornament." In general terms, I will consider argument as a process of communication regarding what Plato calls "the class of doubtful things" (*Phaedrus* 263b), matters in any community—including our classrooms—where varied perspectives are the norm and where reasoned justification is required to substantiate claims to knowledge and secure consent. Seen in this expansive context, argument seems to me an unavoidable topic not just for students and teachers but for all who would find their way through the minefield of "endemic disagreement" that surrounds our most pressing professional tasks as well as our routine, quotidian activities.

Simply put, to relate to one another is, inevitably, to argue. Because antilogical reasoning foregrounds the differences that routinely occur in all discursive situations, it is, I will maintain, well prepared to undertake the tasks of invention and negotiation that are the primary responsibilities of those "good people gifted with speech" who would work through controversy to discovery. It is helpful to remember Kenneth Burke's motto for the *Grammar of Motives*: *ad bellum purificandum*, towards the purification of war. In this book, I will present the case for antilogical reasoning as a model for the purification of dissent, a model that transcends current practices and makes of argument not a regimental march toward victory in our own causes but rather a mutually beneficial relationship in which everyone takes turns in the effort to arrive at prudential judgment.

My own recognition of the value of argument, especially in the classroom, has been slow in coming. I believe I have known intuitively for some time that my classes, as well as my family's dinner table, were more stimulating places when we acknowledged something akin to disagreement within the group. In difference is the seed of delight as well as discovery. I also remember my father telling the story of two rabbis who thought that when they could no longer argue with one another their friendship was at an end. But as a teacher, I was tardy in learning that I was more effective in the role of facilitator and mediator of classroom controversies than I was when I presented myself as the expert, the field-marshal, or—as one of my students put it—"the best student in class."

After more than a decade of experience in the composition classroom, I began to teach professional communication; i.e., writing for business and the professions. In an effort to find assignments appropriate to this emerging discipline, I adopted case-study problems—like those in Barbara Couture and Joan Rymer's *Cases for Technical and Professional Writing* (1985) —as an effective imitation of what was then called "real-world writing." I quickly learned that such assignments were a significant improvement over the standard academic essay, which was typically addressed to the all-knowing teacher and so bore little resemblance to the communicative exchanges that characterize actual professional discourse. At about the same time, I also began teaching a course in the history of rhetorical theory for the growing number of English majors who were planning careers in teaching, professional communication, and information management. As my students and I entered into discussions of Plato and Aristotle, Aspasia and Gorgias, Cicero and Quintilian, Christine de Pizan and George Campbell, it became natural to contemplate the affinities between ancient and modern rhetoric. For me, the first and most dramatic of these affinities was the correspondence between contemporary case-study practice and its Roman equivalent, the exercise declamation and. Here was a situation in which my historical studies could contribute, *mutatis mutandis*, to the improvement of my writing courses. The study of declamation, Quintilian, and more general matters of Roman pedagogy led me to Samuel Bonner, H. I. Marrou, and Donald Lemen Clark. And, inevitably, the study of Roman rhetoric brought me to Cicero and the *De Oratore*, with its dramatic dialogue and wonderful orchestration of *multiplex ratio disputandi*, multiple ideas in dispute.

From my first classroom discussions of the *De Oratore*, I found my own students loved the argumentative ethos of this work, in which Roman students press their teachers, resist inadequate formulations, and enter into disputes with an enthusiasm that makes argument seem exciting and productive rather than necessarily anxious and agonistic. In my pursuit of Cicero's rhetoric and oratory, I turned next to the historical scholarship of Richard McKeon, then to Michael Leff, Thomas Conley, Elaine Fantham, and, most importantly for me, to Thomas O. Sloane. Indeed, it was in Professor Sloane's essays on Cicero, Augustine, Erasmus and English Humanism (1985) that I began to sense that there was something intriguing but undeveloped in the practice of *in utramque partem* (arguing either side of a case), something analogous to what Gerald Graff, Susan Jarratt, and others were saying about the value of dialogue and contention in contemporary pedagogy.[5] It was an exciting moment for me as I realized that theory and pedagogy had, in the early epochs of the rhetorical tradition, been fully allied with one another and that both were energized by the confluence of their interests.

My next step was from Roman pedagogy to Protagoras and Isocrates, backwards in time though deeper into my subject. Happily, my historical curiosity coincided with the explosion of Sophistic studies in the 1980s and early 1990s, when Jarratt, Edward Schiappa, G. B. Kerferd, Jacqueline de Romilly, Sharon Crowley and others did so much to rekindle the Sophistic legacy. Schiappa's book, *Protagoras and Logos*, was particularly helpful as I began to investigate the origins of antilogic and the practice of arguing all sides. More than this I need not add here since the book that follows is, in large measure, a record of my inquiry into these subjects. One final anecdote, however, may help to clarify my own authorial intentions.

Some years after my study of Protagoras and antilogic was in full swing, I attended the Alta Conference on Argumentation and delivered a somewhat over-enthusiastic paean to Protagoras. During the discussion that followed my panel, Ed Schiappa graciously but in earnest asked me why we should recuperate such a sketchy figure as Protagoras, especially as a presence in contemporary classrooms. In other words, "What can Protagoras (or Cicero or Quintilian) do for us now?" (see Schiappa Ch. 1). My response was probably more than a bit inept; but I have continued to think about the question. There are, of course, problems with the appropriation of ancient concepts for contemporary use, none the least of which in this case is that Greek and Roman rhetoric was fashioned in and for slave societies dominated by elites and unfriendly to women, non-citizens, and foreigners alike. In what follows I have tried to acknowledge these problems. But as Quintilian notes, "of that which is good by nature we may surely make good use" (*Institutio Oratoria* 2.10.3). The rest of this book is dedicated to the proposition that there is not only "good use" to be made of antilogic and the pedagogy that follows from it, but that antilogic—with its emphasis on contingency, difference, ambiguity, and tolerance—is especially resonant in our own times. *Many Sides* will attempt to identify just what the "good use" of this recuperated practice may be.

In the chapters that follow, I proceed in historical order from ancient Greece (Part I), to Rome (Part II), to the present (Epilogue). The book also moves generally from the theoretical to the practical, and finally to the pedagogical dimensions of my subject. Chapter 1 is devoted to the origins of antilogic in the philosophy of Protagoras. More specifically, I explore the philosophical ideas at issue in the dispute between Protagoras and Socrates, issues of skepticism, subjectivity, relativism, and the nature of knowledge, all of which follow from the human-measure doctrine and related Protagorean dicta. Because of the unfamiliarity of this material for many readers with primary interests in rhetoric and composition, this first chapter seeks to contextualize the discussion by providing introductory detail on Protagoras himself and on the intellectual milieu of his day. The chapter also approaches ancient Greek philosophy from a decidedly rhetorical perspective. In the process, I hope to convince readers unfamiliar with such terrain that neo-Sophistic argument originates in the powerful ideas that distinguish the Protagorean worldview and that a comprehensive appraisal of our topic begins here. Too often pedagogy is characterized as thin discourse because it is divorced from its theoretical base. This first chapter works to theorize Protagorean principles in a way that will lend substance to the pedagogical discussions that follow.

Chapter 2 focuses on antilogic and the rhetorical extensions of Protagorean theory. Because antilogic remains relatively unknown, I begin with its basic features and with a comparison between antilogic and the related, but antithetical mode of dialectic. I also seek to place the rhetorical innovations of Protagoras against the backdrop of Periclean Athens and Presocratic thought. Chapters 3 and 4 are devoted to the intrinsic features of the subject; i.e., to the pragmatic, ethical, and formal dimensions of antilogic itself. I must admit that these practical and ethical considerations asserted themselves as something of a surprise to me. But as I pursued the origins of my topic, I came to realize that antilogic places a premium on practical action and so could not be divorced from the concept of "utility." Moreover, a full appreciation of the subject demands that we confront Aristotle's charge that Protogoras was prepared to substitute "the worse argument for the better," a charge that has stigmatized Sophism in general for millennia.

Having attended to such matters, it remains for me to address the fundamental question of how the rhetor trained in antilogic actually proceeds in an argumentative situation. Chapter 4 addresses this strictly rhetorical question by first detailing the particular *technai*, or specific habits of practice that distinguished the first exponents of antilogic, and then by examining the more general features that characterize antilogic as a unique rhetorical *praxis*. At this point, I seek to connect the nature and function of antilogical argument to various strands of 20C rhetorical and critical theory. My general motive in these first chapters is to present a comprehensive introduction to and interpretation of antilogic, a portrait that befits the extensive scope of the ideas involved. Michael Billig is surely correct in noting that "Protagoras is one of the most innovative figures of all time" (40). In Part I, readers will have a chance to track these innovations from their source and, in the process, judge their value for themselves.

Part II is devoted to Roman developments of earlier Greek practices; but again, while the material may be ancient, my own interests are as much contemporary as antiquarian. Chapter 5 takes up the shift from antilogic to *controversia*, from theory to practice, from Greek concepts to the Roman invocation of Hellenic ideas. It is under the expert guidance of Cicero that the incipient techniques of the Athenian originators begin to take on a form that we can identify with and adapt. With regard to argumentative practice, Cicero's *De Oratore* is perhaps the most sophisticated elaboration of antilogical methods in the entire rhetorical tradition; so, in Chapter 5, I explore Cicero's approach to argument in detail. And because there is a distinct instructional component to the *De Oratore*, I also begin here to explore the pedagogical implications of antilogic/*controversia*. In Chapter 6, our attention turns directly to pedagogy, to the greatest of all classical teachers, Quintilian, and to the prominent place of *controversia* at all stages of his rhetorical curriculum. As in my examination of Cicero, I close this chapter with an appraisal of pedagogical practices from the *Institutio Oratoria* which seem suitable for adaptation to our own milieu.

This investigation of Roman pedagogy in its ideal form (Chapter 5) and as a practical method (Chapter 6) provides a transition to the Epilogue, a summary assessment of the potential adaptability of an ancient rhetorical tradition to a modern pedagogical environment. In particular, the Epilogue seeks to provide teachers of rhetoric, composition, and argument with an overview of what they would need to actually implement an antilogical pedagogy for themselves. This overview begins with a "philosophical prelude," a synopsis of the Protagorean concepts that motivate antilogical argument. With these principles as a point of theoretical reference, the Epilogue proceeds to consider the dialogical and pragmatic components of antilogic as these contribute, respectively, to rhetorical invention and practical judgment. In brief, dialogical protocols provide rhetors with an innovative approach to the invention of *copia* (the widest possible range of ideas), while antilogical pragmatism offers practical methods for testing options and determining propriety (what is most appropriate under prevailing conditions). In either case, antilogical procedures mark a significant departure from the mainline of current pedagogical practice.

The Epilogue also offers a unique series of pedagogical exercises I have adapted from classical models and refashioned to fit contemporary classrooms. The site of my own classroom experience has been college courses in composition, literature, rhetorical theory, and argumentation. But I see no reason why antilogical pedagogy would not be suitable to any discipline in which conflicting positions condition the search for knowledge. As I note throughout, antilogic is not bound by the traditional protocols of classroom rhetoric, not intent solely on well-rounded discourse; it is also a method of critical reasoning appropriate to any situation attended by diverse positions. As such, the theory and practice of antilogic can be employed as a rational means for working through the unavoidable controversies that condition not just intellectual discourse, but all social interaction. What the ancient tradition of antilogic contributes to the discussion of contemporary argument pedagogy is a well-articulated, thoroughly refined alternative to formal argument, an alternative that places the process of "intellectual contact" with one's dialogical partner(s) ahead of the formal justification of claims. It is this accent on interaction among those engaged in argument that makes antilogic especially responsive to the diversity and ambiguity of our times, cultures, and classrooms.

It is, I believe, appropriate to end with pedagogy. As a teacher of rhetoric and composition, I am admittedly preoccupied with how we prepare our students to engage in controversy. There are many readers who may justifiably feel that we are too exposed already to contention, disagreement, and discord in our daily lives to claim an even larger role for argumentation in our classes. But as Protagoras knew, democratic culture cultivates both difference and dissent, and if we are to respond to such circumstances with the sense of justice (*dike*) and respect for others (*aidos*) that he insists we are capable of, then we must learn to confront our unavoidable differences with civility and even grace. In short, we must learn to approach our arguments with others in ways which honor the basic Protagorean insight that goodness and knowledge are multiform. To return yet again to Dewey, the "essential need" of any culture in which difference is ascendant is to find ways to improve "the methods and conditions of debate, discussion, and persuasion"; such is "*the* problem of the public" (1927, 208, italics mine). Protagoras would applaud the privilege accorded to argument as the cornerstone of public life, and he initiated a rhetoric which, when developed, can address many of our own present needs. I hope I have argued his case well enough to revive discussion of this important, if neglected rhetorical method and to provoke in my readers antilogoi of their own on the potential contribution of this ancient practice to a contemporary rhetorical *paideia*.

Part I

Greek Origins and Organizing Principles

Chapter 1

Protagoras and the Philosophic Origins of Antilogic

"For I do not see the whole of anything; nor do those who promise to show it to us. Of a hundred members and faces that each thing has, I take one, sometimes only to lick it, sometimes to brush the surface, sometimes to pinch it to the bone. I give it a stab, not as wide but as deep as I know how."

Montaigne, "Of Democritus and Heraclitus"

The antilogical theory of argumentation originates with Protagoras of Abdera, the preeminent Sophist of Periclean Athens. Protagoras was the first Sophist to hold public debates, the first professional teacher of advanced studies, a major philosophical and rhetorical influence, a friend and ally of Pericles, and, according to many, the founder of humanistic education (see Schiller, B. Smith, Bouwsma). Despite his stature, however, any effort to reconstruct a Protagorean approach to rhetoric and argumentation begins basically from scratch. As Edward Schiappa notes, the investigation of Protagoras by communication scholars has been "virtually non-existent" (16). That is, while we have a growing volume of commentary on Protagoras, little of this scholarship directly addresses the relation between Protagorean ideas and the rhetorical tradition, and even less attention has been paid to the potential contribution of Protagorean thought to the theory, practice, and pedagogy of *contemporary* discourse.

In response to this situation, *Many Sides* attempts to reconstruct the Protagorean legacy as an alternative current in rhetorical history and to reclaim this legacy as the basis for an innovative, neo-Sophistic conception of argumentation. However, because Protagoras himself and the notion of a Protagorean rhetoric are only vaguely familiar to most scholars/teachers of rhetoric, composition, and argument, it seems best to begin this project with what the Greeks called the *archai*, or starting points. And in the case of Protagoras, the starting points are his two most original and influential ideas: the doctrine of opposing arguments and the allied concept that humanity is the measure of all things.[1]

Any theory of argumentation built upon Protagorean ideas will naturally foreground the two-*logoi* fragment, in which Protagoras is reported to have been the first to claim that "on every issue there are two arguments (*logoi*) opposed to each other" (*DK* 80 A12).[2] Indeed, the concept of opposing positions in argument (*antilogoi*) is, as G. B. Kerferd notes, "the most characteristic feature of the thought of the whole Sophistic period" (1981, 85). It is also the driving concept behind this book. But if we are to understand the full resonance and promise of the opposing-*logoi* concept, it is crucial that we also consider the theoretical framework that supports the unique practice of Protagorean antilogic. So while antilogic will be the centerpiece of my inquiry and will occupy our full attention in the ensuing chapters of Part I, this opening chapter is devoted to the human-measure doctrine, that momentous philosophical theory which F. C. S. Schiller calls "one of the great monuments of Greek genius" and which I take to be the first principle of Protagorean thought (1970, 37). First, however, a brief introduction to Protagoras himself is in order.

Like most of the Sophists, Protagoras was an outsider in Athens. He was born (c. 490 BCE) in the small city of Abdera on the coast of Thrace in the northern Aegean. His arrival in Athens (c. 460) coincides with both the dawn of the Sophistic tradition and the Age of Pericles, in which innovations in drama, politics, philosophy, art, and medicine all contributed to the inauguration of Western culture. Protagoras himself was acquainted with Pericles and was dispatched by him to draft the laws of Thurii, a Greek colony in southern Italy. But Protagoras' primary legacy comes in the intellectual sphere. In the dialogue that Plato names after him (which remains the best portrait we have of the great Sophist), Protagoras declares with pride that he is "a confessed sophist" and educator (*Protagoras* 317b), implying that he was a professional teacher of practical knowledge; i.e., of how to conduct one's public and private life with virtue, and, in particular, how to reason and speak well on all occasions (see Plato's *Gorgias* 449a and *Protagoras* 318-19a; see also Kerferd 1981, 24-41). We also know that Protagoras was the first of the Sophists to charge a fee for his lessons, which makes him a founder of Western pedagogy and the first paid professor (*DK* 80 A2, Philostratus 1.10.4).

Diogenes Laertius lists fourteen books by Protagoras on subjects ranging from argument and debate to government, theology, mathematics, and wrestling (9.55).[3] None of this work survives in anything more than a few isolated sentences, and for these fragments we depend upon the reports of near contemporaries (Plato and Aristotle) and on the work of much later Greek writers notably Diogenes Laertius, Philostratus, and Sextus Empiricus, all figures from the 2-3C CE. So in dealing with the Protagorean canon there is a good deal of speculation as historians and theorists grapple with the provocative, yet gnomic dicta that constitute his surviving corpus. Nor are the details of his life any more precise; e.g., it is unclear if Protagoras died in disgrace or distinction, though his death itself is conventionally set around 420 BCE. What we do know is that Protagoras was a formidable intellectual presence whose challenging, original ideas—on diction, grammar, criticism, and theology as well as philosophy and argument—not only compelled careful scrutiny by Plato, Aristotle, and Isocrates but also continue to reverberate with implications for our own age, as we will see.[4]

1. THE PROTAGOREAN FRAGMENTS AND THE THEORY OF PERSPECTIVISM

The generally accepted version of the human-measure fragment is as follows:

> Of all things the measure is [humanity], of things that are that they are, of things that are not that they are not.
>
> (Sextus, see *DK* 80 B1)

By any standard, the statement is momentous; nor is it too much, I think, to claim for it status as the point of origin, the *fons et origo*, for 2,500 years of Western Humanism.[5] In a purely philosophical context, the fragment gives rise to a host of potential controversies, many of which I will address in the course of this chapter. With regard specifically to matters of discourse, however, the human-measure doctrine can be effectively approached by concentrating on two of its primary attributes, **perspectivism** and **relativism**. This division is admittedly arbitrary because the two topics dovetail. But to take up these attributes in turn should help us to distinguish some of the complex theoretical concepts that undergird antilogical practice.[6] So, in this chapter, I begin by examining the human-measure doctrine and its theory of perspectivism, and then by exploring the long-standing objections to this philosophical program. Once these basic concepts are on the table, I proceed by reviewing the controversy over Protagorean relativism and then by pointing out some of the promising implications of Protagorean theory for rhetoric. You will note, then, an antilogical movement in this chapter as we oscillate between contrasting approaches to this provocative and enduring philosophical doctrine.

With regard to **perspective** or point of view, the human-measure doctrine implies that all experience is based on, measured by, filtered through the perceptual apparatus of the individual (*panton anthropos metron*, or "of all things humanity is the measure"; *DK* 80 A16). It is, of course, possible to argue that *anthropos* should be read as "society at-large," thereby positing cultural knowledge (*nomos*) as the "measure" of all things; and, indeed, Protagoras does place particular emphasis on the role of social groups in the formation of knowledge (see *Protagoras* 322). At present, however, I will follow Plato as well as Cornford, Versenyi, Kerferd, and most contemporary scholars in assuming an individualized rather than a collective agent for the human-measure concept, so that it is the individual percipient or knowing subject who becomes the medium through which all knowledge is manifested and measured (see Piato's *Theaetetus* 152a, 158a, 161d; Kerferd 1981, 86; see also Donovan 37-38, Guthrie 3.188-89, Schiller 1970, 33). In order to more fully appreciate what is at stake in the Protagorean emphasis on personal point of view, it will help if we set this epistemic theory in the context of early Greek thought.

Prior to the arrival of Protagoras in Athens, Greek philosophy had been heavily influenced by Parmenides of Elea, perhaps the most prestigious of the Presocratic philosophers (Kirk 266, *Encyc. of Philosophy* v. 6, 47). In contrast with earlier theories of natural order, such as the Pythagorean table of opposites or the Heraclitean notion of flux, the Eleatic philosophy of Parmenides propounded a monistic vision in which reality or Being (*ousia*) was conceived of as single, indivisible, continuous, changeless, and motionless.[7] Correspondingly, Parmenides regards all apparent manifestations of variety and transformation in the world as an illusion, the "way of seeming" as opposed to the "way of truth." Moreover, according to Parmenides, truth itself can only be approached by logical deduction, unaided by the senses. Conversely, any reliance on opinion or sense perception constitutes a descent into error, into "two headedness" and an unwarranted acceptance of pluralism, division, and contradiction.[8] The human-measure concept, then, with its implication that knowledge always presents itself through the filter of individual perspective and is thus subject to variation and multiplicity, this bedrock Protagorean notion stands in direct opposition to Parmenidean monism. More specifically, for Parmenides, the fixed structure of the world is subject to precise calculations that provide a foundation for all true knowledge; while for Protagoras, no invariant reality is available to human perception and so knowledge about the world exists only within the domain of human conjecture.

This basic, if simplified opposition is central to any consideration of Protagoreanism. For example, when viewed from the Parmenidean perspective, no single event could be conceived of as both A and not-A because Being itself is one and consistent, unable to both exist and not exist simultaneously (see Plato's *Sophist* 237a). In more practical terms, the same wind can not be acknowledged as both cold and not cold at the same time (cf. *Theaetetus* 152b). In this context, truth is bivalent: there are only two potential responses to a truth-claim (true or false) because the world itself is determinate; any middle ground is inadmissible (hence the philosophical principle of the "excluded middle"). But, from the Protagorean perspective, bivalence is inadequate because it fails to comprehend the actual range of human response. If the individual is the measure of all things, then the nature of event A (our perception of the wind) depends not on an objective standard of abstract binaries, but rather on the frame of reference of the individual(s) actually involved in calculating the attributes of the event in question. And, as Protagoras was the first to point out, not all frames of reference for individual measurement are the same. A wind coming off Lake Michigan may be experienced by person B as cool, by person C as warm, by person D as mild, E as blustery, and by person F as without character. Our perception of the wind—and of everything else that is subject to human evaluation—relies, according to Protagoras, upon what we as individuals bring to the experience and how we conduct our assessments of the various phenomena that make up the world. As a result, knowledge is best seen as multivalent, an expression of the variations that distinguish the human community and our diverse ways of knowing.

Correspondingly, when we shift our focus from the realm of sense perception to more distinctly cognitive experience, we encounter the same perspectival differences. The experience of interpretive difference is neatly described by Lewis Carroll's Alice and by Tolstoy in *War and Peace*. In the first instance, Alice is speaking with the Caterpillar about her difficulties in adjusting to Wonderland:

> ". . . being so many different sizes in a day is very confusing."
>
> "It isn't," said the Caterpillar.
>
> "Well, perhaps you haven't found it so yet," said Alice; "but when you have to turn into a chrysalis—you will some day, you know—and then after that into a butterfly, I should think you'll feel it a little queer, won't you?"
>
> "Not a bit," said the Caterpillar.

> "Well, perhaps *your* feelings may be different," said Alice; "all I know is, it would feel very queer to *me*." (italics in text; Ch. 5)

The same principle is at stake for Tolstoy's protagonist, Pierre Bezukov:

> [Pierre] was struck for the first time by the endless variety of men's minds, which prevents a truth from ever presenting itself identically to two persons. Even those members [of his club] who seemed to be on his side understood him in their own way, with limitations and alterations he could not agree to.... (Bk. 2, Ch. 7; 463)

In diametric opposition to Parmenides, the Protagorean principle of perspective will inevitably give rise to individual differences in how any experience is measured. Thus, instead of appearing to be one and indivisible, the human vision of reality—i.e., reality as perceived through the lens of human cognition—will declare itself as it does to both Alice and Count Bezukov: as both *variable* (according to perspective) and *conflicted* (subject to controversy).

The consequences of such a position obviously stand in opposition not simply to Parmenides but also to the dominant tradition of Western philosophy, a tradition which takes its fundamental cue from both Parmenides and Plato and their attempts to ground knowledge on invariant ontological structures and in corresponding principles that are timeless, universal, and immutable. Whereas Parmenidean epistemology assumes that we can discern clearly the nature of reality independent of our perspectival orientation, the Protagorean worldview assumes an interpretive or hermeneutical stance in which our experience of the world is constituted not by (im)mediate exposure to things-as-they-are-in-themselves, but through the mediation of our perceptual horizons and histories. For Protagoras, the single or absolute nature of the wind coming off Lake Michigan is indeterminate (i.e., indeterminate in so far as it is experienced by humans) because it is inevitably "measured" by individuals whose perspectival orientations (or horizons or points of view) are bound to vary. As we will see, this position need not mean that "truth" itself is a fallacy invented by humans as protection against chaos; nor does it deliver us into a world of incommensurable ideas and solipsism, as a great many opponents of Protagoras from Plato to the present have claimed. What we can note at this early stage of our discussion is that the human-measure doctrine resists reliance on ideal, invariant frameworks and on bivalent truth-values, while at the same time it regards all "truth-claims" as the product of an interpretive response to the world and the resultant multiplicity of claims as an ineluctable feature of human inquiry (*multiplex ratio disputandi*). In a phrase, there are many sides to most questions of significance, a reading of things that, instead of ushering in anarchy, allows for the recovery of the world under the aegis of human agency.

We will return to these philosophical issues in greater detail shortly; for now, it is enough to point out that the human-measure doctrine, with its emphasis on a pluralistic rather than monistic view of perception, knowledge, and truth, cannot help but have profound consequences for rhetorical practice. With the move from Parmenides to Protagoras, we mark a momentous intellectual shift from an ontological concern with the nature of things as they are in an abstract, ideal mode to a hermeneutical interest in the various interpretations that any human community is bound to generate in response to shared experience. For Protagoras, the search for knowledge involves us inevitably in the interaction of contending claims. And, as Isocrates (in the next generation of Sophists) was fond of repeating, the adjudication of conflicting interpretations is always a matter of discursive expertise (*Antidosis* 253-56, *Nicocles* 8). Consequently, any theory of discourse based on Protagorean principles will place argumentation and the effort to weigh conflicting claims at the core not only of rhetorical practice but at the center of all efforts to understand the world and orchestrate cultural life. For Protagoras, argument becomes the very soul of the *paideia*, our effort to build a well-reasoned community.

The strict focus on individual perspective implicit in the human-measure doctrine is buttressed by the additional Protagorean theory that there is no transcendental authority available to human scrutiny, no ultimate court of appeal for the adjudication of human claims. This notion is articulated in another Protagorean fragment, a fragment assumed to be the first sentence of a book titled *Peri Theon*, or "On the Gods." Charles Kahn translates the first part of the fragment thus:

> Concerning the gods, I am unable to know, whether they exist or whether they do not exist or what they are like in form.
>
> (in Schiappa 141; cf. Diogenes Laertius 9.51; *DK* 80 B4)[9]

There are a number of interesting historical issues connected with this potential case of ancient agnosticism; but for an inquiry into a Protagorean theory of rhetoric, it is important to note that by effacing the gods as a foundational source for human knowledge, Protagoras eliminates any claim to unimpeachable authority.[10] The gods may either be inaccessible or non-existent, the fragment is inconclusive on this point. But if there is no universal authority available as such, then we operate according to our own lights, without final guarantees and permanently fixed principles. In this context, no proposition can claim to represent the world as it is outside the domain of human reasoning (Parmenides' *Way of Truth*); rather, all arguments are offered as artifacts of human perception subject to contingency and flux. To borrow from Sextus Empiricus—human perception becomes the only *kriterion* for judgment because no transcendental source is available for appeal (*DK* 80 A14; cf. Kerferd 1981, 107-8). And if human perception is by nature variable (as the human-measure and antilogoi fragments imply), then we are faced with diverse (though not irreconcilable) measurements as the inescapable data out of which knowledge and discourse are constructed.

At this point, I should note that by "perception" I mean to connote the full range of our cognitive life; so that individual percipients may be seen as invoking not only the senses, but also memory, intuition, emotion, imagination, reason, or any synthesis of these in an effort to achieve both invention and judgment.[11] The clarification is necessary because Plato makes an extended effort to limit the human-measure doctrine to sense perception only and so to debunk it as a viable epistemic theory. His *Theaetetus* is devoted to the question "what is knowledge?" and begins with "the account given by Protagoras" that things are "to each person as we perceive them to be" (146a-158a). Socrates immediately connects such perception with the sense impressions of the "common herd" who believe that "nothing is real save what they grasp in their hands" (155e), a critique of both the *demos* as a source of knowledge and the concept of knowledge itself as subject to Protagorean conditions. Socrates goes on to argue that if we limit knowledge to human perception we are consigned to a vision of the world that is both unstable (subject to change) and fallible (subject to distortion).[12] If we were to accept as real and true only what is grasped through sense perceptions, we might as well say that a pig, a baboon, or "some other sentient creature still more uncouth" is, in fact, the measure of all things (161c). This sarcasm constitutes a serious misreading of Protagoras; for, as Paul Feyerabend points out in his own dialogue on the human-measure concept, Protagoras nowhere limits the concept of "measuring" in the way Plato implies (1991, 26-27).[13] Once we step away from Plato, we are in a position to recuperate the human-measure doctrine along lines more generally sympathetic with Protagoras' own thought and to reconstruct its function in accordance with more contemporary conceptions of knowledge and truth. I would argue that because there is no absolute arbiter of knowledge in the Protagorean system (as the *peri theon* indicates) and because there are incongruent truth-claims available on all questions of significance (as the two-logoi fragment maintains), the practice of Protagorean "measurement" is best conceived of as a hermeneutical act, a process variously represented as interpretation, evaluation, and judgment.

In direct opposition to Protagoras, Plato posits an epistemology based upon reflection; i.e., on procedures of formal reasoning designed to resolve the "contraries" of sense experience and, by appeal to a reality of pure, transcendental Form, achieve knowledge that is permanent, infallible, and universal. Protagoras abjures access to any such invariant standard by which human judgments can be certified. As the *peri theon* would have it, human perception is the matrix for each effort at understanding, and since human knowing is contingent on the knower, knowledge itself is conditioned by diversity and fallibility. This hermeneutical theory may indeed conflate knowledge and opinion and weave contention into the operations of knowing, but the concept hardly warrants epistemological despair. I will try to indicate as our discussion progresses that the absence of an absolute authority for knowledge actually endorses multiple ways of knowing and invites the convergence of incongruent claims. As such, Protagoreanism becomes what Lorraine Code calls "an enabling rather than constraining position" (1991, 3), a stance that enfranchises *mutliplex ratio disputandi* and *petits recits* instead of insisting upon the authority of an overarching and unquestionable logos.

So far, I have argued that Protagoras inaugurates a cognitive theory in which knowledge is always conditioned by the position and agency of the knowing subject. However familiar, contemporary, or agreeable such a concept may now seem, opposition to Protagorean perspectivism has been, from the beginning, powerful and persistent. Plato responds that unless we purge ourselves of personal bias (via dialectic) and come to know the Truth as it stands apart from individual experience, we cannot claim knowledge at all (*Theaetetus* 186c; cf. Plato's *Timeaus* 51c-d).[14] And Aristotle, with Protagoras clearly in mind, counters that "all things that we know, we know in so far as . . . some attribute belongs to them universally"; i.e., in so far as they are themselves beyond modification by the individual percipient (*Metaphysics* 999a, 27-29). This argument against relativism (and on behalf of universal truths) spearheads the ancient attack on the human-measure doctrine; nor is this critique solely of historical interest. In various forms, many contemporary thinkers continue to express comparable longings for self-evident principles and normative certainties on which to ground the search for knowledge and the procedures of argument (Bizzell 1990, 633; Cherwitz 1990).

Equally persistent and influential as a response to Protagoras has been the charge that the human-measure doctrine is incoherent and indefensible as a philosophical statement; i.e., it propounds a universalist premise (humanity is the measure of *all* things) inconsistent with its own rejection of universal knowledge. As Joseph Margolis points out, the charge of incoherence lodged against Protagoras has been "an abiding feature" in the history of philosophy, a charge which a great many ancient and modern theorists consider both fatal and inescapable (1). It would be contrary to the spirit of antilogic itself if I did not examine my topic in relation to these opposing positions; and so, in the next section, I take up the charge of incoherence, while in Section 3 we turn to the question of relativism.

For the moment, however, it is important to recognize that, despite vigorous, sustained attacks, the philosophical stance that Cornel West calls "perceptual contextualism" has not only persisted but actually come into its own in the modern world (18).[15] Indeed, Margolis argues that the "prevailing philosophical winds" are now blowing more favorably in Protagoras' direction than was ever the case in ancient Greece (5). A sampler of modern variations on the perspectival stance should help complete our initial portrait of Protagoreanism and recast these ancient Sophistic ideas in a more familiar idiom. For example, Emerson (in his essay on "Experience") develops the case for perspectivism this way:

> We have learned that we do not see directly, but mediately, and that we have no means of correcting these colored and distorted lenses which we are, or of computing the amount of their errors . . . . [P]eople forget that it is the eye which makes the horizon . . . . Thus inevitably does the universe wear our color . . . . (156-58)

It was Nietzsche, however, who provided the first detailed explanation of the cognitive operations that give rise to perspectival thought. In his essay "On Truth and Lying in an Extra-Moral Sense" (1873), Nietzsche invokes Protagoras when he notes that in order to understand the world we "treat [people] as the measure of all things," but in doing so we must recognize that individuals do not have "things immediately before [them] as mere objects" (251-52).[16] Human consciousness, claims Nietzsche, is not in contact with "the thing itself" because it registers images, not things (248). Moreover, these things represent themselves to us only as language, i.e., as a system of signs or metaphors that display not "the full and essential" nature of phenomena but only characteristic features of those phenomena that are prominent for the observer (Gilman xiii). Correspondingly, there is no "unrhetorical 'naturalness' of language" that simply mirrors rather than modifies experience, nor is there a single criterion available outside of language that can assess the "right perception" of any particular linguistic sign (Gilman xiii; Nietzsche 252, resp.). Instead, in every instance of thought or speech, we employ a network of tropes that are, by nature, partial, metamorphic, and partisan, a language, in other words, that will always reflect the perspective of the rhetor and beyond which there is no appeal to an absolute (cf. Gilman xiii-xxi).[17]

Another philosophical approach to the role of perspective in knowledge is detailed in Heidegger's theory of "forestructure." As part of a general theory of hermeneutics, Heidegger posits an a priori framework of intentions that all people bring to all interpretive acts: any instance of "appropriation" (or understanding), he writes, "is always done under the guidance of a point of view, which fixes that with regard to which what is understood is to be interpreted. In every case interpretation is grounded in *something we see in advance—in a foresight*" (150). Heidegger separates forestructure into three different, rather complex phases, but the effect is clear enough; every appraisal of every event will be unique, because, as James Kinneavy puts it, "every interpretation, even by the same person, is made from a somewhat different perspective" (3; see Crusius 26-30; cf. Gadamer's notion of prejudice; 1994, 269-77). A more sociological approach to the subject of perspective can be seen in Veblen's theory of "trained incapacity" and Dewey's "occupational psychosis," both of which emphasize the partial perspectives that follow from a world of strict occupational specialization (cf. Polanyi's notion of "tacit knowledge," 49-65). The ideas of Veblen and Dewey are, in turn, incorporated by Kenneth Burke into his own theory of "orientation."

Developed in considerable detail in *Permanence and Change*, Burkean orientation refers to "that bundle of judgments" about present and future matters "in the human sphere" that lend character, meaning, and significance to our ideas and actions (14; see also 5-18).[18] Different orientations, or what he elsewhere calls "frameworks for interpretation," will obviously lead to different conclusions as to what reality is (35). Despite a certain "liquidity" in these orientations as we pass from one framework to another (33), each of us acts as if our own orientation were *imperium in imperio* (an empire within an empire, 26). What Burke, in particular, brings to the perspectival tradition is an insistence on the limitations of individual orientations, all of which, he says, are "self-perpetuating structures" that work to enthrone "the *measures* by which [they] are *measured*" (262; italics mine). In response, he argues that we should do more than simply recognize the multiplicity of available orientations and that we ought to seek out counter-perspectives that aid in the ongoing process of our own "reorientation" (102, 169). This argument in favor of opposing perspectives is deeply Protagorean and replete with rhetorical reverberations, because any effort to find one's way to counter-perspectives will involve transactions with others, most often in language, and so expose us to the ambiguity, difference, and conflict that are inherent in any argumentative exchange.[19] The act that Burke calls "reorientation" asks us not to persist obstinately in "the courage of our convictions," but rather (as Nietzsche also argues) to have the courage to admit, or even solicit, an alternative, an antilogos to our own perspective (in Griswold 202).

Other variations of perspectivism discuss the concept in specific relation to physics, psychology, and sociology (see Cherwitz and Hikins 1984, 250,

n. 6-10).[20] But the basic idea should by now be clear: all knowledge is influenced by the cognitive process of the knowing subject; or, in the original phrase, "humanity is the measure of all things." Indeed, this primary Protagorean concept has become so widely acknowledged that we might well say that the idea of "contextualized knowledge" is a fundamental assumption of most modern worldviews. It is fully compatible with a good deal of modern philosophy (e.g., Heidegger, Gadamer, Derrida, Kuhn) and with most variants of postmodernism: i.e., with approaches to knowledge that encourage cultural differences, that recognize the tacit prejudice conferred by cultural positioning, that posit reason and discourse as functions of diverse conceptual schemes, and that reject foundational principles which claim ontological status. And yet, despite the refashioning of Protagorean Sophism in attractive modern attire, there remains a deeply embedded antipathy to the Protagorean program, an antipathy not simply antiquarian or arcane.

So, while my own study of Protagoras is focused on antilogic and its potential influence on our own argumentative practice, a comprehensive discussion of the subject requires that we at least identify and briefly engage certain points of philosophical opposition. Nor is it simply a matter of balancing the record; other recent writers have done an admirable job of reconstructing the Sophistic legacy in a more positive key (see Barrett 77-81; Jarratt 1991; Kerferd 1981, 4-14; McComiskey 25-38). Instead, I would juxtapose the various arguments over the human-measure doctrine as a way of broadening the base of our inquiry and preparing a more substantive picture of antilogic in its own right. Among those opposing arguments that claim to refute the human-measure fragment are those that raise technical questions about the logical consistency of the doctrine, questions that will occupy us in the next section. Behind these claims of logical fallacy, however, lurk other objections to the human-measure doctrine—objections to its putative relativism and radical subjectivism which we will return to in the third section of this chapter. But even relativism, so often conceived of as a heretical and nihilistic abdication of philosophy, will appear in revised form once we have assayed (exorcised?) the negative terms of the logico-formalist critique.

2. THE CASE AGAINST THE HUMAN-MEASURE DOCTRINE

Under the heading of objections to Protagoras are the related charges that the human-measure doctrine is **self-refuting** and that it violates the philosophical **law of non-contradiction**.[21] These are complex issues with an extensive tradition of philosophical commentary, much of which is outside the scope of this study (see Burnyeat 1976). For our purposes it is enough to summarize the general nature of the charges and outline pertinent features of a response to them.

With regard, then, to the claim of self-refutation, I have already noted the argument that the human-measure doctrine adopts a universal perspective which its own premise (the individual as measure) seems to deny. That is, if truth-claims are relative to the individual (true for A) from what vantage point are we able to assert that a particular claim—namely, the human-measure doctrine—is true for all? Sextus Empiricus, following Democritus, describes the charge in a slightly different way (*DK* 80 A 15): it cannot be argued that all perceptions are true, because this self-same argument (call it Argument T) is invalidated as soon as someone puts forward the opposite position (that all perceptions are not true, or Argument NT). Since Argument T guarantees the right of all interpreters (including those who hold position NT) to their own assessment of the issue, then T cannot be true because there is obviously someone who does not believe it, and to insist upon T in the face of NT would contradict the very principle which the doctrine itself would defend (cf. Democritus [*DK* 68 A113] and Plato's *Apology* 167a-172b; see also Maguire 135-37). Hence, the proposition refutes itself. This kind of objection is called a *peritrope*, a turning-of-the-tables, and Plato puts the peritrope this way: suppose "the world in general" did not believe that humanity is the measure (which, in fact, they don't, says Socrates), then the human-measure doctrine "is more false than true by just so much as the unbelievers outnumber the believers" (*Theaetetus* 170e-171e). I will return and respond to this critique shortly.

The second objection to the human-measure doctrine—the charge that it violates the philosophical law of non-contradiction—overlaps the claim of self-refutation. This second charge can be illustrated as follows: if I say the wind is cold, it is not possible for you to contradict me because the perception of any experience is always relative to the percipient. The doctrine of non-contradiction is formulated in the phrase *ouk estin antilegein* ("it is impossible to contradict"), which is not listed separately as an authentic Protagorean aphorism but which has been attributed to him in some form or another (see *DK* 80 A 1 and 19; Schiappa 134, n.1). In the *Euthydemus*, Socrates connects Protagoras specifically with the notion that it is impossible to "speak against" another person and asserts that this proposition amounts to a claim that "there is no such thing at all as a false opinion" (285e-286d). In brief, the charge is that according to Protagoras contrasting logoi on any given issue are all true and non-contradictory because each follows from a different perspective; i.e., all are true for someone (cf. Kerferd 1981, 90). For Aristotle, the implications of non-contradiction are serious indeed.[22] In Book IV of his *Metaphysics*, Aristotle explores the nature and attributes of Being (*ousia*), and in the process he includes an extensive discussion of the law of non-contradiction. At several points in this discussion he cites Protagoras directly and notes that "if all opinions and appearances are true, all statements must at the same time be true and false," since obviously there are conflicts of opinion on just about everything (1009a 5-15). But if all opinions are both true and false, there is no way to make distinctions, be they important or trivial. Instead, "it is equally possible to affirm and to deny anything of anything," to which he adds that such a position "must be accepted by those who share the views of Protagoras" (1007b 19-22).

For Aristotle, this debate over contradictability hinges on the relation of particular opinions to universal ideas and the role of these "universals" in the formation of knowledge. "If there is nothing apart from individuals," he maintains, "there will be no object of thought, but all things will be the objects of sense, and there will not be knowledge of anything, unless we say that sensation is knowledge" (*Metaphysics* 999b 1-4). That is, sense perception (rather than some more dependable form of cognition) is what follows from contact with particular objects or events, from what he elsewhere calls the "this" and "now" of experience (*Posterior Analytics* 87b 28). But in order to transform these manifold individual experiences into actual knowledge, we must extrapolate from a group of related objects or events a common form, or *eidos* (the "shape" of the sensible thing; *Metaphysics* 1033b 5). Only these universal ideas—ideas that posit an identical and unified shape that inheres within a set of diverse particulars—can claim the status of knowledge. And if non-contradiction makes that extrapolation from particulars to universals impossible (because it denies our ability to distinguish what is and is not true), then the entire enterprise of intellectual inquiry is thrown into doubt. How can we posit that "this is true" and "that is not" if all positions are true for the people who hold them? Under such conditions, the distinction between knowledge and opinion is erased as the former is reduced to the latter. It is little wonder that Aristotle referred to this philosophical issue as "the hardest of all and the most necessary to examine" (999a 24-25).

How Protagoras himself might have responded to all this is, of course, beyond the reach of the fragments. Moreover, the nature of the Protagorean position as characterized by Plato and alluded to by Aristotle is obviously problematic because these reports constitute the interpretations of the opposition. Such, however, is the ironic status of Protagorean thought: from the outset it has been reformulated by those who would refute it. And yet it remains possible to hypothesize a response to the claims against Protagoras based on a position more fully compatible with his own canon, a body of remarkably coherent ideas despite their fragmentary form. To continue to approach Protagoras from the perspective of his ancient adversaries is to operate, as Blake said of Milton, as a member of the devil's party without knowing it. And while such an analogy may be melodramatic, it is nonetheless accurate to claim that the basic assumptions of Protagorean theory and the philosophical epistemology of his harshest ancient critics are fundamentally antithetic.[23] It seems only fair, then, to take stock of these paradigmatic differences in philosophical perspective and to accord Protagoras the benefit of a more positive reconstruction of his position. My intention is to briefly review these opposing philosophical paradigms as a way of establishing a context from which to respond to the two charges on the table.

In this sketch, I adopt—by and large—the terms of analysis employed by Joseph Margolis, a contemporary philosopher particularly attuned to the Protagorean position. For Margolis, comparing "the master tradition" of Plato and Aristotle with the Protagorean alternative begins with a profound difference in metaphysics (88ff). For Plato, as for the Presocratics, reality is manifest in a transcendent order of changeless essences, a domain beyond the flux of the world and yet discernible by human reason. Aristotle locates his own metaphysic at some distance from the Platonic realm of pure Form and includes considerations of sense and substance; but reality itself remains eternal and changeless, while knowledge in its purist form is tethered to a universal framework of invariant principles (see Aristotle's *Metaphysics* 987b 5-9). This outline of traditional metaphysics is, of course, too brief and unqualified; both philosophers are subject to competing interpretations (see Covino 1988; Barnes 1982, 37f). Nonetheless, we can still say with some confidence that for these two patriarchs of Western philosophy the object of knowledge is a timeless, universal structure available to human scrutiny under opportune cognitive conditions.

For Protagoras, however, any conjectures regarding ontological structure take place *within* the orbit of human cognition. Metaphysical order—be it transcendent or abstract—is unavailable and intransparent because we cannot distinguish between the world as constructed by consciousness and the world as independent of ourselves. The Protagorean fragment "on the gods" (the *peri theon*) does not, you will recall, *deny* order, nor does it imply that we are abandoned to randomness only. These are the nightmares of the master tradition. But according to the human-measure doctrine, the world as it appears to us is always a human artifact, an "indissoluble symbiosis," says Margolis, of cognized world and cognizing agent (141). Instead, then, of locating ontological reality in a universal framework that is the point of origin for human knowledge, the Protagorean worldview reverses the process (in a kind of epic chiasmus) and posits human cognition as the starting point of any effort to make the world intelligible.

These basic distinctions in metaphysical/ontological outlook are well known to modern scholars, but there follows from them equally dramatic, though less familiar differences in epistemology and alethic theory (theory about the nature of truth). If, for the master tradition, the world is a fixed structure discernible by human cognition, it follows that knowledge emanates from and is "addressed to" what is immutable in reality, even if the immutable is sometimes obscured by the seeming flux of things (Margolis 78). As Plato puts it, the eternal gives "truth to the objects of knowledge and the power of knowing to the knower" (*Republic* 507d-e). Aristotle divides knowledge into a variety of categories (*Metaphysics* 1025b); but in its most refined form (as scientific knowledge or *apodeixis*), what counts as knowledge is our understanding of the invariant and universal structure of being (*Post. Anal.* 87b 38). For example, Aristotle argues that we know something to be white because "whiteness" possesses an unchanging, universal form (*eidos*) which can itself be known (*Metaphysics* 1007b). And because this knowledge is founded on what is universal (i.e., on what is common to all related particulars), it is exact and invariant: "The object of knowledge in the strict sense," he says, "cannot be otherwise" than it is (*Post. Anal.* 71b 9-16). We can demonstrate, therefore, that a subject does or does not possess an attribute such as whiteness (whether or not X is predicated by whiteness) because, as Guthrie puts it, we can know the abstract *eidos* which informs concrete substance (6.182). As a result, we can also adjudicate any contradictory knowledge claims by reference to what is eternal and uniform. To claim that the same subject can be predicated by contradictory attributes (cold/not cold) at one in the same time is to mistake the vagaries of opinion for knowledge of what "cannot be otherwise." To make this mistake is, as noted, to "share the views of Protagoras."

Aristotle's epistemology is accompanied by an alethic theory of particular interest to our inquiry. If knowledge is anchored in a reality that is determinate, and if contradiction can be resolved by reference to a reality that is uniform and invariant, then truth values may be adequately distributed along strictly bivalent lines. A claim is either true or false, a subject white or not because the reality to which a particular truth-claim refers is, so to speak, monolithic; i.e., it manifests itself in singular, determinate form. To think otherwise, says Aristotle, would be to "predicate of every subject every attribute and the negation of it indifferently" (*Metaphysics* 1007b 29-30). That way, of course, lies chaos. But under the conditions of decidability that govern Aristotelian analytics, we can with confidence articulate alethic claims according to a precise, binary semantics. Any middle ground between the bipolar options of true and false can be justifiably *excluded* from consideration because the reality to which a valid truth-claim corresponds is itself determinate. Truth-claims offered in this context are bipolar because the Truth itself is singular. The axiom of the "excluded middle" is, of course, incommensurate with the pluralism implicit in the human-measure doctrine. A Protagorean response to Aristotle's alethic claims would argue (as Margolis does) that "bipolar truth-values are not always, perhaps not anywhere, adequate to the range of truth claims that may be made of human 'experience'" (75).

However, before we return to Protagoras, one more feature of the Aristotelian method remains to be clarified. If the authority of a fixed ontological structure fosters bivalent truth values, the representation of that structure as a set of abstract concepts endorses, in turn, the exacting procedures of formal logic as the privileged vehicle, the method-of-all-methods for arriving at determinate knowledge. That is, the symbolic ciphers and quasi-mathematical formulae of analytical logic become the indispensable tool or *organon* for careful reasoning because they are congruent with, or "fitted to" the invariant order which their own operations seek to re-present. In its Aristotelian origins, then, formal logic may be characterized as an isomorphic extension of the fixed structures of Aristotle's metaphysics; and, by analogy, the outcome of successful logic will itself be universal, exact, and incorrigible. The core of Aristotle's logical system is, of course, the scientific syllogism, which is based on incontrovertible first-premises and formulated according to precise deductive procedures. When valid, such procedures not only allow for the inferential movement from what is known to what is new, they also produce propositions which are themselves *necessarily* the case (*Prior Analytics* 24b 18). The acquisition of knowledge by syllogistic means proceeds, therefore, from certain basic principles which are themselves axiomatic and beyond scientific proof to judgments about truth and falsehood which carry the imprimatur of certainty. Foremost among the axioms that govern the syllogistic system are, as I noted, the related concepts of non-contradiction (*Metaphysics* 1005b 5-34, 1011b 13-14) and the law of the excluded middle (*Post. Anal.* 71a14, 88 b1). Both also serve, as we will see shortly, as weapons in Aristotle's argument against Protagoras.

There are obviously gaps in my sketch of such a complex subject. In the first place, the primacy of the syllogism in Aristotle's schema has been disputed on the ground that other, less-apodeictic procedures obtain for Aristotle under different epistemic conditions (cf. Ch. 2; see Barnes 1982, 36ff). Nonetheless, Aristotle himself claims that analytical logic is "the best knowledge" (*Nic. Ethics* 1141a 21); and its precise, mathematical purity was assumed for two millennia to be "the sum total of logical truth" (Barnes 1982, 31). Guthrie notes that even in the 20C, much serious thinking remains "cast in an Aristotelian mode" (6.157). Indeed, a good deal of contemporary argument theory continues to appropriate the model of

Aristotelian logic as a regulative norm upon which to pattern many of its own procedures (see Fulkerson 1996, 155-66). Secondly, I have barely alluded to the critique of Aristotle's logic and the extensive tradition of revisionary response (see Lukasiewicz). Such a subject is beyond the scope of this inquiry and my expertise; suffice it to say that we have long since abandoned the universalist metaphysic on which the patterns of formal deductive logic (FDL) are founded and that the applicability of FDL to real world conditions has been questioned at least since C. S. Peirce in the 1870s and 1880s (Margolis 52). Even more striking, however, has been the feminist critique of Aristotelian thought (see Code 1991; Spelman). Andrea Nye, for example, points out that the goal of Aristotelian logic was not "to open the discussion to all viewpoints with the purpose of establishing a consensus, but to establish a model that excluded what was contradictory and irrational" (49-50). Included in this latter category of excluded views were not only philosophical opponents such as Protagoras, but also the voices of women, foreigners, laborers, slaves, or any other group outside an exclusive circle of male property owners. For Nye, syllogistic reasoning is both a monistic and patriarchal ideal that insulates the logician within an artificially circumscribed space where claims can be insisted upon as superior and dominant. "Logic needs no respondent" because it is convinced that its conclusions cannot be otherwise (59).

My purpose in the last few pages has been to identify the epistemological perspective from which the ancient refutation of the human-measure doctrine was mounted. From the vantage point of a fixed metaphysical framework and the epistemological assumptions that follow from it, it makes sense to consider Protagoras an intellectual anarchist and the principal tenets of Protagoreanism a manifesto for epistemological misrule. And, in fact, the charges initiated by Plato and Aristotle have not only established the standard philosophical account of Protagoras, but the power of their critique is still assumed by many to have dealt a decisive blow to the human-measure doctrine. We should now be prepared, however, to approach the charges against Protagoras from a less adversarial perspective, a point of view from which alethic values and logical paradox will appear in quite a different guise.

To return, then, to the charge of self-refutation, the argument is that the human-measure doctrine propounds a theory that is overturned by its own logic. Socrates introduces the critique this way: Protagoras claims that what an individual believes to be the case is so "for him" (*Theaetetus* 160 a-d). The example this time involves wine: if a wine appears to be sweet to A, it is so for him. Yet, when the same wine appears to be dry to someone else (N), human-measure requires that it is so for her. As a result, when A and N drink the same wine and respond to it in their alternative ways, one of the two, says Socrates, must be wrong; and the mistaken party is consequently not the measure of what is, in fact, the case (172 a5-b9). This same basic objection (the peritrope) is shared by virtually all Greek philosophical commentators on the human-measure fragment as well as by many modern ones, including Husserl (cf. Democritus *DK* 68 A114; Sextus *DK* 80 A1; Burnyeat 1976). Socrates, at least, is generous enough to add that if Protagoras were to "pop up his head" from the underworld, he would surely have something to say in response to this charge (172 d2). In lieu of his reappearance, we can offer several possibilities.

First, according to the human-measure fragment, knowledge is always conditioned by the position and agency of the knower. The traditional critique, however, radically transforms the Protagorean position: the perspectival notion of "knowledge *for* A" is supplanted with an emphasis on the object, on "knowledge *of* X" (see Kerferd 1981, 106-7). That is, the peritrope depends upon the reconception of human-measure as a statement about the ontological status of the object of knowledge, the true nature of X. Once the qualification "for A" is dropped (as it is by Plato and Sextus) and statements about the true nature of the object are substituted as the crux of the concept, it becomes possible to argue that either A or N is wrong about X because X itself is determinate (see Vlastos 1956, xiv, n. 29). In the face of someone's error, to argue that each person is the measure of all things is preposterous. The peritrope, therefore, requires a fundamental shift in emphasis away from the cognitive agent and onto the nature of the objects themselves; it rejects the conceptual priority which Protagoras grants to human cognition and replaces it with an overriding interest in determinate knowledge of an invariant order. In sum, the peritrope demands a paradigmatic shift in perspective.

Historically and theoretically, however, what is so striking about the human-measure doctrine is its own refusal of any fixed order and its rejection of the realist epistemology that follows from such an order (i.e., the possibility of determinate, bivalent statements about the "real" nature of things). Instead, the fragment emphasizes the primary role of cognition in the construction of knowledge and the commingling in consciousness of human agency and a world always already marked with the traces of the agent (see Nietzsche on perspectivism above). I will attend to the relativistic implications that follow from the "knowledge for A" hypothesis shortly. For now, I would argue that the peritrope is sound only if one allows the objectivism of the master tradition to dictate the terms of critique. Given a post-positivist belief in the role of human agency in the construction of knowledge, the peritrope dissolves as the bias of a distant and incompatible world view (cf. Untersteiner 28). In the end, the radical reconstruction of the human-measure doctrine required for self-refutation to succeed strikes me (as it did Montaigne) as an argument in bad faith (392-93).[24]

A second approach to the peritrope takes up the distinction between knowledge and opinion. Self-refutation relies on the assumption that we know that X is "necessarily the case" rather than only "appears to be so." But if such certainty is unavailable (as is the case for Protagoras), we have no position from which to claim that X is determinate, that N must be wrong and, therefore, cannot be the measure of things. Alternatively, in arguing that what appears to be so is the case for A, Protagoras acknowledges the potential multiplicity of positions, opinions, *doxai* on matters of controversy. We might well argue that in collapsing the categories of knowledge and opinion Protagoras is offering an observational hypothesis (what Feyerabend calls a "rule of thumb"), an empirical assessment of the way knowledge-claims actually operate in the world (1989, 45). Regardless of this option, critics of Protagoras present human-measure as an epistemological theory that exalts opinion, abandons knowledge, and leads inevitably to naive pluralism—to a leveling of all positions to the same alethic status.

On this point, Protagoras does propose that what appears to be the case is so for the person who holds that view; but he also maintains that we can make distinctions between knowledge-claims that are stronger or weaker with regard to their utility; i.e., with regard to the advantage they hold for those who accept them (*Theaetetus* 172b). If, for example, it is possible for a doctor to convince a patient to substitute a weak *doxa* (this medicine tastes foul) for a strong one (it is good for you), then the substitution of one truth claim for another is in the patient's best interest. Again, we verge on issues of relativism which I will defer for a few more pages. But I would note here that from the Protagorean perspective the conflation of knowledge and opinion does not enfeeble our ability to make distinctions and argue for and against competing claims; it only denies the possibility of absolute claims. Indeed, when approached from a Protagorean perspective, the effort to claim absolute authority for a single logos becomes as much a political act as it is an alethic claim. Which brings us to a final and considerably more radical response to the peritrope.

To the master tradition, alethic claims are made relative to a fixed order; more specifically, the ability to identify an invariant order amidst the flux of things and to demonstrate it in the neutral metalanguage of logic allows, in turn, for the distribution of truth claims along a bivalent axis. The human-measure doctrine, however, challenges the notion that there are fixed grounds and methods for "matching truth claims to truth and falsity *tout court*" (Margolis 4). Instead, Protagorean theory argues (by implication) that the humanized nature of knowledge legitimates a wider conception of truth in response to the full scope of human experience. That is, multivalent truth values untethered to fixed and absolute principles become an alethic option, especially as these pertain to the knowledge required for ordinary living. The human-measure doctrine is not an appraisal of how things are in an essential and abstract form, but rather a "rule of thumb" for assessing how our experience of the world is perceived by individuals under actual conditions. In the *Protagoras*, the eponymous hero claims that in ordinary experience, goodness is "diverse and multiform" (334b). We can infer that truth itself is similarly multivalent. Porphyry (a 3C CE scholar) notes that in a lost text by Protagoras there are arguments against the notion that Being is one (*DK* 80 B2). It is consistent, then, with the Protagorean program to maintain that truth (as it can be grasped by human cognition) is "diverse and multiform" and to reject bivalence and the law of the excluded middle as inadequate alethic standards.

Admittedly, truth claims under the conditions of multivalence will not have the force and privilege of absolute authority. Ideas presented as consistent with fixed principles traditionally claim to be both compelling and sovereign, whereas a multiplicity of competing *doxai* represent the epistemic process as inherently unstable. On the other hand, the existence of multivalent truth values implies that incongruent truth claims are not necessarily antithetical or adversarial. Indeed, from a rhetorical perspective, multivalence is enabling: not only does a diversification of truth values authorize resistance to dogma and hegemony, it also invites the articulation of neglected truth claims and recognizes the partiality of any monological logos. The multivalence of truth also allows us to reply one last time to the peritrope and its argument that "true for A" is a universal claim and so self-refuting (Burnyeat 195). We can now reject the charge as one that compels a response in bivalent terms (X is/is not the case). Rather than repeat the reasoning of previous paragraphs, let me put the matter another way.

It is not clear from the Protagorean canon whether or not a multivalent alethic theory can be extended equally to all branches of knowledge (e.g., is it universal in the sense that it applies to statements about natural phenomena with the same force as it does to human behavior?). Just how far we should extend the notion of "true for A" is a knotty issue, one which I will address briefly towards the end of the chapter. For the present, we might recall, with Aristotle, that rhetoric deals with matters "that might be otherwise" rather than with what "must be the case" (cf. *Rhetoric* 1357a with *Post. Anal.* 71b). From a rhetorical and Protagorean perspective, truth values (as these appear in dialogic arguments rather than in formal propositions) are always multivalent and truth claims are always qualified by the position of the rhetor; the wine is not dry in some absolute sense, the wind is not cold in the abstract, the object is not white because it is predicated by universal whiteness. It is as it is because it is perceived to be that way by A or N, or you or me. Truth in argument, therefore, manifests itself along a spectrum; it is protean rather than bipolar, paratactic rather than propositional. To refuse self-refutation is to reject yes/no and either/or as alethic standards; to accept human-measure is to approach truth as a human construct and to allow for its expression as heteroglossia.

Having developed this "Protagorean" response to self-refutation, I trust I can abbreviate the discussion of non-contradiction without much loss. Along with the laws of identity (whatever is, is) and the excluded middle (everything must either be or not be), the law of non-contradiction is an axiom of formal logic accepted as a necessary condition for valid reasoning leading to demonstrative knowledge. You will recall that Aristotle formulates the law this way: "It is impossible that contrary attributes should belong at the same time to the same subject" (*Metaphysics* 1005b 26-27). The human-measure doctrine constitutes a clear denial of this law (the same wind can be cool for A, not cool for N), and so Protagoras is singled out for reproof (1007b 22, 1009a 22).[25] Plato develops the charge by noting that if all perceptions are "true for A," it is not possible to contradict (*ouk estin antilogein*; *Euthydemus* 285a-86b), and if contradiction is impossible, then we cannot identify falsehood (*Euthydemus* 285a-86b; cf. *Cratylus* 429c; see Brett 158-59, Maguire 136, Kerferd 1981, 88-95).

For the master tradition, the problem is that "true for A" is incompatible with determinate knowledge of the real world. The cherished notion of epistemological objectivity (of knowledge detached from personal orientation and the flux of experience) is thus at stake: Protagoras must be refuted if rational knowledge of the world is to be salvaged. In our own time, the concept of relativized knowledge confronts the scientific conviction that the objects of knowledge may be known as they exist independently of the knower and that a rational methodology is able to represent the real world in formal terms that are incontestable. The claim to certainty, of course, authorizes the elevation of apodeixis and mainstream science to the status of a paradigm for all knowledge. There is, then, a powerful motive behind the negative hermeneutic which would condemn human-measure as irrational and label it permanently as a logical fallacy. Indeed, the conflict between Protagoreanism and various forms of epistemological objectivism presents itself as more than an intellectual disagreement. The vigor, vituperation, and longevity of the debate indicate that the controversy evokes something on the order of a fundamental difference in temperament: the drive to know the world in some absolute, universal, and eternal sense as opposed to an embrace of multiplicity and ambiguity as primary constituents of human experience. Paul Feyerabend, an arch-Protagorean, indicates that these opposing temperaments are characterized by a desire for unity and convergence on the one hand, and a desire for diversity and proliferation on the other (in Rorty 1991, 27). Perhaps the prolonged debate between the two indicates that more is at stake than rational propositions.

Before I turn to Protagoras on his own terms, I would sum up where we now stand. In a nutshell, the charges against the human-measure doctrine are that it constitutes a theoretical paradox, a fallacy that entails its own negation and incapacitates the rational pursuit of objective knowledge. Such charges, as we now know, are framed according to the canons of formal logic. Once we refuse the related notions of a fixed, universal order, of bivalent truth values that follow from that order, and of the superordinate status of formal reasoning, the charges tend to dissipate and an alternative presents itself. According to this alternative, the human-measure doctrine asks us to accept hermeneutical diversity as the self-evident context within which all knowledge claims arise. Given this condition, the longing for perfect knowledge and precise statements about the true nature of things is itself irrational; it not only ignores the contingent and approximate nature of human cognition, it also refuses to acknowledge the legitimate claims of knowledge constructed through discursive exchange, claims which are no less binding because they are also provisional (cf. Bizzell 1990, 661).

Aristotle begins his *Metaphysics* with the assertion that "all [people] desire to know" (980a 22). If knowledge is constituted primarily by universals that have been isolated from the specious mediation of *doxa* and established as abstract forms of an invariant reality, then there is clearly a need to winnow the grain of certain truth from the chaff of possible falsehood. But if we accept the alternative notion that knowledge is constructed within the framework and according to the procedures of human cognition, then contradiction and collective exchange become the context within which we work towards better understanding. In the next section, I turn from the negative reconstruction of the human-measure doctrine to an ameliorative hermeneutic which attempts to assess how the Protagorean program enables this effort at better understanding.

3. PROTAGOREAN RELATIVISM AND THE HUMAN-MEASURE DOCTRINE

Arguments against relativism have an ancient pedigree; they also constitute a leitmotif, a persistent theme in the history of Western philosophy. Plato and Aristotle do not simply attack the human-measure doctrine as logically incoherent, they also seek in their own philosophies to offset the threat of relativism by positing a fixed, universal order which elevates true knowledge above the confusion of variable and conflicting opinion.[26] Descartes similarly rejects "false beliefs" and "doubtful opinions" in favor of "a firm and permanent structure" of universal and immutable ideas (*First Meditation*). Jeffery Bineham describes the Cartesian "anxiety" toward relativism this way:

> Either some ultimate ground for knowledge and action exists, some objective and ahistorical foundation against which claims to know can be measured and the utility of actions ascertained, or we are beset by relativistic skepticism and are unable to speak of knowledge or "justified" action in any meaningful sense. (1990, 44)

If the ancients assume a fixed metaphysical order, Descartes and his Enlightenment heirs rely on a body of "underlying structures," "invariant factors," "determinate patterns" (Rorty 1991, 22). In either case, traditional epistemology is predicated on the accessibility of ultimate reality and on ideas about the real and true which are fixed, eternal, and necessary. In the process, knowledge is insulated against the threat of variability in history, culture, and personal orientation; i.e., against the various forms of contextual and cognitive relativism. Due to the dominance of this tradition, the discussion of relativism has been consistently framed in the language of its opponents, in the context of a falling away from universal and determinate knowledge. Little wonder that Protagoras and the human-measure doctrine have so often been approached as *reductiones ad absurdum* and relativism itself censured as a form of philosophical abuse.

For example, Bacon, in the *Novum Organum* (1620), decries "the false assertion that the sense of man is the measure of things." He writes that "human understanding is like a false mirror which, receiving rays irregularly, distorts and discolours the nature of things by mingling its own nature with it" (1.41; in Bizzell and Herzberg 631). Berkeley, in the early 18C, acknowledges that our sense of the real is derived from perception; but he reiterates the Baconian notion that human perception is subject to distortion, and he appeals instead to a transcendental mind which apprehends the world in absolute, infallible terms (see *Three Dialogues* and *Principles of Human Understanding* 1.24). Kant also acknowledges that mental frameworks condition our representation of the objective world; but the rules that the mind itself obeys are fixed and necessary. Indeed, in *The Critique of Practical Reason* (1790), even moral law is objective and universal. Hegel, in his influential *Lectures on the History of Philosophy* (1830s) explicitly condemns the human-measure doctrine as the source of a destructive relativism which "plunged the Greek world into ruin" (262-63).[27] And Husserl (in the early 20C), while defending the law of contradiction, writes that the relativistic notion that truth is "one thing for one man and the opposite for another must be considered as the purest nonsense" (in Meiland 1977, 569).

In our own time, Karl Popper, in *The Open Society and Its Enemies* (1966), reduces relativism to the arbitrary choice between competing claims, a choice which is unresolvable because for the relativist there are no objective standards of judgment (369). And in 1990, Martha Nussbaum chastises the human-measure fragment as a "Protagorean farrago" based on the debilitating notion that "my-law-is-as-good-as-his" (221-22). So the long campaign and persistent *agon* with relativism continues, with each generation developing variations on a two-part critique: relativism is both a conceptual error which undermines its own premises and a philosophical heresy which destroys the basis of knowledge, undermines education, enfeebles ethics, and opens the floodgates of irrationalism, nihilism, skepticism, solipsism, and anarchy.[28] Clifford Geertz puts the case dramatically:

> What the anti-relativists, self-declared, want us to worry about, and worry about and worry about, as though our very souls depended upon it, is a kind of spiritual entropy, a heat death of the mind, in which everything is as significant, thus as insignificant, as everything else: anything goes, to each his own, you pays your money and you takes your choice, I know what I like, not in the south, *tout comprendre, c'est tout pardonner*. (15)

Of course there is reason to worry: relativism does seem to consign us to intolerable uncertainty, to an uneasiness about truth and to the loss of immutable standards which secure the authority of knowledge. Without such authority, what grounds are there for mounting compelling arguments and making justified choices? What hope is there for harmony when the world is structured so differently by so many? So we continue the search for a silver stake to silence the threat of relativism once and for all.

And yet the specter still walks; indeed, at perhaps no time since Protagoras himself has the defense of relativism been so vigorous. Margolis claims that "there has never been such a protagorean age as our own" (119; cf. Bernstein 1983, 13). Indeed, the catalog of contemporary assumptions favorable to relativism is extensive and would include the unavailability of absolute certainty (Heisenberg) and the undemonstrable nature of metaphysical order (Nietzsche); the prevalence of cultural diversity (Geertz) and variation among different "styles of reasoning" (Code); the changing horizons for knowledge dictated by historical placement (Gadamer, Toulmin); the deconstruction of the self (Foucault); the attack on scientism as a hegemonic paradigm for knowledge (Rorty); the appeal to patterns of reason not dependent on determinate knowledge (Blair and Johnson); and, finally, the belief in the value of norms and standards constructed in the process of social exchange (Scott). Consider also the late 20C interest in *petits recits* over metanarratives, in historicity, contingency, and relationship rather than totality, demonstration, and internal consistency. Consider finally the various levels of sympathy for relativism expressed by many modern thinkers, including Peirce and Dewey; Heidegger and Gadamer; Quine, Putnam, and Goodman; Kuhn and Feyerabend; Foucault and Derrida. In such a climate, there is no "good reason" to assume that relativism is necessarily abhorrent or that knowledge claims that are not absolute are therefore inadequate. We are poised, therefore, to respond to Protagoras in new ways, to reconsider him not as a patriarch but as the precursor of an enabling philosophical position that is surprisingly consistent with a good deal of contemporary thought.

In my response to the charges of logical incoherence, the perspective of Protagoras' opponents was admittedly in the ascendant. To borrow from Hume, at a certain point the master tradition appears to be "in possession of the throne," so that its opponents must "take shelter under her protection" and present their own arguments according to "the patent" of the sovereign.[29] Having responded to the charges of the last section, however, we are now prepared to offer a counterpoint to the negative interpretations of philosophical orthodoxy and present Protagorean relativism in positive terms as a theoretical construct which enfranchises multiplicity and foregrounds free discursive exchange. Regarding this effort, a couple of provisos are in order. First, what follows is not an exercise in historical reconstruction, an attempt to explicate Protagoras' position in the context of his own age. Such an effort is already well underway as part of the contemporary recuperation of Sophism; and while debate continues over Protagoras' stature in the philosophical tradition, readers have a growing list of studies which consider relativism in its own milieu.[30]

Instead, I would acknowledge my own position as a teacher of rhetoric and composition and as a student of contemporary argument. And, as I indicated in the Introduction, this book is addressed primarily to other teachers/students who, like me, are looking for an alternative to traditional forms of argumentation. Therefore, my own examination of the human-measure doctrine is conducted with present utility fully in mind: I would establish the basic Protagorean concepts that serve as the philosophical corollary of antilogical argument and lay out the rationale for a dialogical rhetoric appropriate to our time. Protagoras is a fascinating historical figure and his role in Western philosophical history is a matter of considerable importance, but my own interests are predominantly contemporary.

Second, in my effort to appraise the complex topic of philosophical relativism, I stand on the shoulders of many contemporaries (including

Gordon Kaufman, Joseph Margolis, Lorraine Code, Paul Feyerabend, Jack Meiland, Barry Brummett, Clifford Geertz, and others), all of whom have contributed to the reconstruction of relativism along positive lines. I do not claim to see any further than my sources; what follows in the rest of this chapter is an exercise in integration, a weaving together of threads into a sympathetic revision of Protagorean thought.[31] What I do add is a concern throughout for the effect of these philosophical ideas on rhetorical theory and the practice of argument. And I must admit that for me the view of discourse provided from a Protagorean prospect is expansive, even emancipating. We will turn more directly to rhetorical matters in the coming chapters; for now, it remains to round out this discussion of the human-measure fragment with an assessment of its relativism, an assessment that abandons the dismissive hermeneutic of the master tradition and the criteria of absolute and universal values, an assessment that privileges instead contingent positions, multiple options, unfettered exchange, and a place at the table for all, an assessment based on practical standards and the potential of Protagorean theory to function effectively in practical circumstances.

Our subject is potentially complicated by the many forms of relativism that have developed in the modern era. Feyerabend refers to "the wide distribution of relativisms" (1987, 20); Hillary Putnam speaks of the "plethora of relativistic doctrine being marketed today" (119); and Rom Harre and Michael Krausz distinguish such general categories as semantic, epistemic, ontological, and moral relativism (1996). As the source of subsequent variants, however, the human-measure doctrine stands both before and above such distinctions. When speaking of Protagoras, then, it is possible to describe his relativism in very general terms and to offer a basic exposition. According to the human-measure fragment, knowledge is relative to the knowing subject and to that subject's position in the world. Or, a bit more expansively: all cognitive representations of the world (all our thoughts) are the result of an immanent relationship between the knower and the world and in this relationship the cognitive operations of the knower are primary. That is, what we take to be real and true about the world is conditioned by the perspective of the individual. "Humanity [in the form of the individual percipient] is the measure of all things" because the personalized patterns of cognition are a "preconditon" for knowledge (Margolis 87).[32] In sum, what we know is relative to the way we measure.

We can approach the matter in a slightly different way if we refer to the already familiar distinction between relative and absolute knowledge claims. When the relativist offers a knowledge claim, she will be careful to add the proviso that "X" (the claim) is true "for A" (the person, group, or historical period from whose perspective the claim is valid). To assert that the truth-value of a claim is relative to the claimant (the wine is dry for A) not only allows us to acknowledge the role of the individual in what we now call the "construction of knowledge," it also supports the contention that what is true for one person, one period, one culture, is not necessarily true for another (a point we will return to shortly). On the other hand, when the relativistic qualifier (true *for A*) is dropped, the focus shifts from the knowing subject to the object of knowledge and, more specifically, to the correspondence between formal propositions and the state of affairs that the proposition would represent. As a statement about the world, "X is true" aspires to be both objective (representing the nature of things as they are in themselves) and abstract (removed from connections with specific persons). Jack Meiland points out that an absolute proposition presents a two-term relation: (1) a claim that a particular statement of the case corresponds unequivocally with (2) the true nature of things. In the process, however, such propositions also eliminate the crucial third term: for A (1977; see Code 1982).

Protagorean relativism stands in direct opposition to this effort to substitute the dyad of claim and object (X is true) for its own formula (X is true for A) because the former (which we observed in the *Theaetetus*) effaces the presence of the knowing subject from the production of knowledge. For Protagoras all knowledge is indelibly marked with human presence; consequently, the very notion of abstract knowledge (knowledge somehow separated from concrete persons, times, and places) is itself a paradox. At the core of the human-measure doctrine, then, are the twin notions that (1) all knowledge is relative to persons and that (2) the particular angle of a person's vision is implicit in and inseparable from that person's representation of the world itself. Admittedly, this description does little more than hint at the shape of a complex topic; but it does provide a starting point from which to explore the implications of Protagorean relativism for rhetorical studies.

For example, we can infer from Protagorean relativism that knowledge of this or that event, datum, experience, or idea will manifest itself in multiple forms and in abundant, subtle distinctions. The reasoning here is relatively direct: the human-measure doctrine argues that all knowledge is perspectival (knowledge for A), that it is enmeshed in and arises out of our subjective orientations (those systems of meaning and value by which we order the world). If we add that our subject positions are all uniquely configured because each of us is located somewhat differently in relation to a multitude of historical, social, and psychological factors, then it follows that each of us will understand the world in somewhat different ways. To wit, my understanding of this or that idea or event is bound to differ from the understanding of others because their subject positions will always differ (by varying degrees) from my own. Indeed, the potential differences in the construction of knowledge on a particular topic are theoretically inexhaustible. In itself, this idea is not particularly notable, but on its heels follow some compelling questions; viz., how are contradictory knowledge claims regarding a given topic related to one another? How is agreement possible under relativistic conditions? And how can we claim knowledge at all if truth itself is relative to a myriad of subjective frameworks?

I can begin to respond by offering some comments on the status of subjective knowledge. Discussions of the human-measure fragment inevitably come hand-in-hand with concerns about the radical subjectivity of Protagorean doctrine, about the isolation of individuals in a cocoon of private cognition and the subsequent incompatibility of diverse knowledge claims.[33] In response to this concern, it is crucial to keep in mind the Protagorean notion that while all beliefs are true for those who hold them, not all such beliefs are equal; there are inevitable distinctions between stronger and weaker logoi, between knowledge with normative authority and eccentric ideas that are valid *only* for A. It is the responsibility of wisdom, says Protagoras, to distinguish between sound and unsound claims and to prompt a change from the latter to the former (see *Theaetetus* 166c-167d). Gordon Kaufman, writing in defense of relativism, puts the matter this way:

> It is never valid to regard an opinion as true merely because it emerged in one's experience and thus is pragmatically related to one's own purposes; anyone can make this kind of claim for any belief whatsoever. In pursuit of truth, it is not mere subjective certitude that we seek . . . . (67)

There is a world of difference, then, between the pragmatic recognition that knowledge arises out of the subjective orientation of the knowing subject and the naive conviction that "it's all how you look at it" and "everyone's right in their own house."[34] As Rorty indicates, "(e)xcept for the occasional cooperative freshman," there are very few advocates of naive relativism, of a relativism which would guard the infallibility of every knowing subject in every instance (1982, 166). The Protagorean would point out that not only are some logoi more advantageous and enabling than others, but that being a relativist does not require one to sacrifice all standards of judgment. On the contrary, by identifying the perspectival origins of all claims, relativism sanctions our ability to move about among competing logoi, to consider the merits of all by recourse to dialogical exchange, and to argue in favor of one by appeal to *inter-subjective agreement*. So while relativism does indeed conceive of the universe of knowledge as marked by subjectivity, multiplicity, and flux, such subjectivity is not atomistic, with each position possessing absolute boundaries that mark it off in solipsistic isolation from others. That is, knowledge claims, while issuing from distinct subjective positions, arise within a network of relations and mutual practices that counter-balance the potential for decline into private eccentricity.

We can fill out this line of thought briefly because the basic ideas are familiar to modern readers. Protagoras argues that "humanity is the measure" and that "each of us [has] peculiar perceptions" (*Theaetetus* 166c). I have translated these dicta to mean that knowledge is relative to our subject positions and that we all "think and value" differently from one another. To this basic construct, we can now add that knowledge is not simply relative to individuals because, despite the uniqueness of our positioning, the "systems of meaning and value" that constitute our personal orientations have not developed *sui generis*. Rather, the cognitive practices of the knowing subject are formulated in and shaped by the complex network of relations (historical, familial, cultural, political, economic) that make up our social environment. Kaufman writes that "thought emerges from and is guided by the mass of distinctions and relations" conventional in and made available through one's culture (89); and he cites Ortega y Gasset on this same point:

> . . . my intellectual personality partakes of the diffuse collective life that is the human species up to the present moment. The intellect of the individual is not individual in the sense that it is free to forge its ideas from naught; rather it is, from the outset, shaped by the heritage of the historical collectivity. (50)

Kaufman argues that chief among the influences on the personal orientation of the individual is language, which he describes as

> . . . a vast network of interpretive categories which exists prior to the individual, which is a crucial factor in the coming to self-consciousness of the individual, and which, upon reaching consciousness, the individual uses as the primary set of distinctions and definitions in terms of which he approaches and understands all his experience. (51)

In sum, thinking is relative not simply to the thinker but to the complex social and linguistic systems that contribute to making the thinker who she or he is. Which is not to say that the knowing subject is *determined* by the relations and practices of her social environment; rather, we can argue (with Lorraine Code) that while we rely upon the customs, histories, and knowledge of our various communities, "these factors constitute the stuff out of which knowers, as creators of meaning and as agents, must construct their [own] meanings, purposes, and actions" (1991, 269). The knowing subject, then, is really a social subject whose independent efforts at knowledge construction are always mediated by her relationships with others. In turn, our reliance on shared systems of knowledge allows us to understand how and why, despite the unique location of each knower, inter-subjective agreement is indeed possible and communication is so often successful.

The social grounding of knowledge was not, of course, unknown to the ancients. The basic idea is hinted at in Pindar's aphorism that "custom is king." A generation later, Herodotus (who is contemporary with Protagoras) notes in his *Histories* that customs and values are constructed according to the distinct perspective of a particular group (3.38). Protagoras himself develops the notion in what is often called "The Great Speech" (referred to here as "The Great Myth"), an extended speech in Plato's eponymous dialogue which argues that the *polis* acknowledges certain normative beliefs that have been "distributed to all alike" through "practice and instruction" (*Protagoras* 323d-324e). Among these beliefs is the concept of justice (*dike*) which Protagoras elsewhere indicates is relative to the needs and circumstances of the community (*Theaetetus* 172b). For Protagoras and other Sophists, then, knowledge is relative not only to the "peculiarities" of the knowing subject but also to the socially-constructed frameworks and practices that we all inherit as members of a community.[35]

However, it is not until the advent of historical studies in the 19C that the social component of relativism comes to the foreground. Beginning with Hegel and Dilthey, later in Collingwood and others, studies in the philosophy of history have prompted a recognition that diverse conceptions of truth and reality are equally valid in their own domains and that such conceptions are best understood in relation to the historical situation and epistemological norms of a particular culture.[36] As we move, for example, from Medieval Christianity to Mahayana Buddhism, from Renaissance humanism to logical positivism, standards of meaning and value obviously differ: *autres temps, autre mœurs*. But according to the canons of judgment at work *within a particular culture*, normative claims to knowledge and value are likely to seem both consistent and correct. From the historical perspective, then, knowledge is always knowledge *in situ*; and since cultural situations vary so widely, we are faced with a panorama of epistemological standpoints and alethic positions, all of which claim validity in their own domain. With the advent of historicism, therefore, we return to the topic of epistemic multiplicity, to the theory that there are many ways of knowing, ways which vary with epoch and place as well as with persons. This time, however, the topic is cast in different terms: relativism in its historical form sheds its outdated image as the scapegoat of the master tradition and takes on new force as a framework for understanding and responding to crucial issues of cultural diversity.

As a central feature of contemporary relativism, the historical perspective directs our attention to empirical data that characterize the life and times of a given culture. History's insistence on the temporal and concrete differs notably from the emphasis of universalist philosophies on concepts that apply across synchronic and diachronic boundaries, concepts such as the fundamental nature of humanity, the formal structure of reason, or the fixed patterns of nature. Historical inquiry, on the other hand, is distinguished by its interest in the details and contingencies of contextualized experience (see Rosen 333). In particular, the historical perspective is marked by its attention to the role of the agent in the life of a particular culture. Collingwood writes in *The Idea of History* that the object of historical inquiry is the *res gestae*, the actions of human beings at work in the past (9). Accordingly, history is typically an *ad hominem* argument, an effort to plot the ways in which particular acts of human agency are reflected in the unique development of a culture and its various traditions. We can, therefore, reformulate Macaulay's famous aphorism that history is "philosophy teaching by examples" to indicate that the emphasis of modern historical analysis is clearly on the examples themselves (rather than on immutable ideas), on examining and interpreting the persons and data of the historical moment and their particular roles in a culture's process of development. And—of most importance to my line of thought—because of its attention to the concrete particulars of human experience, historicism also encourages a recognition of the considerable variation in the ways we understand the world, variation which is unmistakable as the student of history moves from culture to culture and era to era.

Cultural pluralism, therefore, becomes a primary tenet in the doctrine of a renewed Protagoreanism. As we know, for the relativist all knowledge emerges from and responds to the epistemological standards and practices of one's time and place.[37] Implicit here is the original Protagorean notion that what is known is mediated by the structuring procedures of the knowing subject. But the manner in which people orient themselves in the world varies under the influence of evolving historical conditions. Protagoras, who had little access to historical variability, would reasonably see synchronic differences in the way the world is "measured" by different individuals. For the historical relativist, however, knowledge emerges in a wider diachronic field; for us to think "relatively" is to challenge the notion that knowledge exists somehow independently of its cultural context and to insist instead on the relation between the knowing subject and the local (rather than universal) standards of reason and judgment at work in the historical moment.[38]

Moreover, historical relativism extends beyond the common-sense appraisal of cultural differences to include a distinct evaluative component in response to the phenomenon of historical/cultural diversity. According to Protagoras, justice appears in different incarnations relative to the norms and histories of particular groups. For the contemporary Protagorean, conceptions of truth and reality are similarly relative to and at least provisionally binding within a particular tradition. That is, given the fact of historical/cultural pluralism, standards of evaluation are best articulated in accordance with the cultural traditions involved. Or, put yet another way, understanding the claims made by members of another culture begins by recognizing the standards of thinking and value at play within and around the knowing subject.

Also implied here is the notion that there is no neutral standard, no pan-cultural perspective, no framework independent of all cultural-historical settings; there are only other frameworks generated under different historical conditions, frameworks with their own unique orientations, horizons, and prejudices (see Brett 144). As Montaigne puts it: "Each man calls barbarism whatever is not his own practice; for indeed it seems we have no other test of truth and reason than the example and pattern of the opinions and customs of the country we live in" (152). As a critical perspective, historicism recognizes the threat of parochialism and insularity expressed by Montaigne; and in its inherent pluralism, the historical perspective contradicts the myopic assumption that the norms of one perspective, one location in history are somehow paradigmatic and so should apply automatically in other historical and cultural contexts. More specifically, historicism recognizes that the thoughts and beliefs of other peoples, places, and times are not an underdetermined space waiting to be inscribed with the insignia of our own way of knowing. What Clifford Geertz calls the "news from elsewhere" is this: though there may well be certain commonalities in the mental lives of various cultures, differences in the way we think and value are ineluctable (1989). Consequently, inquiry in the human sciences conducted under the aegis of history will begin with an expectation of multiplicity.

On this point, contemporary philosophers routinely cite the history of scientific knowledge in support of the notion that knowledge itself is historical, contextual, and mutable: Rorty notes that what is "reasonable" for scientists to believe at one historical moment may not be so at a later age simply because science is constantly in the process of superseding itself (1987); Code writes that if we deny the status of "knowledge" to scientific truths of previous eras, we are "in a difficult position where present knowledge is concerned" (1982, 173; see also Bernstein 1982, Feyerabend 1987, Kuhn). It follows that the application of epistemological/hermeneutic standards derived in one setting to knowledge claims generated in another historical or cultural context requires considerable caution and care. To apply an incompatible standard is too often an attempt to subjugate, to adopt the imperial posture and seek to colonize the "other."

As you might suspect, then, the argument of contemporary relativism is political as well as historical: it asks us to recognize the potential for cultural collision that is particularly acute in our own historical milieu, and it prepares us to negotiate the interactions between and among disparate traditions in useful new ways. What are these new ways? As you might also expect, this book will argue that a reconceived approach to argumentation is the most "useful" response to the challenge of cultural pluralism. After all, argumentation is a *praxis* designed to address difference and conflict. But before we turn directly to rhetoric in general and antilogic in particular, there remain a few last strands to weave into this review of Protagorean relativism.

4. SOME IMPLICATIONS OF PROTAGOREAN PHILOSOPHY FOR RHETORIC

There are important questions about our topic that I have not yet addressed; and while a detailed response to these would take us too far afield, several matters deserve at least brief review. In the last pages of this chapter, I will take up an assortment of issues related to ancient and contemporary relativism with a special interest in their implications for rhetoric and argumentation. To begin, let's take stock: for Protagoras, the knowing subject is the standard or criterion by which knowledge is measured (*DK* 80 A14). To this initial formulation we have added the contemporary caveat that subjects are always located in and responsive to concrete socio-historical conditions, so that thinking emerges in relation to particular norms of truth and value. Such a construct asks us to surrender the notion that knowledge claims correspond directly to an unmediated reality; but to what extent is this surrender synonymous with scepticism, with a general uncertainty about the truth-value of all claims? Sextus Empiricus, a key figure in the ancient discussion of scepticism, writes that for the sceptic "(e)verything is indeterminate [and] inapprehensible," so that nothing can be known for sure (*Outline of Scepticism/Pyrrhonism*, sec. 198-200). But while this condition of intellectual uncertainty (and discomfort) may describe Plato's characterization of Protagoras or may even fit Gorgias, scepticism does not seem to me to be inherent in Protagorean philosophy itself.[39] The unavailability of universally-distributed knowledge claims does not invalidate knowledge itself; it merely places knowledge in relation to the individual and to the orientation, perspective, or framework for interpretation that the individual brings to bear in the course of judgment. There is, admittedly, a certain diffidence in the Protagorean approach to truth claims because—as the antilogical principle has it—all claims are subject to debate. But for Protagoras, debate itself is an affirmative act, and the well-conducted argument is sufficient ground for binding judgments and firm commitments.[40]

But if relativism is not inherently sceptical (in the sense that it doubts our ability to achieve knowledge), one might still wonder about its putative subjectivism. After all, the proposition that something is "true for A" has suggested to many that consciousness is a closed space and that knowledge, in turn, is confined entirely within this subjective cell. Such a conclusion appears to follow from the notion (attributed above to Nietzsche) that we are not in direct contact with the actual stuff of the world, but that what exists in consciousness is always already marked by cognition, already structured in relation to the other contents of consciousness, always the stuff of mind only. What we think about when we think about X (as in "X is true for N") is not the world itself, but our own unique representation of it. Under such conditions, it is only reasonable to ask if relativism is synonymous with a thorough-going subjectivism. Or, alternatively, does it entail anti-realism, the rejection of the external and objective as categories relevant to human cognition?

As a philosophical position, radical subjectivism (or solipsism, from *solus*, alone, and *ipse*, self) would indeed seem to restrict all knowledge claims to the realm of personal consciousness. Descartes, for example, begins his *Meditations* by doubting the existence of natural phenomena (*res extensa*) as anything but the inventions of mind itself (*res cogitans*). But solipsism's prison-house of consciousness is not (despite the claim of some powerful voices) necessarily synonymous with the kind of relativism authorized by Protagoras himself or by modern Protagoreans.[41] Lorraine Code, for example, argues that while each person knows "a somewhat differently structured world . . . the raw material which provides the data for all of these structurings is the same, independently existing, real world." She goes on to say that this raw material "imposes checks upon what can be observed"; consequently, cognitive orientations "which do not demonstrate correspondence to the point of being usable, of allowing the one who holds them to act in the world and to make sense of his or her experience of it, would clearly collapse" (1991, 168). Such a position is compatible with Protagoras' notion that there are qualitative differences in knowledge and that wisdom is manifest in knowledge that is useful or advantageous, knowledge that "acts effectively in the world." Accordingly, it is the successful interaction of individuals with the world, not the Cartesian isolation of the former from the latter, that is the defining feature of the Protagorean subject.[42]

This pragmatic emphasis on the responsiveness of the subject to the flow of things works to subvert the specious separation of the knowing subject from the materiality of the world, even while this dichotomy remains—as Donald Davidson reminds us—deeply embedded in our language and thinking (see "On the Very Idea of a Conceptual Scheme"). For the pragmatist and the Protagorean alike, the knowing subject is not a passive spectator who merely registers the parade of passing impressions upon some eclectic personal apparatus; instead, the subject becomes an active participant who continually responds to encounters with the world of objects and events by commingling the overlapping spheres of agent and other "*within* the aegis" (as Dewey puts it) of particular human experience (see 1970, 23-29, 44-47).[43]

But if the Protagorean subject is in constant interaction with the world, it is nonetheless the case for Protagoras that the world we know is primarily constituted through the cognitive practices of the same human subject.[44] This emphasis on subjectivity and the power of human agency leads us, by implication, to questions of communication; e.g., how do we overcome our unavoidable differences in subjective orientation and communicate effectively, to what extent are these differences (in)commensurable, and how do we evaluate rival perspectives? These matters are the specific subject of the next few chapters; but by way of a temporary response, let me first rehearse what has already been said about intersubjectivity. According to the Protagorean view (see "The Great Myth," *Protagoras* 320c-323a), we do not enter communicative encounters from the sidelines; i.e., with a lack of communal experience and social knowledge. Rather, we share with other members of our various communities certain contours in the shape of our orientations, certain attitudes, ideas, assumptions, language, mores, and practices which overlap and so provide a common ground for the collective scrutiny of experience and the building of shared interpretations. At the very least, many of our ideas, beliefs, and opinions bear a family resemblance with ideas held by others and so are open to comparison, despite attendant differences. The possibility of comparison between/among disparate items plays a key role in building intersubjective agreement and, as such, merits a bit more discussion.[45]

Teachers of composition know that the *topos* of comparison is inseparable from that of contrast. *Comparison* asks us not simply to take note of similarities but also to recognize differences among compared items. It consequently calls attention to the distinctiveness of each unit, to its separateness from others, and to the overall variety of potential forms. Comparability, therefore, should be distinguished from *compatibility*, insofar as the latter connotes the orderly integration of items within a system (cf. *American Heritage Dictionary*). Comparison requires a point-by-point assessment of related elements from two or more discrete entities; compatibility, on the other hand, assumes a neutral framework or order within which distinct items can be integrated. To be incompatible is to be in conflict with some sort of generic standard. This distinction may help us to come to grips with the notoriously slippery notion of *(in)commensurability*.[46]

For if incommensurability is defined as "lacking a common standard on which to base a comparison," it is essentially synonymous with incompatibility and the appeal to universals (*American Heritage Dictionary*). But the issue is more complicated.

Incommensurability has been the subject of heated debate by philosophers since it was employed by Kuhn in his discussion of the relation among diverse scientific paradigms: "what differentiated these various schools [of science] was not one or another failure of method—they were all 'scientific'—but what we shall come to call their incommensurable ways of seeing the world and practicing science in it" (4; cf. 103 and 148-50).[47] For some of Kuhn's critics, however, incommensurability implies the inability to communicate across boundaries and is the price we pay for the surrender of universal standards. To abandon "neutral" or objective standards is to accept the notion that the disparate paradigms of experimental science and tribal magic (both of which attempt to comprehend and control the natural world) possess inherent standards and may each be valid relative to those standards. For the anti-relativist, such a position replaces the search for Truth with an ascent on Babel in which the cacophony of incommensurables reduces all to irrationalism. On the other hand, Richard Bernstein argues that what Kuhn is actually concerned with is the difference between compatibility and commensuration (1983, 82ff). According to the former, we can only make sense of two dissimilar paradigms by reference to "something permanent and determinate that stays the same in all such cross-paradigmatic comparison" (85). But from the relativist perspective, compatibility is not the only standard by which we determine what makes sense. For Protagoras—and for Kuhn (as I read him)—commensurable ideas need only be comparable, and since there is typically some point of overlap in our concepts, standards, problems, methods, it is possible for us examine the relationship between discrete items by means of individual distinctions, by careful comparison and contrast rather than by logical appeal to an overarching framework.[48]

For the rhetorician, commensurability based on comparison provides a framework for making sense of intersubjective communication. Despite wide variations in subjective orientations, it remains possible for practical judgment to identify points of comparison among seemingly incommensurate or incongruent claims. The fact that individuals are routinely at variance with one another, that we approach problems from such different interpretive angles does not in itself diminish the possibility of comparison leading to cooperative understanding and action. Indeed, we might well claim that the goal of argument—at least argument in its antilogical form—is to confront the specter of Babel and contribute to the *sensus communis* by incorporating opposition into the formation of knowledge and the process of judgment. In other words, antilogic aspires to "a productive encounter with difference" by approaching difference itself as a source of discovery (Dassenbrock 247). In this context, intersubjective agreement is the basis of knowledge and is secured through rhetoric, through our ability to compare differences and communicate meanings.

A final proviso remains. Faced with the enormous scope and variety of contemporary research and scholarship, the modern Protagorean does well to acknowledge that relativism is not equally at play in all sectors of inquiry (Margolis 11-13; Code 1982). For example, in the physical sciences precise measurement of empirical phenomena can produce propositions of limited ambiguity (water boils at 100° centigrade at sea level). In such circumstances, the constitutive power of subjectivity is tempered by the determinate nature of the materials out of which claims are constructed. In human relations and cultural matters, however, the data to be had are inherently ambiguous and thus alterable according to the interpretive practice of the subject. Since there will always be variation in these interpretations, what counts as knowledge in human affairs follows from some form of agreement (what Rorty calls "solidarity") rather than from a putative correspondence with the "real". There are, of course, precedents for this kind of categorical division; e.g., in Aristotle's separation of *apodeixis* (scientific demonstration) from *phronesis* (practical reasoning) and in the 19C German distinction between the degree of certainty available to *Naturwissenshaften* (natural sciences) and *Geistewissenchaften* (the historical or human sciences).[49] In our own time, the division of knowledge takes form most often as a bifurcation into "social" and "technical" domains (see Farrell 1976).

All such distinctions recognize that standards of knowledge are field-dependent. At the same time, they allow us to address the extent to which relativism obtains in particular sectors of inquiry. For example, we can say that technical knowledge (i.e., the product of inquiry in science, technology, and the skilled crafts) seeks to minimize its reliance on the knowing subject by adopting strict inquiry procedures and a neutral metalanguage (preferably mathematical) for communicating results. The assumption here is that the "truth" about the object of investigation lies in an invariant set of principles/relations which are embodied within the phenomena themselves. Careful inquiry procedures seek to quantify these factors, with the resulting data providing a warrant for propositions and predictions about the nature and function of the object of study.[50] In brief, technical knowledge aspires to comprehend a state of affairs outside and beyond the influence of the knowing subject. And yet, the objectivity of technical knowledge—its avowed independence from time, place, and persons—is obviously a matter of dispute. Once discrete data are set within meaningful relationships, they are re-humanized, re-embodied in a framework of human construction, and these frameworks or paradigms are always relative to the people who develop and use them. Hence, I. A. Richards points out that "[t]he world—so far from being a solid matter of fact—is rather a fabric of conventions" (36). R. G. Collingwood is even more to the point:

> The natural world which human scientists can study by observation and experiment is an anthropocentric world; it consists only of those natural processes whose time-phase and space-range are within the limits of our observation. (in Code 1982, 168; cf. n. 20)

The crux of the issue for us is not the substantiality of an independently existing material world; rather, I would argue that while all knowledge involves the contribution of the knowing subject, technical inquiry works assiduously to reduce the potential for subjectivity to intervene and alter the nature of the object of study (Farrell 1976, 611). Relativism is thus operating in circumscribed fashion, though it is never reduced to zero degree.

With regard to social knowledge, however, it is hardly contestable today to maintain that the form and content of what is known in this domain is thoroughly influenced by the place, purpose, and process of the knower. Given the familiarity of this position, perhaps I can say something new about the topic if I shift terms and adopt Lorraine Code's distinction between "knowing that" and "knowing about" (1982, 158-60; cf. 1991, 37f). Roughly analogous to technical knowledge, "knowing that" confers conceptual priority on the object of study and its ontological status; it is focused on the substantive, the nominal, the true nature of X as X exists in itself and outside history. Conversely, "knowing about" acknowledges the hermeneutical process of the knowing subject; it is concerned with human agency, human language, and the fabric of history, motives, and implications that encompass and connect human events. I "know that" the United States Senate voted to acquit President Clinton of the two impeachment charges in votes of 45-55 and 50-50 on February 12, 1999. But I "know about" the event in a very different way, a way undoubtedly distinct from the way that you have made sense of the same complex experience. As noted, Paul Feyerabend makes a useful division between (a) knowledge that seeks "convergence" in the uniform and certain and (b) knowledge which runs to "proliferation," to multiplicity and ambiguity (1978, 9-10). There is, of course, a rhetoric of the first category, a discourse practice for articulating claims to convergence; but rhetorical theory has traditionally dealt with knowledge of the second kind, "knowledge about" the contingent, the multifaceted, the ambiguous and mutable. Such knowledge is more relative than its technical counterpart. Moreover, as Feyerabend points out, it is this kind of knowledge that interested Protagoras and to which the human-measure fragment is addressed.[51]

In the next few chapters, I will concentrate almost exclusively on antilogic, its various features and its unique contribution to argumentative practice. Readers primarily interested in rhetorical matters may feel that the philosophical survey of this chapter has delayed a bit too long this turn to the central topic of *Many Sides*. The guiding assumption of this chapter, however, is that antilogic is theoretically of-a-piece with the philosophy of the human-measure doctrine and that a careful appraisal of the former is requisite if we are to understand "the full resonance and promise" of the latter. Hopefully, my philosophical précis will provide the vantage required for a comprehensive view of the subject. It is at this point in our inquiry that attention shifts decisively to the rhetorical; so it is perhaps appropriate to begin this transition and close out the chapter with a brief reprise of the common framework which supports both the philosophical and rhetorical extensions of the Protagorean program.

So, this is where we are: the human-measure fragment outlines or adumbrates a common-sense hypothesis on the way human knowledge actually operates in the world. It argues that what counts as knowledge for human beings follows from the cognitive practice of knowing subjects, that culture and history are immanent in this practice, and that, consequently, there are a great many ways of knowing or "measuring" our experience of things. As a result, what is "true" for A is not necessarily so for N, nor is the opinion of one necessarily false if the two happen to differ. Such a position not only acknowledges that the world is multifaceted and dynamic, it also calls into question any dogmatic move toward singularity or totality since the claim of supremacy invariably depends upon obscuring and overcoming difference.

Under these conditions of inevitable multiplicity, meanings and standards are appropriately established through intersubjective exchange, an exchange managed in large part through the offices of rhetoric. Argument, in particular, plays a crucial role in the process of symbolic exchange through its efforts to transform the "clash of adverse opinions" into cooperatively-derived principles and standards (Mill). For the Protagorean, these principles of belief and standards of judgment are no less valid or binding because they are arrived at through argumentative exchange rather than deduced from universals. To be a relativist, then, is not to be bereft of standards; it is a commitment to standards won through exchange and reviewed as needed. Socrates says of Protagoras that the Sophist believes that "the just and unjust . . . are not things that exist by nature, with their own essential beings, but that the public decision becomes true at the moment when it is made and for such time as the decision stands" (*Theaetetus*

172b).[52] Moreover, Protagoras routinely represents the argument of opposing beliefs as an opportunity to identify the social good rather than as an opening for personal preference (cf. *Theaetetus* 167a). In its Protagorean form, then, relativism is neither the recipe for chaos nor the surrender of standards that its detractors claim; instead, it is an enabling doctrine that calls for direct democracy in the construction of meaning. It is functionally synonymous with the practice of argument and the effort to posit standards and justify belief under conditions of multiplicity and change.

If the philosopher is a lover of great thoughts, the rhetorician is passionate about the thinking process itself. In Protagoras, the two roles are inseparable: in order to find "the one set of thoughts [that are] better than the other," one must be willing to argue (*Theaetetus* 167b). And, in a world without universal guarantees, everything must be argued. The road to wisdom, as it turns out, is mapped by the discourse of opposing logoi. Just how Protagoras thought we might best proceed along this road is the subject of the next chapters.

Chapter 2

Protagorean Practice and the Nature of Antilogic

> *"For this is a very true presupposition: that men are in agreement about nothing, I mean even the most gifted and ablest scholars, not even that the sky is over our head."*
>
> *"So that, since equal reasons are found on both sides of the same subject, it may be the easier to suspend judgment on each side."*
>
> Montaigne, "Apology for Raymond Sebond"

In *The Romance of the Rose*, Umberto Eco imagines finding a lost part of Aristotle's *Poetics*, a treatise on comedy with momentous implications for what we value in art and culture. The recovery of the book that begins with the human-measure fragment would be no less momentous. After all, the first sentence alone initiates a radical departure from the canonical notion that truth and knowledge are by nature universal and invariant; while the alternative titles for the full text—*Truth* and *Refutations*—pose the intriguing possibility that the former is somehow congruent with the latter (*DK* 80 B1). As we actually have them, however, the Protagorean fragments are not much more than epigrams; they suggest rather than define, but they suggest much. Early in the 20C, the British pragmatist F. C. S. Schiller argued that the fragments "compress the largest quantum of vital meaning into the most compact form" (1903, 33). Compact, provocative, and open to elaboration.[1] Historical reconstruction must, of course, concentrate on what is given in the historical record; but the persistent vitality of the Protagorean fragments also invites us to contemplate what was lost, to fill in the enormous lacunae according to the light of our own time and place (cf. Schiappa 64-85). In other words, the extant form of the human-measure doctrine operates as what Rorty calls a philosophy of "edification," an enabling source whose emergent theory is continually enriched by new contributions (1979, 357-60).

My own response to this enabling doctrine has been to cast the human-measure fragment as a theory of knowledge in which the knower and the known exist in symbiotic relation. That is, what counts as knowledge is contingent upon the interpretive processes of the knowing subject ("humanity is the measure"), while, in turn, the interpretations of the subject are themselves informed and corrected by the stuff of the world and the shared views of others. More simply, knowledge claims are relative to persons, and persons are "placed" in history and culture. It is but a short step from this relativist emphasis on personal knowledge to the recognition that, in any collection of individuals, divergent perspectives on the same topic are bound to follow from the more-or-less different positions that individuals inhabit. In other words, when the human-measure doctrine is extended beyond the single subject and distributed throughout the inherently diverse human community, what counts as knowledge will inevitably be attended by multiple and contradictory views (*multiplex ratio disputandi*). Sharon Crowley neatly summarizes the Protagorean perspective on social knowledge when she notes that "disagreement is endemic to the human situation" (i). Nor can our disagreements be easily dismissed; for under Protagorean conditions, contradiction will not be resolved by traditional appeals to either individual authority or universal principles. Instead, with the acknowledgement that there will always be a diversity of views on any given topic, disagreement becomes the substratum of human reasoning, the material with which thinking must grapple and out of which knowledge is constructed.

To put the case yet another way, a particular idea or logos may be relatively true (true for A), but for it or any logos to have currency in the world (i.e., to be valid for others as well), it must be examined alongside alternative positions, the antilogoi which naturally arise in the process of open, social exchange. Only by comparing the widest variety of positions and counter-pointing each against objections raised by another can we arrive at ideas agreed upon as sound. Such, in short, is the epistemological narrative of the human-measure doctrine: knowledge (the measurement of things) begins with the unique perception of the knowing subject, but the acceptability of any proposition offered to others depends upon the conduct of argument and, in particular, the collation and comparison of opposing views. Protagorean doctrine, therefore, theorizes argument as central to the prospects of human knowledge and posits antilogic—or the juxtaposition of opposing logoi—as argument's appropriate protocol. As it turns out, the search for *truth* and the act of *refutation* are parts of the same process.

That Protagoras accorded argument a central role in his own life and thought is corroborated by various historical details. Diogenes Laertius tells us that Protagoras wrote a book called *The Art of Debating* (now lost) and that he was the first person to conduct the public debates that became so popular in 5C Athens and were so closely connected with Sophism itself (*DK* 80 A1; see *Protagoras* 335a). We also believe that he employed yet another text, titled *Contrary Arguments,* to teach his students how to speak on both sides of a case (Guthrie 3.181-82). He may even have been the originator of the dialectical method of reasoning by "catechism of question and answer" that we now associate with Socrates (Kerferd 1981, 60). And, in Plutarch's "Pericles," we hear that the great Athenian leader once spent the whole day with Protagoras debating "the most correct judgment" in a complex case of accidental homicide (*DK* 80 A10). So, even if we discount some of the biographical record regarding Protagoras as hazy and undependable, we can still with some confidence connect Protagoras with the theoretical origins of argument theory and also with the actual practice of *antilogike techne*, the art of antilogic.

G. B. Kerferd, whose commentary on antilogic has become the critical standard, refers to *antilogike techne* as "the most characteristic feature of the thought of the whole Sophistic period" (1981, 85). And yet, while the Sophistic tradition has been enthusiastically reclaimed by a continually growing number of modern scholars, antilogic, though occasionally referred to, remains for the most part undeveloped. My own intention is to extend the analysis of antilogic beyond the abbreviated comments of Kerferd, Schiappa, de Romilly, and others, and to reclaim its distinctive practice not simply as a neglected feature of the neo-Sophistic revival but also as an instructive precedent for a dialogical approach to contemporary argumentation. The dialogics of antilogic is a topic that will emerge in the course of this book, as we move from its origins in Protagorean theory, to its refinement by Cicero and its pedagogy in Quintilian.[2] But the full resonance of this alternative tradition of argumentation resides ultimately in the relation between antilogical practice and the theoretical/epistemological matrix with which it was originally connected. For that reason, I rely, as noted, on the vocabulary of Protagorean relativism developed in Chapter 1 as a vehicle for articulating what I believe has been left unsaid about the art of antilogic. Put another way, the living doctrine of the human-measure statement provides an "edifying" framework for exploring the rhetorical implications of the two-logoi fragment, implications that have remained for the most part inchoate because they have been disconnected from their theoretical base. The goal of this chapter and the next two, then, is to reclaim antilogic as the pragmatic extension of the human-measure doctrine and to employ this connection as an aid in developing the neglected potential of antilogical practice.

The relationship between the human-measure doctrine and the art of antilogic has not gone unnoticed. At the turn of the last century, Schiller, an enthusiastic supporter of Protagoras, devoted much of his *Essays on Humanism* (1903) to an interpretation of the human-measure fragment as an "empirical" account not only of the relativism of individual perception but also of the conceptual "ambiguity" that attends the inevitable diversity of personal judgments (35). Schiller argues that Protagoras meets this threat of epistemological ambiguity by invoking the offices of rhetoric as the means by which we make "*distinctions of value* among individual perceptions" and, in the process, transform "the mass of subjective judgments" into "collective assessments arrived at in society at-large."[3] Similarly, in the 1960s, Laszlo Versenyi argues that the philosophical insight of *homo mensura* leads directly to "the formal invention" of antilogic. In his words, "antilogical arguments are rooted in and exhibit Protagorean relativism" (1963, 21-22). More recently, Stanley Fish writes that for Protagoras and the Sophists truth is "a contingent affair [which] assumes a different shape in light of different local urgencies and the convictions associated with them." In this context, rhetoric moves

> from the disreputable periphery to the necessary center: for if the highest truth for any man is what he believes it to be (*Theaetetus* 152a), the skill which produces belief and therefore establishes what, in a particular time and particular place, is true, is the skill essential to the building and maintaining of a civilized society. (1989, 480-81)

It is rhetoric, then, that negotiates the passage from "endemic" difference to shared belief, though Fish does not seem to recognize the specific role of antilogic to this process.

There are, of course, others who have commented in passing on the connection between Protagorean philosophy and antilogical practice.[4] But this commentary seldom pauses to consider Protagorean relativism in significant detail or to fully contemplate the continuity between this philosophy and its rhetorical complement. In consequence, the actual operations of antilogic have remained what they are for Kerferd, a preliminary stage in the development of dialectics (1981, 67; cf. Solmsen 1975, 244). My own exposition, however, would develop a more ambitious profile for antilogic, a role in keeping with the progressive humanism of Protagorean thought and, in particular, with its willing engagement in the construction of knowledge under the aegis of disagreement and flux. I begin with a brief analysis of the two-logoi fragment itself and an outline of its basic features as a distinct argumentative practice. After this extended introduction, I next compare antilogic to its argumentative cousins, eristic and dialectic, a comparison that should allow me to more fully develop the cardinal features of my subject. Finally, this chapter closes with an assessment of Protagorean antilogic in the context of 5C Athenian culture, most notably its relation to the discourse of Athenian democracy and the maxims of Heraclitus, Protagoras' most notable predecessor. In turn, a review of the historical/social grounds for antilogical practice should prepare us for more specific analyses of its pragmatic and ethical dimensions in Chapter 3 and its formal features in Chapter 4. Throughout it all, I argue that antilogic is best approached as an extension of the human-measure doctrine, a pragmatic response to the exuberant relativism of the Protagorean worldview.

1. ANTILOGIC TRANSLATED AND DEFINED

The core concept of the two-logoi doctrine is effectively paraphrased by Billig as "a method of argumentation by which contrary positions are examined in relation to one another" (Billig 45; cf. Kerferd 1981, 63). I cite this paraphrase first because, if anything, we are in this case even further removed from the actual language of Protagoras, his *ipsissima verba*, than we were from the human-measure fragment. In the latter case, we have the relatively contemporary paraphrases of Plato; but for antilogic, we rely on Diogenes Laertius again, who, in the 3C CE, was removed from Protagoras by more than six centuries. Nonetheless, Schiappa thinks Diogenes is basically faithful to the source in this case when he writes that Protagoras

> was the first to say that on every issue there are two arguments (*logoi*) opposed to each other on everything (*pantos pragmatos*). (*DK* 80 A1)

Other ancient variants are as follows: "Every argument has an opposite argument" (Clement of Alexandria, 2C CE) and "one can argue equally well on either side of any question" (Seneca the Younger, 1C BCE; see *DK* 80 A20). The import of these variants is basically consistent with Diogenes, and so I will assume his rendition as the base text of the two-logoi fragment and begin this analysis with its two key terms: *logos* and *pragma*.[5]

The range of meanings for logos is broad and various, including the following variations: logos is (a) reasoning, thinking, or accounting for; (b) speech, discourse, or even specific statements; and (c) the organizing principles, formulae, or laws of the world itself. G. B. Kerferd, from whom I draw these distinctions, writes that "(w)hat we are confronted with is not strictly speaking one word with a number of different meanings, but rather a word with a range of applications all of which relate to a single starting point" (1981, 83; cf. Havelock 192ff). In the case of Protagoras' fragment, this "focal" notion is to be found in the concept of "reasoned speech," so that both Kerferd (84) and Michael J. O'Brien (the English translator of Diels-Kranz) render logos in this instance as simply "argument." In response, antilogic (from *antilogike*, as in *Phaedrus* 261c-e; cf. *Phaedo* 90b-c) becomes arguments in opposition to one another, and *dissoi logoi* (the presumed name of a 4C BCE treatise based on Protagorean ideas) becomes divided, doubled, or "contrasting" arguments (Robinson), or simply "debate" (Freeman 471).[6] The other key term in the fragment, *pragma* is typically translated as "things," but to best understand its implications for antilogic, some further clarification is in order.

Behind this particular fragment stands an extensive tradition of Presocratic thought that focuses on the oppositional, contradictory nature of the "natural world," or *phusis*. And from among the Presocratics, we can single out Heraclitus, for whom dyadic opposition in nature is a cardinal principle and who routinely calls attention to the ubiquity of opposites (e.g. night/day, love/hate, strong/weak, etc.). The Heraclitean theory of opposites provides essential background to the two-logoi fragment, and I will return to it later in this chapter. For purposes of general translation, however, it is enough to say at this point that in Protagoras' hands, the Heraclitean philosophy of opposites is appropriated from the realm of nature and applied to the realm of discourse; as there is day and night, so there is logos and anti-logos (cf. Untersteiner 35). Such an appropriation dictates a corresponding shift in our conception of *pragma*. Consequently, most 20C translators of the fragment tend to render *pragma* not as "things," which carries the weight of the Presocratic reference to the physical world, but as "experience" (Untersteiner 19) or even "reality" (Robinson 90, n. 74). *Pan pragma,* then, becomes not just "every thing," but "all subjects," as in this line from Euripides: "On every subject, it would be possible to set up a debate of double arguments" (*dissoi logoi*; in de Romilly 1992, 77).[7] Or we might just as well use the terms "issue" or "question" borrowed from O'Brien's translations of Clement and Seneca cited above. With the help of these translations, we arrive at this basic rendering of the fragment: "Protagoras was the first to say that there are two opposing arguments (logoi) concerning every subject."

With this translation as our guide, we should be able to identify, in a preliminary way, some of the primary characteristics of antilogic. We can begin by defining the root concept as a rhetorical strategy for dealing with "the class of doubtful things" (Plato) or "things which might be otherwise" (Aristotle); i.e., topics of discussion in which certainty is indeterminate and controversy hence immanent. Like any other form of argumentation, antilogic is a *techne* designed to address the quarrelsome subjects that arise in the process of social discourse. As a specifically Protagorean response to controversy, however, antilogic is distinguished by the conscious effort to set contrasting ideas or positions side by side for the purpose of mutual comparison; i.e., in order to comprehend any "doubtful thing," each logos must be complemented with its anti-logos rather than approached on its own. The juxtaposition of opposing logoi is most commonly mounted by two or more individuals, but antilogic might just as well be conducted by a single person who would contemplate the relative merits of opposing views on any contentious topic. In either case, antilogical discourse is founded on the reciprocal relationship between/among multiple participants or positions in controversy, all of whom contribute distinctive perspectives to the problem at hand.

At this point, we should distinguish argumentation (the interaction between opposing parties in contention) from "an argument," insofar as the latter is construed as a discursive artifact designed to assert a justified belief. More exactly, "argument as artifact" is intended to stand alone, an autonomous structure that demonstrates its validity through the formal congruence of its parts, as in the case of a syllogism or even a Toulmin-based claim. As such, an argument is something we *make* rather than an something we *have*, a distinction I will return to later (see O'Keefe 1977, 121). In contrast, *argumentation* always involves a relationship between two or more claims in opposition, though of course the nature of that relationship is variable. In the Aristotelian model of argumentation, communication flows predominantly in one direction, as the rhetor attempts to secure the compliance of an audience who may be anywhere from neutral to negative towards the rhetor's claim. Similarly, there is the interaction that takes place in Socratic dialogues such as the *Meno* or *Phaedrus* in which Socrates, despite protests to the contrary, guides the discussion between participants in the direction of his own thinking. What distinguishes antilogic as a mode of argumentation is the bilateral relationship of the opposing arguments: i.e., the lines of influence in antilogic are meant to flow in both directions. Eric Havelock writes that according to Protagorean method both sides must speak, both must listen, both can ask questions, and both must "wait while the answer is given" (210). Of course, judgment may go in one direction or another, but not until both sides render their own positions, review the alternatives, and respond to the juxtaposition. So rather than "standing alone," antilogical claims proceed side by side, mutually conditioning one another in a process Bakhtin refers to as "answerability," the exposure of each to the presence of the other (Clark and Holquist 63-94; G. Clark 9). There is in antilogic, then, an implicit commitment to dialogic exchange that distinguishes it in almost every way from the unilateral emphasis of traditional debate.

Admittedly, dialogue has become somewhat diluted as a rhetorical term of art. But at bottom, the defining feature of dialogue is its two-sidedness, its alternation between/among opposing logoi. There is, of course, a powerful tendency to resolve the tension of contraries by supplanting it with something solitary. When we say that the goal of argument is the discovery of "truth" or the achievement of consensus, we acknowledge an intention to pass beyond dialogue, to replace the two or more voices in contention with some *tertium quid*, some new logos that transcends prior divisions or that occupies a middle space between interlocutors. *E pluribus unum*; in which case, the dialogue of contraries is resolved in favor of the singular, difference is overcome, monologue triumphs. Antilogic, on the other hand, resists the effort to resolve disagreement by erasing difference too quickly; there are always other antilogoi, other oppositions that follow from the oscillations of dialogue, another side to the same story waiting to be told. By insisting on the perpetual two-sidedness of dialogue, antilogic not only asks us to recognize that the other is always present, but, more importantly, it also compels us to acknowledge that "alterity" is not to be subordinated, that the antilogoi of the interlocutors are "real and true" for those who hold them and should be seen as such. Antilogical argument, then, is not a dyadic preliminary to some kind of synthesis, nor is it intent on the end of dialogue. Instead, it is first and foremost an encounter with the autonomous other, an encounter that offers each side the chance to see beyond the boundary of its own logocentricity, but only so long as we remain in dialogue.

If dialogue is the heart of the thing and understanding is to be sought in the interaction of two or more sides, then the role of any single utterance or text is superseded by an emphasis on context, on the relation among the logoi in play and the forces that condition their contact. Such contextuality distinguishes antilogic once again from "formal" argument. Whereas the latter tends to isolate ideas from the intricacies of their situation for the purpose of generalization and clarity, antilogic is immersed in the distinct and fully historical world out of which the discourse develops and in the psychological nature of the living human beings involved. Hence, the arguments we offer are always more than "the accumulation of evidence and inference sufficient to establish a claim"; they are claims which follow from the particular position of their author and which exist within a landscape of opposing alternatives (Ehninger 1970, 103). Or, to put the matter in the terms of the previous chapter, the arguments of our dialogical partners are formulated in and shaped by the complex network of relations that surround and contribute to the collective life of the individual. To come to grips with what is "true for N" requires that we restrain a natural attachment to our own logos and seek to understand the socially generated categories that inform the "other side" and allow it to makes sense. In a phrase, antilogic asks each side to understand "where the other is coming from."

The prevalence of the personal in antilogical relations leads, in turn, to the notion that Protagorean argument is a form of practical reasoning, or *phronesis*, a method for managing actual controversies among specific people rather than speculating about the universal and impersonal. The historical context for this pragmatism is clear enough: Protagoras abandons Eleatic theories about essential oppositions in nature and avoids reliance on "first principles" that govern formal reasoning. Instead, the focus of the human-measure doctrine is on the existential status of human knowledge and on the multiple claims that emanate from within the moving stream of human relations. Under such conditions, we are faced not only with "endemic disagreement" but also with the crucial yet quotidian problem of discriminating among a legion of conflicting claims. In the absence of appeals to fixed and universal principles, how do we make the transition from individual perception to social judgments? Or, more importantly, according to what standard does an individual claim establish itself as superior to others and so warrant at least provisional agreement?[8]

The Protagorean response to these questions is largely pragmatic in its emphasis on the practical consequence of particular claims. More exactly, we make judgments about the merits of individual assertions based on collective assessments of how an idea might actually function in a particular context and, more specifically, what common benefits it might promote. For instance, in the *Theaetetus*, when asked by Socrates to expand on the human-measure doctrine, Protagoras argues that while one set of thoughts may not be "in any way truer" than another set, we can still discriminate among competing claims based on what is most *beneficial* or *useful* in any particular circumstance (167a-c). He develops the case for utility by reference to the practice of the physician or the husbandman, both of whom prefer one treatment to others according to what is most likely to promote beneficial effects. Similarly, the Sophist in debate will be guided by the relative potential of each logos to result in definite instances of human benefit. In this way, Protagoras surrenders any purchase on fixed and absolute standards while at the same time salvaging the concept of value under the heading of the common good (what is conducive to the best interests of those involved). Given the practicality of such evaluative standards, we can characterize antilogic as an instrumental procedure for calibrating the relative utility of conflicting claims as these present themselves within a particular context (cf. de Romilly 1995, 192; Mailloux 11-13). Put another way, antilogic is a means for responding to controversy not by appeal to a timeless reality or to hypostatic values, but rather in terms of the prevailing interests that condition local decisions about reasonable belief and practical behavior.

These matters are complex and I will take them up in detail in the next chapter; but, for the purpose of this abbreviated preview, the pragmatism of antilogic can be rendered a bit more fully by pursuing the analogy with the practice of medicine.

As noted, Protagoras himself suggests that the physician and the Sophist both appreciate the relative nature of particular claims (what seems healthy to A may not be the case for N) and both are intent on practical benefits (what will make for physical and social well being, respectively) (*Theaetetus* 167b). Similarly, the author of *The Art of Medicine*—a Hippocratic treatise contemporary with Protagoras—argues that the art or *techne* of medicine should not proceed by deduction from "empty postulates"; instead, the good doctor will focus on the constitution of a particular patient and the nature of this patient's physical environment (Hornblower 947-48). Versenyi neatly articulates the Hippocratic rationale for this attention to concrete circumstance: "What agrees with one man, in one situation, may not agree with him, or other men, in other situations" (1963, 33). By analogy, then, Protagoras can argue that just as a particular medicine is only suitable under certain conditions, so a particular logos or claim is helpful (or harmful) relative to its utility within a particular community. In sum, antilogic—like medical diagnosis—is a practical art expressly intended to calculate the relative merits of opposing claims and to suggest the position most appropriate to the well being and best interests of these people, at this time, in this setting.

When this practical interest in the common good is considered from a political perspective, antilogic can be further identified as a potential aid to democratic practice. To be sure, we cannot claim an essential political nature for any rhetorical *techne* since all are subject to misuse (see *Rhetoric* 1355b; cf. *Euthydemos*). Nonetheless, it is reasonable enough to assert that neo-Protagorean antilogic—given its insistence on negotiating meaning within an expansive field of competing preferences—would encourage democratic participation in the production of knowledge and thereby promote a significant reorganization in the structure of human relations.[9] Consider, for example, that the human-measure doctrine acknowledges the potential contribution of each knowing subject to the determination of particular claims (see *Protagoras* 322c-d). But if knowledge claims are to accrue social currency, they must be negotiated with others; antilogic, therefore, insists on the need to counterpoint dominant logoi with those voices that have not been heard, that one cannot imagine on one's own, the constituencies not yet at the table, the quiet student in the back row. Consider, in turn, that while conventional beliefs (*nomoi*) are the consequence of prior deliberations and while dominant voices are inclined to control these deliberations, the introduction of the heterodox and the hitherto-unheard opens up the possibility for the transformation rather than the reproduction of convention. In this way, argument intervenes in the world and seeks to influence both the production of knowledge and the politics of discursive exchange (who says what to whom and when). The logic of antilogic, in particular, leads to a deeply layered or "thick" conception of public culture in which alternating and alternative voices engage in the constant revaluation of social norms based on continual challenges to reigning beliefs.

Consider, in addition, that the Protagorean denial of transcendental appeals is also a denial of a priori claims to authority and privilege. In fact, it is the express aim of antilogic to subject all views or perspectives to challenge by equally enfranchised alternatives, so that critique is distributed universally. No single notion that seeks general approval can assume for itself unchallenged obedience, nor can a reigning assumption adopt the mantle of an edict since every claim, when set in a truly democratic field, will be accompanied by alternatives that question its status. Of course, it is at the moment of decision-making, when ideas to be executed in the realm of human affairs are actually determined, that issues of power become most prominent (see Crosswhite 202-16). As a pragmatic *techne* oriented toward concrete benefits, antilogic would not defer decisions indefinitely through an infinite series of discrete challenges. Instead, antilogic would make the inquiry process that precedes judgment as inclusive as possible, incorporating not only all relevant positions but also all available challenges. Eric Havelock points out that any participant in Protagorean discourse could initiate a critique of existing *nomoi*, the conventions that govern social knowledge (Havelock Ch. 8; see Jarratt 1991, 53). Similarly, a contemporary antilogic would encourage the comprehensive exploration of diverse and contradictory logoi as the appropriate prelude to democratic judgment. In this context, "justice for all" depends upon our ability to interact with difference.

So, if we acknowledge that all rhetorical practice is politically marked, then antilogic is on the side of radical egalitarianism: we are all endowed with something to say, with the "truth" as it appears to us, and in the process of struggling towards a shared or consensual view, it becomes an "inalienable right" for all positions to have the floor, if not to carry the day. And yet, it is, of course, one thing to have something to say and quite another to be able to say it to effect. That is, how are we to represent ourselves to best advantage in the open forum of antilogical exchange? The question leads us from politics to pedagogy, from democratic practice to organized efforts that prepare students to become citizens. It is pedagogy, then, that assumes pride of place as the final and perhaps most telling feature in this introductory outline of our topic; for in its pedagogy antilogic has its basic features summarized and its fundamental promise most clearly on display.

"I admit to being a Sophist," says Protagoras, "and an educator" (*Protagoras* 317b). Education was central to Sophism; to be a Sophist was to teach and, in particular, to teach what we now call discourse studies.[10] However, like so much else about the Protagorean legacy, we are left to reconstruct the nature of this teaching from related evidence and inferential leaps. And, as ever with Protagoras, the process of reconstruction begins with the human-measure doctrine—with its theory of knowledge as conditioned by multiplicity and flux, with its particular interest in the relationship of contending perspectives, and with its overriding concern for practical responses to controversial situations. Given this theoretical context, it is little wonder that rhetoric is the organizing element in the Protagorean curriculum, the *logon techne* (art of reasoning) that prepares students to negotiate the shifting ground of human controversy and manage the endemic disagreements of social life. In brief, Protagorean education—like Protagorean philosophy—is fundamentally pragmatic, focused on argumentation in the interest of practical conduct and at odds with the dialectical pursuit of ideal definitions characteristic of Plato's Academy.

For centuries prior to the Sophists, Greek education had been essentially military. Early in the 5C, however, it becomes increasingly cultural and academic, with training not only in sport but also in music and poetry, arithmetic and astronomy. Nonetheless, according to H. I. Marrou, the French historian of "education in antiquity," the emphasis of even this new training was more artistic than intellectual (36-45). Consequently, when the Sophists introduce a form of instruction "wider in scope, more ambitious and more effective than any previous system," the result, writes Marrou, is "a veritable revolution" (48, 58). This instruction was practical in focus (based on the specific practices of antilogical *techne*) and utilitarian in aim (intended to cultivate effective skills for use in concrete circumstances). Protagoras himself points out that when a student comes to him, the student will learn

> [t]he proper care of his personal affairs, so that he may best manage his own household, and also of the state's affairs, so as to become a real power in the city, both as a speaker and a man of action. (*Protagoras* 318e-319a)

James Jarrett glosses this passage by noting that Protagoras was principally interested in nurturing "the sound judgment required to act effectively in both private and civil affairs" (34).

This pragmatic agenda was supported by methods of instruction that were similarly original; instead of contemplating the wisdom of "divinely inspired poetry" (Schiappa 162) or dissecting the "barren subtleties" of abstract philosophy (Isocrates' *Antidosis* 262-69), Protagoras' pupils nurtured sound judgment through direct engagement with concrete rhetorical situations. Such engagement often involved the critical evaluation of literary models, which Protagoras considered an exercise in practical reason "transferred to the realm of poetry" (*Protagoras* 339; cf. Quintilian 2.20.5). For the most part, however, students of the Sophists were engaged in the production of their own discourse, in writing and speaking arguments they had drafted themselves, often in imitation of models, but nonetheless the result of their own composition (Marrou 50-51). In sum, instruction proceeds through rhetorical practice itself rather than by recourse to theory or precepts about such practice (de Romilly 1992, 55).

Of course, the elevation of practical training over theoretical discussion is an extension of Protagorean pragmatism, with its emphasis on the concrete and contextual and its empirical interest in how things actually happen.[11] Instead of focusing on the abstract features of rhetorical genres (cf. Aristotle's *Rhetoric*), students in training with Protagoras (or Quintilian) would compose speeches on opposing sides of a topic (*DK* 80A), present these in student debates, and discuss the alternatives in small groups (Kerferd 1984, 34; Schiappa 162). Such activities immerse the students in local detail, requiring them to observe the kairotic demands of the particular moment and identify appropriate rhetorical strategies in response. Knowledge may thus be said to flow from direct experience, as students inductively generate norms for rhetorical behavior based on discussion of and practice with particular situations. Eight centuries later, Augustine will endorse this same pedagogy when he points out that we "learn to be orators not by studying the rules of oratory, but by reading and listening to the orations of orators, and, in so far as it is possible, by imitating them" (*De Doctrina* 4.5). In fact, the pedagogy of practical experience persists through the Renaissance and into at least the 19C (Murphy, 1990). If longevity is any indication, then Protagorean/Sophistic instruction—focused as it was on particular contexts and appropriate choices, on the critical analysis of existing models and the imitation of what was best in them—proved remarkably effective.

And yet to isolate practical aims and antilogical methods from their socio/political context would be to limit what Burke calls the "circumference" of this pedagogy (1969, 77-85). That is, antilogical pedagogy, like all prescriptions for classroom practice, offers a theory of how groups work and as such is suffused with assumptions about authority, identity, responsibility, and value. Protogoras outlines his own social "values" in the "Great Speech," the mythological narrative in Plato's eponymous dialogue in which Protagoras connects the origins of human culture with our capacity for justice (*dike*) and respect for others (*aidos),* virtues which allow us to mediate differences and cooperate in collective action (*Protagoras* 320b-23a; see Atwill 149-55). These virtues, however, are potentialities only and depend upon good teaching to be realized (323c-e). Rhetorical instruction, therefore, serves first and foremost to develop our latent capacity for responding to social differences with an adequate measure of civility. And, in the political environment of late 5C Athens, where all citizens were at least part-time politicians, such capacity constituted the price of entry into the *polis* itself (Murray 244-48). Training in argumentation is thus preparatory to social and political engagement, to a life in which values are "externalized" in the process of argumentative exchange. The antilogical classroom, in particular, is a gymnasium for training in the art of *concordia discors* (harmony amidst discord), for argument conducted under the aegis of *dike* and *aidos*.

A few pages above, I described antilogical practice as "radically egalitarian," opposed to the dominance of prior status, insistent that all claims be challenged and all rhetors be invited to speak. "We can not be certain," writes Vico, "that we have dealt with something in all its essential respects if not all the possible pertinent questions have been asked" (in Mooney 134). In giving voice to "all pertinent" logoi, antilogic creates an opportunity not only for conventionally "weaker" positions to be heard, but, in the juxtaposition of probabilities, for the dominant order to be challenged and even overturned if the alternative case can be made to the satisfaction of those involved. In the classroom, this invocation of multiple voices creates unusual possibilities for critical independence and conceptual growth. Admittedly, these social conditions were hardly reflected in Protagoras' own circle, where the pupils were drawn from what James Jarrett calls the "democracy of the wealthy" (62). Nonetheless, the internal ethic of antilogic requires that alternatives be heard and so extends a special invitation to voices in opposition, to students, perspectives, and logoi that are, for whatever reason, subordinate. Like the *dike* and *aidos* of the Protagorean myth, however, the full promise of this pedagogy is a potentiality only, a *techne* for instruction in democratic thinking that awaits actualization in classrooms and cultures capable of rising to its considerable challenge. The social dimensions of antilogical practice are the subjects of the next two chapters; in the Epilogue I return to the notion of antilogical pedagogy as a potential contributor, *mutatis mutandis*, to contemporary instruction in argumentation. It may well be, as Joseph Margolis argues, that we are in a better position than any of our predecessors to explore the potential of

Protagoreanism, especially in the classroom. For now, I will close out this section by calling attention to a neglected feature of this potential.

In the last few pages, I have tried to indicate the technical and practical orientation of antilogical pedagogy, its focus on specific argumentative strategies for the sake of improved intervention in "private and civic affairs." This attention to educational method and purpose tends to overlook, however, the benefits of antilogical practice for the individual student. Granted, personal development is a modern rather than an ancient pedagogical interest; nonetheless, a final note on "Sophistic argument for the modern student" is in order.

Marrou asserts (perhaps unfairly) that the aim of Socratic education is the "submission" of the student to the claims of the Absolute (58).[12] In contrast, I would argue that among the aims of neo-Protagorean pedagogy is the liberation of the knowing subject from the claims of unchallenged assumptions. That is, human beings may be the measure of all things, but our efforts to achieve intersubjective understandings are nonetheless encumbered by a variety of cognitive habits, not the least of which are the unacknowledged partialities and dogmatically-held beliefs that we all carry with us. The static tendency of existing belief strains against the demands of argument, which is predicated on the possibility that the parties involved are willing to change in response to alternative logoi. To resolve this tension, antilogic would encourage the emancipation of the rhetor from the imperious demands of the "my-side bias," would overcome frozen allegiances by engaging in dialogue with justifiable alternatives.[13] More specifically, the protocols of antilogic require that all positions be examined in relation to their opposite numbers, that the authenticity of the other side be understood and respected, and that the construction of one's own position involve some response to the collision of alternatives. Consequently, when conducted according to Protagorean principles, argumentation involves more than the incidental relation of parallel positions, of pro and con side by side. Rather, antilogic requires an intermingling of alternatives, a sojourn in the "parlour of discourse," where each position advanced is the subject of critique and each critique, in turn, receives critical response. We may choose to reject or revise, accept or accommodate the arguments of the other side; but once we enter into an antilogical contract, we cannot remain unmoved, for involvement in antilogical exchange in itself serves to extend the circumference of understanding. According to the logic of antilogic, then, emancipation and discovery go hand in hand.

When the emancipatory benefits to the individual student are grafted onto the practical/political interests of antilogical pedagogy, this teaching finally achieves the holistic scope that distinguishes the educational legacy of the Sophists (Jaeger, de Romilly, Marrou, Bolgar 29). The pedagogy of antilogic assumes that interacting with rather than excluding opposing logoi from the process of critical reasoning will liberalize a student's thinking by opening up genuine alternatives. At the same time, the ability to address questions of difference and achieve acceptable resolutions allows for consensual action, for the skills of argument to become an instrument in pursuit of the public good. This merger of public and private interests is a hallmark of Sophistic teaching, an extension into the classroom of what Renato Barilli calls rhetoric's "vocation for fullness" (viii). In a Protagorean pedagogy, such fullness is achieved only when free individuals engage in dialogical exchange as an "indispensable preliminary" to wise public action.[14] Seen thus, antilogic is not simply a pedagogical *techne* for acquiring personal knowledge; it is a discursive means for transforming the "endemic disagreements" among individuals into the contracts that constitute community, for organizing diversity in the interest of the general welfare. This optimistic scenario, however, is not the only way to look at the antilogical heritage. To complement my own neo-Protagorean idealism, I turn next to alternative forms of argumentation opposed to antilogic, its own anti-logoi.

2. ANTILOGIC, ERISTIC, AND DIALECTIC

In due course, I will fill out the various features of antilogical argument that I have only sketched in so far. For now, if am to follow the principles of antilogic myself, it is only fitting that I acknowledge and respond to alternative perspectives on my subject. As Borges puts it, "[a] book which does not contain its counter-book is considered incomplete" (13). In particular, the critical reader might justly object that the flattering portrait of antilogic that I have drawn to date is too one-sided, too monological, an act of advocacy rather than a critique. Or, the reader might argue that my initial claims for antilogic propound an obvious paradox by aligning argument with goodwill and so ignoring the natural enmity that attends refutation (see Plato's *Lysis* 216a). Other charges are readily available. One might cite instances of argumentation in which an antilogos is raised only as a convenient cipher by a crafty rhetor who would eventually overcome all opposition (see Demosthenes' "First Phillipic"). Or antilogic might as easily be employed by the thorough-going sceptic to prove that there is nothing secure from attack in any argument whatsoever (see Plato's *Euthydemus* 272a).

These and similar charges have been mounted against antilogic since antiquity, most notably by opponents who would identify it with eristic, the desire to prevail in argument at all cost (Kerferd 1981, 62-63; Nehamas 3-6).

Nor is such criticism unfounded; for if approached simply as a *techne*, a rhetorical tool for turning the tables and converting alternatives into justification for one's own stand, antilogic will indeed decline into eristic, an argumentative peritrope employed solely for persuasive success. For instance, recall Aristotle's charge that Protagoras promoted the unscrupulous substitution of worse arguments for better ones (*Rhetoric* 1452a 20-26; cf. Plato's *Apology* 19b). Seen thus, antilogic, becomes an "art of deception" that disguises ruthless self-interest as acceptable reasoning. This debased form of antilogic is on full display in Aristophanes' *Clouds*, in which a young student of rhetoric, Pheidippides, employs antilogical sleight-of-hand to confuse his father, obscure his own delinquency, and ultimately achieve a complete reversal of common sense. And in the *Euthydemus*, Plato similarly satirizes two sham Sophists who once taught martial arts and who now practice linguistic quibbling (*logomachy*) as a means of throwing their opponents off balance.

Any effort to recuperate antilogic must certainly respond to such corruptions and counter-arguments. And I intend to do so in detail throughout *Many Sides*. At this point, however, the appropriate next step is to follow the lead of G. B. Kerferd and juxtapose antilogic with its nearest neighbors (see 1981, 59-67). That is, how can we distinguish antilogic from eristic, its most direct contrary, and, alternatively, from dialectic, a form of argumentation that shares much with antilogic yet also differs fundamentally from it. But whereas Kerferd undertakes his analysis of Sophistic argumentation with Plato as his guide ("what then is Plato's attitude toward this method of antilogic?" 63), I will briefly review the same subjects from the perspective of the neo-Protagorean views outlined in Chapter 1.[15] Operating from this vantage point should allow me not only to distinguish antilogic from alternative forms of ancient argumentation but also to meet the charge that antilogic is simply derivative of some other, more distinct form (Nehamas 9-11; *Encyc. of Philosophy* 2.387).

To begin with eristic: the word itself is derived from *eris*, literally meaning strife or wrangling and poetically figured as the sister of Ares, the god of war and warlike frenzy (e.g. *Iliad* 4.440-45; Liddell and Scott; cf. Hesiod's *Theogony*). It makes sense, then, that the defining feature of eristic is not simply the effort to prevail in argument but, more exactly, to obliterate all competing positions. Its tactics include the histrionic display of verbal ingenuity, the use of fallacious premises, and an aggressive form of question and answer intended to overwhelm one's opponents by sheer force of contention (cf. Plato's *Sophist* 231e). Not surprisingly, ancient eristic is routinely condemned both for its attitude and objectives (Isocrates' *Antidosis* 258-60; *Phaedo 90b-c*; *Meno* 75 c-e; cf. *Gorgias* 458a-d). According to this critique, eristic is unreservedly agonistic, the embodiment of the commonplace that "argument is war." In contrast, Protagorean antilogic—as conceived of in *Many Sides*—entails the emancipation of all parties from dogmatic preconceptions and the subsequent negotiation of mutually acceptable benefits. In a phrase, antilogic would foster cooperative engagement rather than the unabashed competition of eristic.

Even more distinctive is the difference in the respective assumptions regarding the relation of argument to knowledge. As Gilbert Ryle points out, the eristic preoccupation with victory displaces any commitment by the rhetor to the truth-value of a proposition. Indeed, the eristic rhetor willingly employs outrageous *adoxa* (fallacious opinions) as premises for argumentative showmanship (74; cf. Grote 388). On the other hand, antilogic is intent upon the progressive development of justifiable beliefs and so will examine, in turn, alternative accounts in search of ever-more comprehensive understanding. Consequently, Protagorean antilogic will view dialogical exchange among people with varied conceptions of the world as itself a victory because the circumference of what is known is potentially enlarged by the addition of each argument to the pool of resources. Seen thus, the eristic fixation on self-interest runs directly counter to the objectives of antilogic.

Admittedly, any *logon techne* may decline into misuse or perversion (*catachresis*) based on the purposes that propel individual rhetors (see *Rhetoric* 1355b 10). And, undoubtedly, the post-Periclean rejection of Sophistic practice was, to a significant degree, the result of excesses represented by eristic (de Romilly 1992, 78-92; Farrar 99ff). Moreover, this condemnation persists in the long tradition within Platonic studies of treating antilogic and eristic as "interchangeable" (Kerferd 1984, 62). But to continue to conflate the two not only ignores valid distinctions, it also perpetuates an anachronistic effort to debunk anything that hints of Sophistic relativism, an effort more than a little eristic itself, albeit of a subtle kind. My own motives dictate different tactics: having distinguished antilogic by contrast with eristic, I turn now to the comparison with dialectic, a companion form whose differences with antilogic reflect fundamental distinctions.

The term *dialektike techne*, or the "art of conversation," is derived from *dialegomai*, meaning to converse or debate, and implies the back and forth nature of argumentative exchange (Liddell and Scott). For many readers, the process is most vividly represented in Platonic dialogues like the *Meno* or *Parmenides* in which Socrates and his interlocutors seek out and test the truth of some concept through conversational exchange. Hence Plato's description of dialectic as "the coping stone" of philosophy, the ultimate method for arriving at truth (*Republic* 534e). Typically, dialectical inquiry is focused on the universal nature of its subject, its unchanging essence, a topic

I will return to shortly (Aristotle's *Topics* 8.14, *Republic* 454a-c). It is more difficult to say "how" such inquiry operates because dialectic has developed along multiple lines. Plato's Socrates tends (in the early dialogues) to proceed through conversational interrogation of his dialectical partners, often approaching the topic through distinctions and divisions (*diaireseis*), often by direct challenge to initial propositions (the *elenchus*, to which I will also return), regularly through a sequence of questions and answers (see *Cratylus* 390c). Like his teacher, Aristotle is concerned with universally valid propositions or laws (*koinai archai*) which are atemporal and impersonal in nature and generally accepted by authority. However, Aristotle approaches these propositions as purely logical constructs, replacing the Platonic connection between ideas and ontology with an interest in self-contained formal methods that eventually results in the syllogism. In the process, Platonic metaphysics gives way to "the methodology of argumentation," with the *Topic*s as a training manual in the proper structure for dialectical propositions and the available strategies for the thrust and parry of dialectical exchange (Solmsen 1968, 67; cf. Ryle 69, Grote 390). Other variations of dialectic proliferate, such as the refutation of unsubstantiated hypotheses (as in the *Gorgias*), or the Hegelean analysis of the historical process by which one idea passes into its contrary, or the socialization of Hegel's dialectic in the Marxist struggle of opposing material interests. But as Gadamer points out, all variants have in common the reasoned inquiry into contradictory propositions and so focus on the examination of opposing logoi (1976, 5).

This concern for opposition marks a distinct similarity with antilogic. Jean Dietz Moss writes that dialectic is fundamentally distinguished by its desire to "demonstrate contraries" (116), a process we might describe more technically as the ability to set up contradictory predicates of the same thing (cf. *Phaedrus* 261d). Aristotle corroborates this idea in the *Topics* by noting that dialectic prepares its practitioners not only to understand the epistemological usefulness of contrasting arguments but also to practice the complex strategies of arguing on opposing sides of a case (1.2, 8.4-5; cf. *Rhetoric* 1355a). Unlike the apodeictic (or demonstrative) sciences, which are based on verifiable data or self-evident axioms, dialectical discourse is characterized by opposites because it is based on *endoxa*, or opinion, albeit authoritative or accredited opinion (Grote 384). As a result, equally legitimate, though opposing opinions are the very substance of dialectic, a perspective which seems—to this reader, with my biases—remarkably Protagorean in character (cf. Kerferd 1981, 63). We can develop these similarities, and begin to establish some distinctions as well, by referring to Kenneth Burke's comprehensive catalog of definitions for dialectic in *A Grammar of Motives*.

Burke writes that dialectic encompasses the following (I have added numbers for convenience of reference):

> 1) reasoning from opinion; 2) the discovery of truth by the give and take of converse and redefinition; 3) the art of disputation; 4) the process of interaction between the verbal and the non-verbal; 5) the competition of cooperation, the cooperation of competition; 6) the spilling of terms out of terms, as the dialectician proceeds to make explicit the conclusions implicit in key terms or propositions used as a generating principle . . . .; 7) the internal dialogue of thought . . . .; 8) or any development . . . got by the interplay of various factors that mutually modify one another . . . .; 9) or the placement of one thought or thing in terms of its opposite . . . .; 10) or so putting questions to nature that nature can give an unequivocal answer. (403)

The majority of these definitions (#1, 2, 3, 5, 7, 8, 9) could apply to both dialectic and antilogic, which is why de Romilly refers to Protagoras' rhetoric as "first and foremost a dialectic" (1992, 75). For example, the "interplay" of opposites leading to "mutual modification" and intellectual "development" (#8) aptly describes the method and goals of both dialectic and antilogic; while the juxtaposition of opposites (#9) is a good shorthand definition of antilogic in particular (cf. Billig 45). Indeed, it may be the strong family resemblance between the two that accounts for the historical "absorption" of antilogic by dialectic in a process that begins with Aristotle (see *Topics* 1.2, 101b3) and is all but complete in the procedures of medieval disputation (see Moss 127; de Romilly 1992, 90). But if there are points at which the two practices "merge" (Burke's term), there are also points of difference which allow us to identify (by *diairesis*) important distinctions between the two (1969, 403-06).

First and foremost, we can say that dialectic, in opposition to antilogic, is directed toward the investigation of "supremely abstract" concepts (*Encyc. of Philosophy* 2.385). In Plato, this tendency toward abstraction involves a commitment to the Platonic Forms, to inquiry into the nature of Justice, Knowledge, Beauty, and other abstract norms of excellence, or *arete*. Whereas Sophistic reasoning dwells (according to Plato) on "the humors and desires" of public opinion, dialectic is conversely preoccupied with "the real nature of things;" i.e., with universal definitions that transcend verbal categories and comprehend a subject's participation in essential Form (*Republic* 532a-b; Aristotle's *Metaphysics* 13.4, 1078b; Solmsen 1968, 49). So, for example, Protagorean inquiry into aesthetic issues will naturally produce what Plato calls a "multiplicity of beautiful things" (i.e., multiple perspectives on the same aesthetic artifact); dialectic, on the other hand, is intent on the one true reality, the nature of "the Beautiful itself" (*Republic* 493e-94a). In Burke's catalog, this commitment to transcendent abstraction is implied in dialectic's attempt to "discover truth" (#2), to spin out key terms and propositions (#6), and to arrive at "unequivocal answers" (#10), all of which indicate that dialectic is not dialogue as commonly encountered in the social sphere. Rather, dialectic is an idealized discourse, typically conducted by an authoritative voice, addressed to an audience of experts, and in pursuit of knowledge in hypostatic form, knowledge that has been removed from its context in the world. In the process, dialectic tends to make formal entities out of fluid, operative relations and to atomize information into discrete and separable units, all for the purpose of generalization and precision. Dialectic, then, is the pursuit of knowledge for its own sake, an inquiry that looks beyond the specific players involved in the cross-fire of dialogue, players who are themselves the principal focus of antilogical attention (cf. Aristotle's *Metaphysics* 1087b 25-32).

Second, because of the abstract vs. situational natures of dialectic and antilogic, respectively, there are potential differences in the relationship between interlocutors (see Burke's #5: the competition of cooperation, the cooperation of competition). Dialectic typically involves the one-on-one exchange of single individuals; e.g., think of Socrates' three, serial interlocutors in the *Gorgias*. Despite this direct relationship, dialectic is (as noted above) predominantly concerned with appeals to reason rather than to persons. In the *Cratylus*, Plato discusses dialectic by analogy with weaving (388a-90d), but the weaving of dialectic involves the warp and woof of ideas in their formal purity, shorn of their context in individual difference and social variation, ideas organized and presented for the sake of systematic proof. Given this impersonality, dialectic is unlikely to allow the rhetorical interest in conciliating one's audience to disturb its primary effort to get at the essence of its subject. And in the absence of an inherent demand for goodwill, dialectical argument can easily take shape as an intellectual duel, a contest in which refuting one's partners or forcing them into inconsistency is part of the process of pursing the truth. Indeed, Solmsen warns that dialectic must be careful not to "encroach on the province of its neighbor, eristics" (1968, 67). This potential contentiousness is illustrated in the practice of the elenchus, one of dialectic's most distinctive features.

Grote describes the elenchus as the cross-examination of one's dialectical partner by rapid-fire questions which solicit abbreviated responses (382; see *Protagoras* 332a-333e). Gregory Vlastos, in a brilliant article on the subject, adds that the principal goal of the Socratic elenchus is moral truth. But he also acknowledges the aggressive means to that end, pointing out that the cognate verb of "*elenchus*" means not only to test or refute, but also to censure and reproach (1983, 28). According to Vlastos, the standard structure of the elenchus is as follows: Socrates' dialectical partner asserts a thesis that Socrates considers false, Socrates then presents alternative premises to which the partner agrees but which contradict the original claim, at which point Socrates can assert that the interlocutor's original claim has been proven false (38-39). The primary goal of this process may not be Socrates' own triumph, nor even the humbling of the interlocutor, but the practical outcome is the debunking of the partner's claims and not the contemplation of the *multiplex ratio disputandi* that motivates antilogic.

Socrates, of course, believes that he is doing his dialectical partners a favor by disabusing them of falsehood (*Apology* 29e); but the cooperative impulse of the elenchus is clearly subordinate to the drive toward conceptual correctness and conformity to procedure as dictated by the questioner. Alternatively, antilogical argument is more fluid and less predictable because it revolves around individual differences and responds to the bilateral interests of both participants. It assumes that all opening positions are "partial" and that everyone has both a right to speak and a responsibility to listen (see Chapter 4). Like dialectic, antilogic must be on its guard against crossing the border into eristic; but as I argue throughout *Many Sides*, antilogic in its neo-Protagorean form is not intent on demonstrating the ascendancy of one thesis over others. Rather, antilogic seeks coherence among opposing beliefs sufficient to justify provisional conclusions and sanction consensual action. As a result, antilogic will make a significant effort to achieve "cooperation in competition" and comity in conflict. Burke's motto for *A Grammar of Motives* is *ad bellum purificandum*, toward the purification of war. Antilogic will attempt to purify the argumentative process by respecting the basic principle that an idea is "true for N" and deserves to be heard.[16] Dialectic, on the other hand, is clearly more interested in ideas than in people; and the Socratic elenchus, in particular, seems willing to sacrifice workable relationships for conceptual precision.

Finally, dialectic, in its effort to systematically demonstrate and firmly establish the truth (i.e., to provide "an unequivocal answer" [#10]), runs counter to the perpetual openness of antilogic. In this context, "openness" denotes the acknowledgement that no claim is absolutely determinate because alternative ideas will invariably arise with the change of time and circumstance. To the extent that dialectic aspires to transform disagreement into exact and imperturbable conclusions, it seeks ultimately to close the conversation.[17] Burke refers to this process as "transcendence," the resolution of local differences through the invocation of a crystallizing principle always and everywhere in force (1969 420-40). Antilogic, on the other hand, places a "priority" on questions over answers because the perpetual presence of antilogoi keeps the "questionability" of a topic always in play (cf. Gadamer 1994, 363). Insofar as dialectic is ultimately oriented toward universal answers (and ignoring the tendency of Socratic dialogue to end in *aporia,* or irresolution), it is clearly distinct from antilogic. The objective of the former is an unassailable truth, while the paradoxical *telos* of the latter is an open-ended process that finds only interim resolution in the utility of proximate judgments.

In summary, dialectic is an intellectual discipline for coming to terms with the contradictory arguments (logoi) that surround complex issues by abstracting them from the pluralistic, contentious, chaotic exchange that conditions public discourse and subjecting them to systematic, formal analysis. In this shift from the phenomenal to the noumenal, from participation-in to speculation-about, from the actual *polis* to the ideal state, dialectic undertakes the contemplation of argument at a level that is self-contained and analogous to a mathematical demonstration (Havelock 197; see *Theaetetus* 167a). Antilogic, on the other hand, is a *praxis*: both a theoretical means for coming to terms with the confusion of logoi that surround us and a practical intervention into the realm of actual decision making for the purpose of assisting collective action (Clark 20). In brief, dialectic would apprehend the controversies of experience for the purpose of contemplation, while antilogic would understand what it can of the endemic disagreements of social experience for the purpose of action. And action in this case means making knowledge effective, making it function within the actual experience of the community. Kerferd writes that for Plato "antilogic is the first step on the path that leads to dialectic" (1981, 67; see *Phaedo* 101e). In true Sophistic fashion, I would reverse that progression and say that dialectic—the contemplation of the necessary contradictions of the world in universal terms—is the prelude to antilogic and the effort to understand and act upon the actual complexities of social life.[18]

3. ANTILOGIC IN ITS ORIGINAL CONTEXT

Having outlined the basic features of antilogic and having compared it with its relatives, we can round out this introductory appraisal by adopting a more historical perspective. Assuming that all intellectual constructs show signs of their origins and that origins offer particular insights, I return now briefly to Periclean Athens and to those features of the times that most significantly mark the evolving practice of antilogical exchange (cf. Farrar 10). In particular, the question of the moment is "what was the inspiration for the form of argumentation first theorized by Protagoras?" I have already discussed philosophical antecedents in Chapter 1; I turn this time to the general milieu of 5C Athens and the hypothesis that antilogic develops out of the cultural practices of emergent democracy.

As noted earlier, *dissoi logoi* are natural offspring of any group discussion. Cicero writes that "it is almost impossible" for any serious discussion not to give rise to contending arguments (*De Oratore* 2.291). The point to make about Athens during Protagoras' residence there (ca. 460-430, a span that almost exactly coincides with the rule of Pericles) is that this period afforded unprecedented incentives for public discussion and, in turn, treasured the prospect of "contending with words."[19] Herodotus (a contemporary of Protagoras) places the origins of Athenian democracy in the early 6C BCE with the decline of archaic tyranny (5.61ff). With the reforms of Solon (also early 6C), the concept of the *polis* begins to take shape as a community defined by citizens themselves rather than dictated by social and economic tradition. In time, members of the *polis* included all citizens of Athens, all of whom had the right to contribute to the aims and practices of the community through various avenues of political engagement. Following further reforms by Cleisthenes (late 6C), not only is the domination of political life by aristocratic elites eclipsed, but—in an astonishingly short time—democratic practices develop to the point that social order is construed as the responsibility of all citizens and all citizens approach this responsibility as political, if not social and economic equals (Farrar 22).

Of course, the direct democracy of the Athenian *polis* extended only to native Athenian males, or about 10% of the population. And, of course, Athens was a slave society, which excluded women and non-citizens from the *polis*, exclusions that leave their own marks on Greek cultural history (see Fantham 1994, Jarratt and Ong, Nye 7-82). But those who were citizens were expected to do their duty in the assembly, on councils and juries, and in administrative offices where policies and decisions were openly debated and decision-making was by majority. With the institution of pay for service to the *polis*, political influence reaches the economic margins and political views become especially diverse. By the time of Protagoras, any citizen could propose legislation and speak in the assembly (the *ekklesia*), and any wronged citizen could appeal for justice and present his own case to a jury of peers (the *dikasteria*). In the Athenian public sphere, then, rhetorical exchange becomes the essential agency of the community's business. Those citizens who could argue well could shape political opinion, while all citizens were routinely exposed not just to endemic disagreement but also to the crafted antilogies (alternative logoi) of public debate. In this environment, Sophistic training in argumentation becomes what Jacqueline de Romilly aptly calls "a *techne* for citizens" (1992, 33).

Add to these circumstances the fact that the Age of Pericles (461-29) was a period of constant turmoil, bracketed by the first Peloponnesian War (461-51) and the plague of Athens (430), and one can easily imagine the variety and vehemence of the argumentative discourse with which the citizens of

Athens were routinely engaged. In such a situation, Protagoras' insight into the dialogic nature of argumentation might well be described (to borrow Gadamer's phrase) as "an effect of history" (1994, 265ff). More specifically, we can reasonably infer that the conceptualization of antilogic follows not only from Protagorean philosophy but also from the empirical observation of the forensic, deliberative, and social discourse of the time (B. Smith 202).[20] We already know that Protagoras himself was a regular participant in the discourse competitions that became so popular in mid-century Athens. We also know that he was commissioned to devise the laws for the Panhellenic community of Thurii (*DK* 80 A1) and that he describes his own teaching as preparation for citizenship (*Protagoras* 319a). So Protagoras was himself professionally immersed in the argumentative climate that conditioned Athenian public life. Moreover, the Athenian preoccupation with argumentation extends well beyond the *ekklesia* and the classroom. Think, for example, of the general excitement with which the debate between Socrates and the Sophists is greeted at the outset of the *Protagoras* (309a-311a), or the mock-forensic performances of Gorgias' "Helen" and "Palamedes" (*DK* 82 B11 and 11a), or the internal argument of Prodicus' "Choice of Heracles" (*DK* 84 B2), or the antilogical speeches of Antiphon (*DK* 87 B1). In addition to these literary arguments, there are also the many paired, antithetical speeches in the drama of Euripides and the history of Thucydides.[21] In the latter, we find the following list of antilogies which surfaced during the Peloponnesian war, a list that nicely illustrates the Athenian tendency to think in antitheses:

> Reckless audacity came to be regarded as the courage of a loyal ally; prudent hesitation, specious cowardice; moderation was held to be a cloak for unmanliness, ability to see all sides of a question, inaptness to act on any. Frantic violence became the attribute of manliness; cautious plotting, a justifiable means of self-defense. (3.82)

The Athens of Pericles, then, provided ubiquitous evidence that, given a certain level of social and discursive freedom, human beings incessantly exercise their capacity to contradict, to map out the differences between their own "measurement" of an issue and the perspectives advanced by others. To Protagoras, therefore, a kind of natural antilogic must have seemed the habitual practice of Athens.

As a theory of origins, the empirical hypothesis finds support in the work of social psychologist Michael Billig, who writes that while the scientist must be cautious in accepting universal theories of behavior, there does seem to be a "two-sidedness" to our thought, so that human beings seldom approach any issue from a single point of view. Indeed, "our species" is so fully characterized by this inclination to contrary views that "(a)ny psychological theory of thought which omits the role of the antilogos will therefore be incomplete" (49). It was the genius of Protagoras, adds Billig, not only to recognize the two-sidedness of human thought but also to provide a philosophical framework for tracing its implications *and* a theory of discourse that applies this theory to natural practice (39-50). If such an achievement, built as it may have been on direct observation, seems humble in comparison with the metaphysical sublimity of Plato or the architectonic thoroughness of Aristotle, recall that the procedures of Protagorean argument nonetheless constitute the initial method and enduring practice of democratic discourse. Seen thus, they become (as Eric Havelock argues) an invention of "no small historical importance" (219).

Of course, the tendency toward antilogic that becomes so prominent in Periclean Athens is not itself without pedigree. In Homer, there is a distinct tension between two contrasting features of human excellence, or *arete*. On the one hand, the heroic code of action encourages the military, competitive, selfish drive to "always be first and best" (*Iliad* 6.208; C. Kahn 1979, 12); and on the other, there is a social, cooperative tendency expressed in the term *sophrosyne*, or restrained reason. The voyage of the *Odyssey*, which begins with the senseless sacking of the Kikoneans (9.39ff) and reaches its apogee in the eloquent humility of Odysseus among the Phaeacians (Bks. 9-14), can be read as a meditation on the appropriate balance between the opposing virtues of heroic valor and common sense. Charles Kahn writes that "both views, the selfish and the social conception of *arete*, and the deep tension between the two, were there in the moral bloodstream of the Greeks long before philosophy appeared on the scene" (14).

Just how ingrained such contraries were in the early Greek mind is clarified by G. E. R. Lloyd's study of "polarity and analogy" as types of argument common to the earliest expressions of Greek speculative thought. Lloyd writes that few who study Presocratic philosophy "can fail to be struck by the recurrent appeal to *pairs of opposites* of various sorts both in cosmological doctrines and in the accounts of particular natural phenomena" (7). In fact, Aristotle, in his *Physics*, indicates that "all" his predecessors invoked contraries as a way of structuring their discussion of nature (188b, 27-30). Lloyd explains this universal appeal to contraries by noting that (1) "many prominent phenomena in nature exhibit a certain duality" (e.g. hot/cold, summer/winter); (2) these dualities often acquire "added significance as the symbolic manifestations of fundamental religious or spiritual categories" (e.g. good/evil, yin/yang); and (3) "opposites provide a simple framework of reference by means of which complex phenomena of all sorts may be described and classified" (80). To this last point, he adds elsewhere that opposites provide "abstract clarity and . . . apparent comprehensiveness" and so help to resolve the most complex problems into a simple, workable schema (66-67). Among the many Presocratic thinkers that he considers, Lloyd distinguishes Heraclitus as "exceptional" for his emphasis on the "interdependence" and "unity" of opposites (17); and because this concept of alliance-in-opposition looks forward so directly to Protagoras, Heraclitus deserves some attention here as a significant antecedent to the conceptualization of antilogic.[22]

Most readers who are already familiar with Heraclitus (a late 6C BCE native of Ephesus in Asia Minor) will no doubt connect him with the doctrine of perpetual flux, or *panta rhei* ("all things flow"). This concept is expressed in his famous aphorism that "one cannot step into the same river twice" (*DK* 22 B91).[23] Parmenides and the Eleatic philosophers would, of course, reject this radical emphasis on change as an illusion, a *fable convenue* brought about by a false and unphilosophical reliance on our sense experience of the material world (M. West 1991, 137). Nonetheless, Plato indicates that this paradoxical notion of permanent flux was held by "a whole series of philosophers," including Protagoras (*Theaetetus* 152e). For many of the early Greek natural philosophers (especially the Milesian cosmologists Anaximander and Anaximenes), the process of change is structured by an unremitting oscillation between opposite poles (day/night, summer/winter), so that the world becomes "a cycle of elemental interchange" between antithetical powers (C. Kahn 80; see *DK* 12 B1). Or as Heraclitus puts it in one of his approximately 130 fragments, "cold warms up, warm cools off, moist parches, dry dampens" (*DK* 22 B126). What distinguishes Heraclitus' theory of dynamic interaction between opposites, however, is the notion that opposites are, in fact, held together in a common unity. So, in another fragment we hear that "from discord comes the fairest harmony" and that "the counter thrust brings together" (*DK* 22 B8). Even in the flowing river, there is a permanence that is the necessary counterpart of change: "as they step into the *same* rivers, other and still other waters flow upon them" (*DK* 22 B12, italics mine); or, conversely, "in changing, things find repose" (*DK* 22 B84a). This paradoxical conjunction of antitheses extends not just to nature, but also to the human world, in which such dualisms as sleeping and waking, youth and age, even life and death are brought together in what Kahn describes as a "total unity within which all opposing principles . . . are reconciled" (21).[24]

It is important to recognize, however, that within this Heraclitean "unity of opposites" reconciliation does not eliminate opposition, as it would in a three-stage dialectical system where antitheses surrender to synthesis. Rather, opposites are seen here as dyadic coordinates of a unified system, a system that takes its essential form from the omnipresence of tension and conflict between its polar elements. From the Heraclitean perspective, then, life presents itself as "an ontological passage from contrary to contrary," a passage in which the dynamism that comes from division never resolves itself in the stasis of merger (Wheelwright 6; see Burke 1969, 403-20). In illustration, we might return to the fragment "out of discord comes the fairest harmony" which is derived from a reference to Heraclitus by Aristotle and which can be translated more literally as "from tones at variance comes perfect attunement" (*harmonia*), and to which is added "all things come to pass through conflict" (*DK* 22 B8; Wheelwright 90; *Nicomachean Ethics* 8.1, 1154b4; C. Kahn 63). But how is it that "a thing at variance with itself" can produce *harmonia*? In fragment *DK* 22 B51, Heraclitus cites as symbols of this harmonious tension first the lyre, whose strings might easily produce discord unless they are coordinated (*harmonia*), and second the bow, which requires the archer to pull the arms in opposite directions to produce the singular flight of the arrow. In a brilliant analysis of these images, Charles Kahn points out that *harmonia*, the skillful fitting together of opposites, is "the mediating concept" that renders intelligible the pattern of polarities that characterizes the Heraclitean worldview (199-200). Once armed with this "key to the system," it is easier to make sense out of the enigmatic concepts that we find throughout the Heraclitean corpus. I cite just a few examples:

> [T]he nature of all things was constructed according to conflict. (*DK* 22 A22)
>
> War is the father of all . . . . (*DK* 22 B53)
>
> Conflict is Justice. (*DK* 22 B80)
>
> The way up and the way down are one and the same. (*DK* 22 B60)

Inherent in all these statements is a deconstruction of the *prima facie* interpretation of conflict and a reconception of opposition as a positive, productive force. This transformation of conflict into a potential source of *harmonia* will, in turn, motivate the Protagorean approach to antilogic. For just as Heraclitean cosmology is animated by opposites in harmony, Protagorean humanism seeks the coalescence of diverse views in the context of continuous challenge. That is, Protagorean doctrine places every logos in relation to opposing positions; but, at the same time, the practice of antilogic seeks to ameliorate the resulting tension and transform social practice by reconciling conflict with cohesion, commonality with difference. Just how this reconciliation proceeds is a subject for Chapters 3 and 4; for now, the Heraclitean worldview provides a helpful framework for thinking about the "harmony in conflict" that is central to antilogical practice.

But before I turn my attention from this catalog of antilogic's antecedents, one addition to the list of "influences" deserves reemphasis. Mario Untersteiner claims that Protagoras "found already in existence a serious problem, more or less clearly formulated . . ." (25). The "problem" referred to is one of ubiquitous opposition, of two-sideness in all matters, from the cosmos and nature to politics and moral values. I have attempted to highlight this tendency toward opposites in the existential practice of Athenian social and literary discourse and in the philosophical concerns of Heraclitus. In fact, as G. E. R. Lloyd points out, there are many other Presocratic theories on the interplay of opposites (15-71). But Untersteiner does, I think, overstate the case in claiming that Protagoras happens upon a pre-existing intellectual formula; and we risk withholding from Protagoras his due measure of originality if we forget the claim made in Chapter 1 that the human-measure doctrine—with its emphasis on individual perspective, on the elimination of absolute guarantees for knowledge claims, and on the resultant pluralism of any discursive situation—if we forget that this complex of ideas serves to connect antilogic as a theory of discourse with a larger frame of philosophical reference.[25]

It can be argued that the concept of antilogic occupies pride of place as the first principle of Protagorean doctrine since it is based on empirical observation of existential fact, and that the more theoretical principles of the human-measure and "on-the-Gods" fragments are deduced from that observation. But from my point of view, such a chronology or arrangement is misleading and fails to recognize the primary status of the human-measure doctrine. The arrangement of this book in itself stakes a claim for the human-measure doctrine as a general philosophy of human experience and perception, a philosophy which is remarkably prescient in its emphasis on the act of interpretation as a primary constituent of our effort to understand the world. Through this insight into "measurement" as the medium of human understanding, Protagoras establishes a context for knowledge in which universals are eschewed and the experience of interpreting/evaluating/measuring is placed at the center of things. Regardless, then, of the other powerful influences "already in existence" in 5C Greece, influences which fill out the historical context of antilogic, we would be remiss if we didn't also assert that the human-measure doctrine maps out the primary principles of the Protagorean worldview and that antilogic works to ground these same principles within the domain of discursive practice.

Chapter 3

Pragmatism, Ethics, and the Function of Antilogic

> "*See how reason provides plausibility to different actions. It is a two-handled pot, that can be grasped by the left or the right.*"
>
> Montaigne, "Apology for Raymond Sebond"

> "*At every opposition we do not consider whether it is just, but, right or wrong, how we can get rid of it. Instead of stretching out our arms to it, we stretch out our claws.*"
>
> Montaigne, "Of the Art of Discussion"

In Chapter 1, we explored the philosophical base of antilogic and the fragmentary Protagorean texts; in Chapter 2, we reviewed the general features of antilogic itself, along with its nearest relatives, potential origins, and original milieu. With this background in place, we can now concentrate on particular features of our subject that define its distinctive contribution to the rhetorical tradition and account for its continued relevance. To this end, I have chosen three constituent elements of antilogic that characterize its unique nature and promise. In this chapter, I take up the first two: the underlying **pragmatism** and controlling **ethics** of antilogical argumentation. I hope to show that antilogic, unlike dialectic, will naturally combine questions of knowledge with matters of conduct, so its procedures of inquiry are invariably oriented toward practical action. In addition, there is an obvious conflict between my claim in the last chapter for antilogic's egalitarian politics and the traditional charge that Protagoras would substitute worse arguments with better ones, a conflict that requires us to attend to the ethics of antilogical exchange and to the place of moral conduct in the Protagorean social order. In the next chapter, I follow these studies in pragmatism and ethics with an investigation of more familiar rhetorical territory: namely, the **formal elements** that distinguish antilogic and constitute the norms of its unique practice. My goal is to round out Part I by detailing the cardinal features of my subject and, in so doing, provide a comprehensive overview of this provocative, if unfamiliar approach to argumentation.

1. THE PRAGMATIC DIMENSIONS OF ANTILOGIC

Throughout *Many Sides*, I have attempted to render the past in a form useable for contemporary practice. With the present topic, there is a natural kinship between past and present. Readers will recall that Protagorean antilogic transforms Presocratic theories on the dualism of nature and the cosmos by applying the Heraclitean concept of related contraries directly to logos; i.e., to language, reason, and argument.[1] In so doing, Protagoras locates all consideration of the "real and true" in direct relation to human linguistic faculties. In Aristotelian terms, discursive exchange becomes the "formal cause" of what we know about the world, the way in which the material of experience takes shape for knowing subjects.

Given such notions, it is hardly surprising that a host of contemporaries—including Harold Bloom, Stanley Fish, Steven Mailloux, Joseph Margolis, Richard Rorty, Cornel West and others—have acknowledged Protagoras as the first philosopher of "linguistic consciousness."[2] If we accept Gustav Bergmann's now famous comment that "all linguistic philosophies talk about the world by means of talking about language," then Protagorean theory is clearly an early effort to talk about the world by contemplating the way in which logos serves to mediate experience and give form to what we know (in Rorty 1967, 8). We can recognize in Protagoras, then, an ancestor to language-based epistemologies even if Protagoras' own influence has, until recently, been modest at best. What has not been well recognized in contemporary commentary is the fundamental role of antilogic in this prototype of the "linguistic turn." In response to this neglect, it is worth dwelling for the moment on the idea that while antilogic indeed foregrounds the linguistic basis of human understanding, it refines this notion by imagining the epistemic process as a constant oscillation between contending logoi. This refinement yields other, productive continuities between Protagoreanism and modern thought.

For example, as opposing forces of nature, contraries remain relatively distinct.[3] In the medium of language, however, the rigid separation of ideas and things tends to dissolve. Language is not simply a tool used by speakers to give discursive form to this or that object or datum; rather, it is an encompassing space within which multiple voices coalesce in the process of understanding, a space where the boundaries of self and other (as well as signifier and signified) are mediated and transformed. Heraclitus claimed that fire erases all distinctions. Discourse may not have quite the same power to erase boundaries, but antilogic would nonetheless operate as a kind of crucible within which various representations of the world (Protagorean "measurements") mix and change form as a result of contact and contention (cf. Bineham 1995, 1). This conception of argument, of course, goes beyond a general anticipation of linguistic consciousness and looks forward directly to those modern philosophies that place the dialogical "mixing" of voices at the core of human experience: most particularly, Bakhtinian dialogism, philosophical hermeneutics, Buber's existential anthropology, and American pragmatism. I will take up various theoretical affinities in due course; for now, we can press the ancient/modern relationship a step further by taking a clue specifically from pragmatism and pursuing the hypothesis that the "linguistic turn" performed by Protagoras in the 5C BCE is a precursor to the notion that language in general (argument in particular) is a form of action in its own right, an effort to make discourse productive not simply as an agent of "intellectual consequences" but also a as means for creating "tangible effects in the public sphere" (Downing 1995, 185). If we would comprehend the full dimensions of antilogic, it is time to consider what it means to introduce a theory of argument that is also a theory of action.

What, then, do we mean when we say that something is practical or that someone is pragmatic? Since we are dealing here with the very origins of these concepts in Western history, perhaps it is best to start with etymology. The English term "pragmatic" (as in a concern for practical affairs and consequences) is of relatively modern usage: William James writes in 1902 of a "thoroughly 'pragmatic' view of religion [that] has been taken as a matter of course by common men" (*OED* 2265). But the root of the term comes from the ancient Greek *pragma*: a deed, a thing done and, in particular, an act of public business or private affairs. Similarly, the ancient Greek *praktikos* implies "fit for action or business" (*Liddell and Scott* 581-82). We can follow this etymological trail a step further if we return to the human-measure fragment and the notion that "humanity is the measure of all things." The word for "things" in this case is not *pragmata* (as it is in the two-logoi fragment) but *chremata*, which Laszlo Versenyi tells us connotes things that we use, or deal with, or events that in some way affect us (1963, 11-13; see Schiappa 117). In an age dominated by the metaphysical speculations of Parmenides and the abstract theorizing of the 6C natural philosophers, Protagoras' choice of the term *chremata* "announces a practical program for philosophy," and "recalls" humanity back to "a world of practical application" (Versenyi 1963 12-13).[4] Protagorean thought, therefore, represents a turn away from arcane theorizing about causal powers and formal abstractions and toward a frame of reference in which human relations, attitudes, opinions, and actions are "decisive" (ibid. 12). So, in a strictly Protagorean context, to be pragmatic is to attend to the business of humanity; or, to extend this base definition slightly, Protagoraean argument seeks to negotiate useful courses of action by executing appropriate judgment in the affairs of society.

Such a definition is not far removed from Aristotle's famous description of practical reason (*phronesis*) articulated in the *Nicomachean Ethics* and developed in both the *Politics* and the *Rhetoric*. Most succinctly, *phronesis* is "a true and reasoned state or capacity to act with regard to human goods" (*Nic. Ethics* 1140b 4-5). Unfortunately, scholarship surrounding the meaning and function of *phronesis* as a central tenet of Aristotelian ethics is distinguished by its complexity, controversy, and—for this reader—a notable lack of clarity.[5] Nonetheless, we can glean from Aristotle some help in understanding the pragmatism of antilogic. According to the complicated typology of Aristotle's ethics, there are five "virtues" or "excellences" by which the soul possesses knowledge: art (*poiesis*), scientific knowledge (*episteme*), practical wisdom (*phronesis*), philosophical wisdom (*sophia*), and instinctual understanding (*nous*) (*Nic. Ethics* 1139b 15-17). But in the discussion of *phronesis* in Book 6 of the *Nicomachean Ethics*, these mental categories are distilled into three basic forms: theoretical wisdom, which is concerned with knowledge for its own sake; the various arts (e.g., sculpture and medicine), which are concerned with making something; and practical wisdom, which is concerned with action. Hence, we have the standard division of Aristotelian knowledge into theoretical, productive, and practical forms, all of which must be informed by *orthos logos* (right reason) to function effectively.

Perhaps most helpful for our study is the distinction between theoretical and practical wisdom, since these two constitute what Aristotle calls the two "intellectual virtues" (ibid. 1103a 406). The function of both these intellectual faculties is the attainment of truth (1139b 12), but they have very different domains and methods. The aim of theoretical wisdom is to understand the universal and invariable ("what is not capable of being otherwise"), i.e., knowledge as we might construe it in the realm of theology, mathematics, or physics (1139b 21; Hardie 222). Its constituent features are *nous*, or the instinctive grasp of or insight into the first principles from which reasoning begins (see *Posterior Analytics* 71a 1), and *episteme*, the ability to understand and demonstrate the truth of universals in terms of their causes (*Nic. Ethics* 1139b 31-3). It follows that in Aristotle's conception of theoretical wisdom, argument finds its appropriate function (*ergon*) in the logical explanation of universal concepts (ibid. 1140b31-41a8).

On the other hand, Aristotelian *phronesis* is characterized by its concern for truth "in agreement with right desire" (ibid. 1139b 30; i.e., a desire that shows itself in our choices about courses of action in particular situations; cf. 1142a 23-30). *Phronesis* (or practical wisdom), is consequently as much a matter of ethics and character as it is of intellect, since moral virtue informs our ability to make rational choices about actions to be performed. I will return to this important connection between *phronesis* and ethos in the next section; for now, I would continue my outline by noting that Aristotle posits as the two aspects of practical wisdom (corresponding roughly to the *nous* and *episteme* of theoretical wisdom) the ability to make rational choices (*prohairesis*) and the ability to deliberate effectively (*bouleusis*) about what these choices should be. He writes:

> Now it is thought to be the work of a [person] of practical wisdom [the *phronemos*] to be able to deliberate well about what is good and expedient for himself, not in some particular respect . . . but about what sorts of thing conduce to the good life in general. (*Nic. Ethics* 1140b 25-28)

This ability to deliberate well (*euboulia*) manifests itself in both the political skill of a Pericles and the more routine demands of "household management" (1140a 7-11). Seen thus, deliberation is concerned with finding the appropriate means to an end, or alternatively, in justifying the nature of one's choice. If we assume, then, that the *phronemos* begins with circumstances or difficulties (*aporia*) that require response and with the choice of an objective to be achieved by one's response, then deliberation is the process of considering the means for accomplishing one's objective (Broadie 228-42). Whereas the argument of *sophia*, or theoretical wisdom, addresses the notion of "why" something is as it is, the argument of *phronesis* is devoted to "how" an objective (or desire) can be effected and justified. Or more succinctly, our proairetic ability to choose well (i.e., to choose in accordance with moral excellence) "fixes" the target and our practical wisdom allows us to determine how best to get there (*Nic. Ethics* 1144a 5-7).[6] Taken together, the ability to choose a goal worth pursuing and to deliberate effectively on the means of this pursuit, these two features of *phronesis* prepare the way for *eupraxis* or good action, which Aristotle sees as the end of *phronesis* itself (ibid. 1140b 7). Since philosophical wisdom itself "moves nothing" (1139a 36), it falls to *phronesis* to perform the critical role of guiding all action, large or small, connected with human welfare (ibid. 1140b 21-22).

My synopsis of Aristotelian *phronesis* is highly simplified and somewhat too schematic, even for Aristotle. But it is intended only to provide a workable vocabulary for application to the analogous pragmatism of

Protagoras, not as an account of Aristotle's *organon*, whose interpreters are legion. Given our more modest goal, we can now say, in the first place, that antilogic, like *phronesis*, is a deliberative process committed to right reason (*orthos logos*) rather than an intuitive insight into first principles (like the *nous* of philosophical wisdom). Second, antilogic seeks practical rather than theoretical knowledge (in the Aristotelian sense) for the simple reason that its basis in Protagorean philosophy eliminates the epistemic appeal to universals or absolutes and so consigns its investigations to the indeterminate realm of human experience. Third, antilogic, like *phronesis*, operates outside the scope of art (*poiesis*) or science (*episteme*) and so functions without the strictures of exact rules, a point I will return to repeatedly (Broadie 191). Fourth, while antilogic tends to culminate, as does Aristotelian *euboulia* (deliberation), in reasoned choice, that choice cannot be approximated until after the dialogical encounter of alternative positions. Unlike Aristotelian *euboulia*, then, which begins with an argument already established in one's "choice" (or resolution), antilogic is a method of inquiry into the options for choice itself, not an a posteriori explanation of a choice already made, a distinction of some significance for rhetoric and rhetorical history.[7] And finally, if antilogic is not the justification of future activity to be undertaken after due deliberation, then antilogic is itself—to borrow from Friere—a form of social action "informed by reflection"—i.e., a practical attempt to intervene in some unsettled circumstance (*aporia*) and to execute the human capacity for right reason by first and foremost drawing attention to the multiplicity that surrounds the issue in question (Friere 1970, 100).

So, like *phronesis*, antilogic is a "capacity to act" with regard to situational demands, but this capacity involves more than the ability to implement a resolution or actualize a judgment that is itself the culmination of the deliberative process. Nor is it simply the "calculation" of appropriate action to follow (*Nic. Ethics* 1140a 29), or "correctness of thinking" in the deliberative process (1142b 14), or "a quality of mind" concerned with right action (1143b 21). Instead antilogic is a *praxis*, an activity whose fulfillment lies in its own effective execution, which in this case means that the equitable exchange of contrasting logoi is an end in itself, since no single logos can hope to transcend its own partiality and since the elimination of dialogue would destroy what Bakhtin calls "the very sphere in which the word [logos] lives" (in Bialostoski 1995, 86). This is not to say that a single, determinate choice is not possible in antilogic; rather, any choice that necessity requires is not an "end" in itself since, according to the norms of antilogic, there is no single argument that might not benefit from further interrogation by contraries. Consequently, antilogic is best seen as a mode of conduct, an ongoing, self-critical activity and not a speculative preliminary to action. These and the foregoing distinctions regarding antilogic press the Aristotelian analogy about as far as is serviceable; to develop Protagorean pragmatism further, we had best trade in an ancient lexicon for a modern one.

Richard Bernstein, in *Praxis and Action* (1971), writes that the modern tradition of philosophical pragmatism, and especially the pragmatism of C. S. Peirce and John Dewey, is best approached as a critique of "the spectator theory" of knowledge that descends from Descartes (174).[8] According to the Cartesian paradigm, the individual is a fixed observer, separated from the world around him or her but capable of knowledge about that world through the adoption of rigorous neutrality and the exercise of rational method. Such method, it is argued, can overcome the vagueness, confusion, and bias of common thought and establish a firm foundation for legitimate knowledge in "clear and distinct" ideas arrived at through neutral, objective cognition (see *Discourse on Method*, Part II). Richard Rorty refers to the conception of knowledge outlined by Descartes (and advanced by Locke and Kant) as an "ocular metaphor," a theory in which the eye of the mind inspects and comprehends a world of matter that is separate from and outside of the mind itself (1979, 3-69; quote, 13). For our purposes, the critical feature of the spectator theory is the notion that reality consists of cognitive and material substance (*res cogitans* and *res extensa*) and that these two are mutually exclusive. In such a conception, thought and action are opposing categories, while theory (a term derived from the Greek verb "to look at" and so etymologically aligned with spectatorism) requires a separation from practice to operate effectively. It is just this categorical split between mind and matter, thought and action that Peirce chose to challenge.[9]

In an 1868 series of essays on the "categories" of phenomena, Peirce writes that "there are always two ways of describing the same experience," and that "consciousness *itself* is two-sided, so it has two varieties, namely action, where the modification of other things is more prominent than their reaction to us, and perception, where their effect on us is overwhelmingly greater than our effect on them" (in Bernstein1971, 180). As a response to Descartes, it is notable that the "two-sidedness" represented here by the actions of an individual agent on the one hand, and by the perception of brute fact or "things" on the other, that this seeming dichotomy is comprehended within the unifying realm of "consciousness."[10] But the kinship of Peirce's statement with Presocratic concepts is also striking. In the first place, Peirce's comment that "there are two ways of describing the same experience" is a very close (though I assume an unwitting) approximation of the two-logoi fragment.[11] And second, his deconstruction of the agent/things dichotomy reflects both the Heraclitean unity of opposites and the Protagorean notion that "things" are always seen in relation to humanity. These parallels aside, what Peirce refers to in the passage above as "consciousness" he also describes as "experience"; and, for the pragmatic tradition, this emphasis on experience functions as a comprehensive critique of Cartesianism and an alternative theory of knowledge. From within the Cartesian tradition and especially from the perspective of empiricism, the concept of experience tends to be conflated, as Bernstein says, with "the notion that what experience forces upon us has absolute authority" (1971,181). But pragmatism will shift the emphasis in the subject/object dichotomy back toward the agent and, in the process, experience takes on a new meaning. Indeed, for the pragmatist, experience—and the allied notion that the individual is an active participant in the process of experience—becomes the "key of understanding all aspects of human life, including human inquiry and knowledge" (Bernstein 177).

Building on the framework of Peirce's insights into the nature of experience, human agency, and the norms of inquiry, Dewey makes the "reconstruction of experience" the central motif of his own philosophy. In "The Need for a Recovery of Philosophy," he advances a series of distinctions between the Cartesian and the pragmatic conceptions of experience. He first notes that experience is not "a knowledge affair," i.e., the contemplation of matter by mind. Instead, it is "an affair of the intercourse of a living being with its physical and social environment" (1970, 23). In the process of this intercourse, the world that we "traditionally" consider objective (the world of natural phenomena) "enters into the actions and sufferings of [people] and undergoes modifications through their responses" (ibid. 23). The idea here replicates Peirce's (and Protagoras') notion that the agent will always influence the experience of interaction between individuals and the environment. From the pragmatic point of view, then, the spectator-mind, enthroned in the enclave of its own cognition, is a misrepresentation of experience, an illusory perspective instituted in large part by the commitment of the Cartesian/empirical tradition to "particularizing" distinctions (like self and other) which are not appropriate in the vital context of experience (ibid. 23). An experience, says Dewey, is "pregnant with connections" and is better conceived of as a constantly shifting process of overlapping influences, a process that is (from the perspective of one's actual life) holistic (ibid. 23-29). As Bernstein explains, subject and object become, for Dewey, "names for changing functional distinctions" that exist *within* the aegis of experience (1971, 205; cf. Peirce on agents and "things" above). In addition, Dewey writes that experience is "characterized by projection, by reaching forward into the unknown," by our connection with the future (1970, 23). We don't simply approach the world as a given; we also seek to manipulate it, to transform it, to bring it into accord with our own desires. By virtue of our "experience," then, we exhibit ourselves as active, vital beings whose interaction with the world is invested with intention and purpose.

Dewey concludes his pragmatic reconstruction of experience by noting that if the human experience of interaction with the world is characterized by intentionality, it is also distinguished by reflection. "In the traditional notion experience and thought are antithetical terms," he writes, "(b)ut experience taken free of the restrictions imposed by the older concept, is full of inference. There is, apparently, no conscious experience without inference; reflection is native and constant" (1970, 23). To the degree that our experience is informed by this reflective spirit, it becomes "intelligent" (34), and so the boundary between thought and action, like that between subject and object, is dissolved in the matrix of experience. The term that Peirce and Dewey use most consistently to describe our intelligent interaction with the world is "inquiry." But this form of inquiry has little relation to the "objective" speculation of Cartesian tradition. For Dewey and Peirce, inquiry is a mode of conduct in which our habits of inference—our ability to move from one idea to another and to build from initial reflection towards knowledge—are guided by procedures and norms over which we can exercise a considerable measure of self-control (Bernstein 1971, 188-89).

So, though our experience is certainly conditioned by the nature of the world with which we interact, our ability to reconstruct experience through our procedures of inquiry allows us to act decisively in pursuit of our own desires. Dewey writes elsewhere that inquiry arises "when there is something seriously the matter, some trouble, due to active discordance, dissentiency, conflict . . ." (in Bernstein 1971, 204-5). Correspondingly, a specific procedure for inquiry is *a deliberative action* designed to meet that discord through the intervention of human agency for the purpose of achieving some imagined human good or objective. What distinguishes antilogic as a method of inquiry from other argumentative practices is its absolute commitment to interaction with other interlocutors, to the experience of give-and-take with opposing positions. In many every-day (i.e., uncontrolled) arguments, we all know that the experience of interaction is often either aimless or may actually even aggravate conflict. And in any formal argument (in which contact with contesting logoi is postponed until after a choice or decision has been made), the concept of interaction is predominantly theoretical because a single, dominant logos has been elevated above other challengers. The purpose of antilogic, however, is not just to tolerate difference but to establish what Dewey calls "a purified medium of action" for the conduct of inquiry, a privileged space inside of which opposition is welcomed and transformed by the norms of antilogical *praxis* into a source of strength and invention (*Democracy and Education* 1916, 24). Because it insists that there are always at least two sides to every question, antilogic seeks to locate the action of inquiry within the experience of opposing voices—not only as a way of overcoming the inherent partiality of individual inference and transcending the boundaries between opposing positions, but also because such experience is a good way to get things done.

Once again, then, antilogic is not the prelude to an action to be undertaken after a choice has been made; rather, it is the active construction, through dialogue with others, of new possibilities for the resolution of uncertainty and the fulfillment of desire. In short, it is not just the capacity to act, but action itself "in regard to human goods." It is *eupraxia*. At the end of "The Need for a Recovery of Philosophy," Dewey writes that philosophy "recovers itself when it ceases to be a device for dealing with the problems of philosophers and becomes a method, cultivated by philosophers, for dealing with the problems of [people]" (66-67). Because antilogic encourages and organizes the experience of interaction among diverse people and logoi, and because it changes with the constant situational fluctuations that condition debate, it is certainly more than a device, or *techne*. Its power comes from the fact that it is a *praxis* explicitly focused on acting constructively in response to the endemic disagreements that constitute human social experience.

A "pragmatic" conception of antilogic, then, reveals an instrumental procedure that responds to the occurrence of some disruption or problem in the state of human affairs by guiding inquiry in search of understanding. The *summum bonum* of this particular procedure is the interaction between critical agents who would substitute as their methodology the give-and-take of dialogue for either the linearity of formal logic or a dogmatic insistence on unimpeachable first principles. Unfortunately, however, no examples of Protagorean argument exist, despite the fact that Diogenes Laertius includes a reference to two books of *Opposing Arguments* in the Protagorean canon (v.2, Ch. 9, 55; *DK* 80 A1). But there are some clues about the original practice of antilogic in the biographical record, and there are relevant examples of 5C and 4C Athenian argument that are undoubtedly informed by Protagorean theory. A brief look at these should allow us to substantiate the practical function of antilogic in its original form.

The most interesting biographical incident is contained in the passage referred to toward the end of Chapter 2 as the Great Myth, the narrative that Plato attributes to Protagoras in the dialogue named after him (see *Protagoras* 320c-322d). As you may recall from that earlier discussion of the myth, at this point in the dialogue Protagoras is defending his own teaching. Just before he begins his mythic tale, he stresses the practicality of his teaching by noting that his subject is

> the proper care of [his pupils'] personal affairs, so that [they] may manage [their] own households, and also of the state's affairs, so as to become a real power in the city, both as a speaker and a [person] of action. (318e-319a)

Socrates then suggests that the proper name for this subject is "the art of politics" (*politike techne*) or "making good citizens"; but Protagoras' pedagogical "subject," with its domestic and civic components, might just as well be called "practical wisdom" (cf. *phronesis*: *Nic. Ethics* Bk. 6, 1140b 8-11). Protagoras is said to have written a work titled *On the Original Condition of Mankind*, and his myth in Plato is an account of origins in which humanity first receives from Prometheus the gift of language and the various crafts of survival (housing, clothing, and agriculture) (*DK* 80 A1). The initial emphasis, then, is on the utilitarian means for rudimentary survival. But these resources are not enough to insure permanent security because "when they gathered in communities [the people] injured one another for want of political skill, and so scattered again and continued to be devoured" (*Protagoras* 322b). The unique ability (in comparison to other animals) to congregate in large groups and form cities was praised later by Isocrates; but, according to Jacqueline de Romilly, it is only Protagoras among the early Greek thinkers who identified the practical need for social skills (cf. Isocrates in the *Nicocles* 6 and *Antidosis* 254; de Romilly 1992, 164). Accordingly, in the Protagorean myth, the gods not only confer "technical skill" on humanity, but when *techne* alone proves inadequate for survival, Zeus also sends to the people "the qualities of respect for others [*aidos*] and a sense of justice [*dike*] so as to bring order into our cities and create a bond of friendship and union" (322c).[12]

Laszlo Versenyi writes that the emphasis throughout this myth is on practical excellence in living rather than on abstract-theoretical pursuits (1963, 23-30). In this instance, justice and respect become the foundation that allow for the creation of social harmony in an environment that Protagoras believes (as we know from the human-measure and antilogic fragments) is characterized by the omnipresence of disagreement. Not long after the narration of the myth, Protagoras remarks to Socrates that "goodness" is especially "diverse and multiform" (334b). Under such circumstances, when the human perception of any virtue or action is bound to involve conflicting views, justice and respect express themselves in the ability to mediate potential discord and to arrive at common agreement (*homonoia*) in particular circumstances. Eric Havelock describes the myth as extolling a "methodology of social judgment . . . leading to practical conclusions" (1957, 192). Just how that methodology will operate will have to wait until we address the norms of antilogical practice in the next chapter;

but from the point of view of our present focus on pragmatism, the Protagorean myth draws an image of political wisdom as a practical virtue, a means by which the wise person (or Sophist) can invoke justice and respect in an effort to overcome conflict and find out what works to create harmony in our private and civic lives.

The practical orientation of Protagorean rhetoric is also on display in a fascinating incident first mentioned in Plutarch's *Lives* (*DK* 80 A10; "Pericles" 36). The Greek biographer (c. 45-123 CE) alludes to a conversation between Protagoras and Pericles that spans an entire day and in which the two discuss the case of a pentathlete who, when throwing his javelin, unintentionally kills a bystander. The question is who is at fault: the athlete who threw the javelin, the director of the games who is responsible for the grounds, or (according to an arcane point of Athenian law) the javelin itself (see Untersteiner 1954, 31-32). And while we don't have the actual debate between these two great men, we can be confident that the issue was argued antilogically. Our principal evidence comes from Antiphon, the first Athenian logographer (or speechwriter) and a contemporary of Protagoras who, in his *Tetralogies*, provides hypothetical examples of forensic antilogic. In these mock-speeches, the prosecution and defense both get to speak twice (hence "tetra-logy"), with each ensuing speech responding to and, typically, reversing the claims of the previous speaker. The second of Antiphon's tetralogies (*DK* 87 BII) is devoted to what we might call "the case of the fatal javelin," and in it Antiphon brilliantly lays out a contrasting set of arguments based on motive and probability that are put forward by the opposing sides.

There are, of course, other famous examples of this kind of dialectical exchange in the non-dramatic literature of the 5C. In Parmenides' poem, for example, arguments for the Way of Truth and the Way of Seeming are placed in rather lopsided opposition; and in Prodicus' "Choice of Heracles" (*DK* 84 B2), the goddesses of Virtue and Vice argue in defense of the benefits they each have to confer. But the abstract, philosophical nature of these two debates contrasts starkly with the concrete, practical nature of antilogic as employed in the case of the javelin. In Antiphon, the emphasis is not on the dominance of one element in a dialectical pair (truth over seeming, certain knowledge over relativism); rather, we oscillate between perspectives, with each speaker presenting the evidence from a different point of view and, as the dialogue develops, with each side answering the alternative logos with an anti-logos in response. At issue in Antiphon is a specific forensic decision concerning the fate of those involved; antilogic thus becomes a pragmatic means for laying out contradictory claims and preparing for a practical decision based on which claim seems to best fit the available evidence.

The appeal of antilogic is similarly pragmatic in Thucydides and the *Dissoi Logoi*. The *Peloponnesian War* of Thucydides (another contemporary of Protagoras and a possible student of the Sophist Prodicus) is studded with paired speeches in which opposing diplomats and statesmen speak for and against a particular cause in a manner much like the competitive debates pioneered by Protagoras (see, for example, the Mitylene debate between Cleon and Diodotus, 3.37-47; see also Rankin 98-121). In the famous Melian dialogue of Book 5, regarding the island of Melios and its independence from or alliance with Athens (5.84-113), the contrasting speeches are not merely epideictic exercises. The pleading of the Milean orators and their stern Athenian counterparts involves the life and death of thousands of people, lives which the ruthless Athenians ultimately extract in response to their opponent's audacious appeal for independence. Antilogic, in this case, allows Thucydides not only to detail the complexity and discover the exact issues in conflict, but also to assess the situation in its totality and to critique the hubris of the arrogant Athenians from a position more comprehensive than that of either party (see 5.105).

In the *Dissoi Logoi* (that 4C treatise by an epigone of Protagoras), virtually all arguments (even those on good and bad, truth and falsehood, and other abstract topics) are directly tied to specific cases (Robinson 51). Chapter 2, for example, examines contrasting arguments on what is "seemly" and "shameful" and does so by assuming a kind of cultural relativism similar to that of Herodotus, the "father of history" (Robinson 105-115; see Herodotus 3.38). The author of the *Dissoi Logoi* argues that what is seemly, or appropriate, or fitting is not the same for everyone (in Robinson 2.18; cf. 1.17). What conditions our decisions about the seemly is *kairos*, or the opportune moment; and any theory of argument that revolves around *kairos* (as this text surely does) may be said to be pragmatic in the sense that it privileges considerations of the particular occasion. There is, then, an inherent focus on the particularity and nuance of the specific case in antilogic, a focus that aims not just at talking over things in general but at getting things done in practice; and—in this practice—antilogic is of a piece with sophism in general.

You will recall that the Sophists were advocates of a practical education and a pragmatic approach to public discourse. As teachers, they specialized in practical training (i.e., oratory) that prepared their pupils for public life; they were the first to charge a fee for their training because they were confident their teaching was useful; and the success of Protagoras and Gorgias is some testimony to the fact that their pedagogy paid off. As professors of oratory, the Sophists invariably emphasize practical issues, as Protagoras does when he tells Socrates that the subject of his teaching is *politike techne*, a realm where the measure of value is determined by what is useful, advantageous, and practicable for the people involved (*Protagoras* 334a-c). Jacqueline de Romilly (whose own distinguished analysis of the Sophistic project is deeply informed by the antilogical spirit) writes that the Sophists were the inventors of a practice designed to be of use within the "existential framework" of daily problems rather than in the Parmenidean/Platonic context of idealized, theoretical issues (1992, 237-38). Antilogic is the pre-eminent expression of that practice and remains committed to the instruction of students who, to paraphrase Homer, would be both orators and "people of action" (*Iliad* 9.443).

2. THE ETHICAL DIMENSIONS OF ANTILOGIC

If Protagorean antilogic has, from the beginning, taken on the role of pragmatic alternative to classical dialectic, it has also from the beginning been stigmatized as a source of rhetorical corruption. After all, if one can and, indeed, should argue with equal determination on either side of a proposition, then one's attachment to the truth must be tenuous at best. Or, if one insists on making the weaker case appear the stronger, what besides duplicity and self-interest can be one's motive? What, for example, could prompt Gorgias to take up the apparently indefensible case of Helen of Troy besides a desire either to deceive his audience through the audacious power of antithetical reasoning or to aggrandize himself through the elaborate excesses of rhetorical ornament (cf. Billig 47)? The general response to these ethical concerns has long been the same: the practice of antilogic has little interest in paving the way for sound judgment by reasonable people but instead has had abiding attraction for the opportunist who seeks personal advantage.

During Protagoras' lifetime, the conservative playwright Aristophanes argued (in *The Clouds*, 423 BCE) that Sophistic pedagogy in general and antilogic in particular had little regard for the truth and led to moral corruption. Not long after, Socrates himself was charged not simply with the corruption of Athenian youth but more specifically with the ability to make "the weaker argument defeat the stronger" (*Apology* 18b). This last charge against Socrates is profoundly ironic since Socrates himself equates the Protagorean "art of contradiction" with insincerity, dissimulation, and demagoguery (*Sophist* 268b-d). But in the main, the charge has stuck: the mid-20C historian W. K. C. Guthrie cites the "promise" of Protagoras to make the weaker seem the better case and comments that not only does this doctrine contain "the very essence of the Sophistical teaching" but that, more importantly, it leads to "moral and political chaos" (3.337 and 3.187, resp.). Alexander Sesonske claims that this same "promise" was the core of Plato's complaint against the Sophists and adds that Plato's attack was both well-founded and conclusive enough "to convict the Sophists for all succeeding ages" (72). Such commentary is routine enough to substantiate Schiappa's contention that the sources we have relied upon to understand Protagorean ethics have been almost uniformly hostile to the great Sophist himself (107).

In the face of this tradition, the time has come to confront the long-standing but ill-informed charge of ethical corruption. The rest of this chapter, then, will focus on and respond to the traditional condemnation of Protagorean immorality, first in a summary of the ancient case for antilogic's corruption and then with a review of the Protagorean perspective on ethical conduct. But, in fact, this discussion will continue well beyond the confines of this chapter: Roman developments in rhetorical ethics will require commentary in Chapter 5, while the consideration of a contemporary pedagogy in the Epilogue will again focus attention of ethos, ethics, and the appropriate conduct of argument. In the rest of this chapter, I hope to establish the basic context for this continuing conversation.

Unlike the other great Protagorean fragments that we have examined (the human-measure, on the gods, and two-logoi fragments), the source of the concept now under consideration cannot be traced directly to the canon of Protagoras himself. The textual origin of Protagoras' "promise" is Aristotle's *Rhetoric*, and the context of this first surviving statement is particularly pejorative. Towards the end of Book II, Aristotle takes up the discussion of fallacious enthymemes; and in the course of discussing false arguments from probability, he writes that "this sort of argument illustrates what is meant by making the worse argument seem the better," to which he immediately adds "(h)ence people were right in objecting to the training Protagoras undertook to give them" (1402a22-25; the Rhys Roberts translation). Translations, of course, differ: Lane Cooper writes that false arguments from probability illustrate "making the *worse* appear the *better cause*" (in Schiappa 103), rather than Rhys Roberts' "making the *worse argument* seem the *better*" (italics mine).[13] On the other hand, George Kennedy's 1991 translation is "to make the *weaker* seem the *better cause*" (210), and Kennedy encloses the whole phrase in quotations, which gives the impression that the concept is a well-known condemnation of Protagoras and unjust argument. The difference between the various contraries implied in these translations is, as Schiappa argues, a matter of importance for any appraisal of Protagorean ethics (see 104-110).

The ancient Greek words for this opposition are *kreitto* and *hetto*, translated above as "better" or "worse," terms with distinct ethical implications. But Schiappa points out that "from the time of Homer to that of Plato one finds passages which document the use of *kreitto* and *hetto* as paired terms meaning 'stronger' or 'weaker'" (see the *Iliad* 16.722 and Plato's *Timaeus 57a*; Schiappa 106). My own Liddell-Scott lexicon lists "stronger," "mightier," "better," and "braver" for *kreitton*, and "weaker," "inferior," and "yielding" for *hetton*. The translations of the source passage from the *Rhetoric* (1402a) in Rosamund Sprague's *The Older Sophists* and Kathleen Freeman's *Presocratic Philosophers* both adopt the "weaker/stronger" phrasing, as in "making the weaker argument stronger" (*DK* 80 A21 and B6; Sprague 21; Freeman 350).[14] So, clearly, there is a choice between the ethically charged "better/worse" and the "stronger/weaker" alternative, though the pejorative option has tended to hold sway.

One reason for the prominence of this hostile translation is undoubtedly the negative context in the Aristotelian source. But well before Aristotle's *Rhetoric*, Aristophanes' *Clouds* conducts an inquisition against antilogic that is considerably more damning than Aristotle. *The Clouds* is one of the better known comedies of this master dramatist, and in it Aristophanes attacks the whole enterprise of Sophistic education which, as he presents it, is guilty of charging unscrupulous fees for teaching the young how to "take up a bad cause and make it triumph over the right" (B. Rogers 263). In Aristophanes' terms, there are "two logics," one better, one worse, the latter of which is the Sophists' specialty and allows one "to talk unjustly and—prevail" (l. 115).[15] The play itself revolves around Strepsiades, a rich Athenian who has been recently plunged into poverty by his son, Phidippides, and who, upon hearing that the Sophists teach a new logic that can confound one's creditors and so alleviate debt, enrolls in the "thinking-house" run by Socrates, who is here associated with the Sophists and ridiculed with them (see l. 94, 243-44). In the midst of this schoolroom satire, Aristophanes presents a debate between Just and Unjust Logic (see ll. 890-1100), characters formed directly in response to the stronger/weaker concept that was universally connected with Protagoras.[16] In this *agon*, Unjust Logic introduces herself by noting that "there never was Justice or Truth" and that she intends to prevail in debate at all cost (l. 901); whereas Just Logic announces herself as an advocate of such traditional values as honor, chastity, silence, militant patriotism, and respect for one's elders (ll. 961-99). Unjust Logic responds that she is the original proponent of *antilexia*, i.e., of how

> . . . old established rules and laws might contradicted be.
> And this, as you may guess, is worth a thousand pounds to me,
> To take the feebler cause, and yet to win the disputation. (ll. 1041-43)

The agenda of antilogic could hardly be less noble, as the students of sophism seem intent on propounding the very worst (i.e., most immoral) arguments imaginable. In due course, Strepsiades sends his son to the thinking school to study the *techne* of unjust logic, which Phidippides learns only too well. First the son employs antilogic to overturn the claims of his father's creditors, and then, in an unexpected but hilarious reversal, he invokes the ability to substitute an unjust argument for a just one by demonstrating that instead of a father having the right to punish his own son, it is actually the son who commands the right to punish his father, a right that Phidippides promptly exercises by beating Strepsiades. At which point, the chorus comments that

> . . . the lad has learned the way
> All justice to gainsay.
> Be it what or where it may;
> That he'll trump up any tale,
> Right or wrong and so prevail. (ll. 1311-14)

From Aristophanes' perspective, then, "better" and "worse" are reasonable terms for the two-logoi, since the two in this case represent a contest between good and evil in which mental and verbal gymnastics triumph over truth and tradition. But the kind of immoral eristic that Aristophanes presents as antilogic is not just disreputable; it is dangerous. For some time after his beating, Strepsiades remains persuaded by the twisted reasoning of his profligate son, indicating that false arguments can indeed convince willing audiences (ll. 1437-39). And even after he comes to his senses (i.e., after the son threatens to attack his mother), Strepsiades is so maddened by the perversity of Sophistic training that burning down the thinking-school seems the only adequate response.

Greek tragedy is built upon a reversal (*peripeteia*) in the fortunes of a noble hero. Antilogic, as conceived of by Aristophanes, seems curiously intent on a similar reversal in the social sphere, as the fortunes of Just Logic and honorable values are attacked by deceptive oratory and replaced by a frame of reference that denies all value but self-interest. To teach antilogic is, thus, to prepare the ground for a *peripeteia* in the fundamental standards of society, to open the door to a tragedy in which morality vanishes because what is best is replaced by what is worst. In comedy, of course, such reversals tend to be momentary and we laugh as all those duped by the sophistry of the thinking school are lampooned and persecuted. But the potential for tragedy is there if Athenians allow their sons to be exposed to the corrupting force of Sophistic pedagogy.

The dogmatic ethics of *The Clouds* delineates a strict polarity between the conflicting logoi. Here and throughout much of the history of rhetorical theory, the assumption is that we approach any controversy with the presumption that one position is clearly superior to other(s) (see Whately's *Elements of Rhetoric* 1.3.2.). Like the epistemological bi-valence we encountered in Chapter 1 (X is either true or not true), moral polarities run counter to the Protagorean emphasis on openness, an openness that would entertain all contending logoi in an effort to uncover ideas that dogmatism obscures and to prepare the way for equitable judgment. Indeed, as antilogic develops historically, it will take as a matter of its own ethical doctrine that, in response to probable claims to knowledge, the search for *orthos logos* requires both the suspension of judgment and the serious consideration of reasonable claims made by those with whom we disagree. And yet, intellectual neutrality and openness, noble as they may be, are not exactly native to argument. Instead, the competing logoi of most arguments tend to assume some ranking in relation to one another. The rigid ethics of "better and worse" is one way, the pejorative way, of describing this ranking. From such a perspective, antilogic appears to be intent on perversity, on the insinuation of an admittedly weaker argument in the place of a better one if this reversal serves the rhetor (cf. Schiappa 107). Fortunately, however, not all classical accounts of Protagorean argument adopt the perspective of "better and worse." As we turn to other sources of information about Protagorean discourse, and especially to Protagoras' own discussion of knowledge and values in the *Protagoras* and *Theaetetus*, we can begin to see antilogic through an alternative lens that reveals a very different vision of ethical action.

As a preface to our appraisal of Protagorean ethics, it is worth noting that the ethical standing of Protagoras himself is never in question. Socrates, who is vehemently opposed to both Protagorean doctrine and pedagogy, nonetheless refers repeatedly to Protagoras with respect and at one point calls him a "teacher of virtue" (*Meno* 91a).[17] Recall also that in the mythic narrative of the *Protagoras*, the great Sophist lays out a chronology of human development that does not stop with the Promethean gifts of fire and art but contains a second stage original to Protagoras in which Zeus confers on all human beings not just virtue (*arete*) but the specifically social virtues of justice (*dike*) and respect for others (*aidos*) (322c-d). Such a scenario hardly seems the work of a theorist who ignored ethics in pursuit of unscrupulous practices and personal advantage. More to the point is Protagoras' own characterization of his profession. He tells Socrates that as a Sophist and teacher his expressed intent is to help others advance "on the road to virtue" (Protagoras 328a-b; cf. 349a) and that during each day of instruction his pupils will "make progress toward a better state," a state in which they can effectively "manage" their own affairs and those of the community (318a-319a). In other words, Protagoras is in the business of teaching both prudential conduct in everyday affairs and political achievement, for which public honor is requisite (cf. Levi 284-88). There is in the Protagorean *paideia*, then, an original attempt to interpret virtue pragmatically; i.e., to cultivate the individual's moral judgment (*euboulia*) in the service of the public good (*Protagoras* 325c-328d). We will return momentarily to the Protagorean integration of politics and ethics as well as to the special role of rhetoric in making virtue an active participant in culture. But for now, it should be clear that Protagoras, no less than Socrates, saw himself (and was seen by his contemporaries) as a professor of knowledge with a strong moral cast.

Directly following the Great Myth, Protagoras returns to the subject that had given rise to his narrative; i.e., whether or not virtue can be taught (see *Protagoras* 302b). He tells Socrates that while it is generally believed that everyone shares in some degree the capacity for political wisdom (323a), in fact, such a capacity is not "innate or automatic" but rather "acquired through care and practice and instruction" (323c-d). Laszlo Versenyi, one of Protagoras' most perceptive modern readers, refers to this capacity for justice and respect as an "endowment," a "talent" to be developed, a "potentiality" that proper instruction can cultivate (1963, 25).[18] According to Protagoras, society takes this training in civic propriety very seriously. From a child's infancy and throughout its life, a host of teachers—nurses and tutors, parents and professors—are actively engaged in promoting a sense of "justice, moderation, and the holiness of life" (324d-26e). These lessons don't always take hold, just as not all offspring of musicians are able to master music (327a-c). But the cultivation, training, and development of our potential for civic virtue are of the utmost importance because "it is to our advantage that our neighbors should be just and virtuous, and therefore everyone gladly talks about it [virtue] to everyone else and instructs [them] in justice and the law" (327b). Given the ubiquity of such instruction in Athenian culture, Protagoras would consider it "a wonder" if virtue were not teachable (326e).

The foregoing passage serves to substantiate Protagoras' commitment to political virtue as a binding force in human life. Far from presenting an immoral ethic in which self-interest reigns supreme, Protagoras outlines what de Romilly calls a "new morality" in which "individual[s] must behave well towards others because [they are] bound to them" (1992, 211). Versenyi goes so far as to say that a culture founded on Protagorean ethics would be "the highest form of human development" (1963, 24). But what exactly is so new and distinctive about morality in the Protagorean *polis*? A quick glance back at the myth may help.

When Zeus hands out that second round of divine gifts (*dike* and *aidos*), he refers to them as "the ordering principles of cities," though, as we have seen, he clearly intends these virtues (*aretai*) to operate as the governing standards of personal behavior as well as the binding force of collective life (322c). Indeed, according to the argument of the Great Myth, the community's survival is dependent on the widest-possible distribution of virtue, so that the maintenance of political order is indistinguishable from the cultivation of moral citizens. Or, alternatively, because the individual requires the community to survive, it is a matter of personal prudence to comprehend and promote justice, respect, and moderation (*sophrosyne*; 324e-325a). The relation between public and private *arete* is therefore reciprocal; i.e., while personal benefits obviously follow from a public commitment to democracy and individual freedom, involvement in the direct, participatory democracy envisioned by Protagoras also solicits from its citizens an engagement with the individual *aretai* on which culture depends. While our subject at present is not political theory, politics and ethics are here inextricable. It is worth noting that according to Cynthia Farrar, Protagoras is the first to offer a "coherent analysis of democratic man"; i.e., how communities achieve social order under conditions of maximum individual autonomy (1988, 2, 77). And yet, this early version of the social contract is not without its theoretical problems, undue idealism, and lamentable lack of detail. Nonetheless, in the brief hints we do have, we can with confidence posit ethical conduct and moral training as necessary conditions for Protagorean culture and community. As Michael Nill sums up the case, the function of the Protagorean city is "to create good men" (1985, 7).

Against the backdrop of this firm moral mandate, other ethics-related issues arise. In the first place, any casual glance at the actual values at play in society will reveal that there is considerable variation in the degree to which individuals manifest their "talent" for virtue. If, as Protagoras argues, our capacity for virtue can be influenced by instruction, then it is a matter of critical importance who is responsible for education and how this responsibility is acquitted. Appropriate pedagogy, as it turns out, is the lynchpin of both ethical and political order. And yet, potential problems arise with the introduction of moral education into a learning environment conditioned by epistemological relativism. For if knowledge is relative to the location and perspective of the knowing subject, then by what authority does any teacher claim the right to understand and insist upon her version of the "good"? Or, to expand the question slightly, can the multiplicity that characterizes Protagorean philosophy be coordinated with ethical standards intended to ensure social unity? Or, more starkly, do the philosophical and pedagogical pillars of the Protagorean program contradict one another? Plato clearly thought that such was the case, but Protagoras obviously disagrees; and the way the great Sophist approaches this apparent paradox should help us appreciate more fully the ethics of the Protagorean *paideia*.

Here again, then, is that primary tenet of Protagorean philosophy: "humanity is the measure of all things, of things that are, that they are, of things that are not, that they are not" (*DK* 80 A1). In the *Theaetetus*,

Socrates paraphrases the human-measure doctrine to say that "what everyone believes as a result of perception is indeed true for [them]," to which he adds:

> . . . if, just as no one is better entitled to be a better judge of what another experiences, so no one is better entitled to consider whether what one thinks is true or false . . . then, my friend, where is the wisdom of Protagoras, to justify his setting up to teach others and to be handsomely paid for it, and where is our comparative ignorance or the need for us to go and sit at his feet, when each of us is . . . the measure of [our] own wisdom? (161d-e)

The clear intent of Socrates' question is to challenge the compatibility of Protagorean philosophy and pedagogy, to expose the wisdom of Protagoras as "an empty conceit," and to reveal the impossibility of any instruction based on relativistic principles (Versenyi 1963, 28). Protagoras' response to these charges is to be found in the "apologia" which Plato creates for the great Sophist and inserts into the *Theaetetus* (166c-68c). According to Protagoras (or rather, according to Plato's reconstruction of Socrates' account of Protagoras), [19]

> . . . (e)ach one of us is the measure of what is and what is not, but there is all the difference in the world between one [person] and another just in the very fact that what is and appears to one is different from what is and appears to the other. And as for wisdom and the wise [person], I am very far from saying that they do not exist. (166d)

The focus of Protagorean theory is not, it will be remembered, on absolute determinations of the Real and True that follow from either the logical or scientific demand for certainty. Rather, the emphasis in Protagoras is on the primacy of individual perception and the effort to identify the logos that proves most effective in response to particular situations. Or, to put the matter somewhat differently, what we know about the world results from the contention of opposing positions and not from accurate demonstrations of the way things are in themselves.

So, at this point in the apologia, the inherent relativism of the human-measure doctrine remains intact. But who is it, then, who can claim wisdom and honestly adopt the mantle of instruction? Protagoras continues:

> By a wise [one] I mean precisely [one] who can change any one of us, when what is bad appears and is to [us], and make what is good appear and be to [us] . . . . [L]et me explain still more clearly what I mean . . . . To the sick [person] his food appears sour and is so; to

> the healthy [person] it is and appears the opposite. Now there is no call to represent either of the two as wiser—that cannot be—nor are the sick [ones] to be pronounced unwise because [these persons] think as they do, or the healthy [ones] wise because they think differently. What is wanted is a change to the opposite condition, because the other state is better.
>
> And so too in education, a change has to be effected from the worse condition to the better; only, whereas the physician produces a change by means of drugs, the Sophist does it by discourse. (166d-67a)

The analogy between Sophistic education and medical practice is fascinating because the two arts (rhetoric and medicine) were developing at the same time and there are clear Sophistic influences at work in the Hippocratic corpus.[20] Protagoras obviously feels that the medical parallel has a particular bearing on the present point, and he continues this line of analogical reasoning by noting that just as physicians have to do with the health of the body, the husbandman is concerned with the health of plants:

> For I assert that husbandmen too, when plants are sickly and have depraved sensations, substitute for these [other] sensations that are sound and healthy, and moreover that wise and honest public speakers substitute in the community sound for unsound views of what is right . . . . On the same principle the Sophist(s), since they can in the same manner guide their pupils in the way they should go, are wise and worth a considerable fee to [their students] when their education is complete. In this way it is true both that some [people] are wiser than others and that no one thinks falsely, and you, whether you like it or not, must put up with being a measure . . . . (167c-d)

The last phrase (that we all must "put up" with being a measure) has always struck me as a basic motto for both the burden and benefits of humanism. But the point at hand is that Protagorean education is focused, *a fortiori*, on the progressive development of the students' moral, prudential, and political capacity, their ability to move each day from a weaker to a stronger capacity for *euboulia* (*Protagoras* 318c - 319a). The means by which teachers (and politicians) achieve this "change to a better condition" is, naturally, "by discourse," about which I will say more shortly. But in answer to the relativism/pedagogy dilemma, we can now say that while students, like citizens, have the right to their own opinions, the teacher brings a wisdom that is fundamentally pragmatic and rhetorical: an authority based, on the one hand, on the critical ability to discern the most productive course of action in each particular case and, on the other, in the knowledge of how best to encourage this capacity for sound judgment in others.

But we must take yet another step before we reach the crux of the matter. To the relativist, the ultimate question always deals with standards of judgment: exactly how does the "wise and honest speaker" or the ethical Sophist discriminate among opposing logoi in a particular case, between the sound and the unsound, the strong and the weak, between opposing cases for goodness? The question is crucial since, in the process of actual debate and according to Protagorean theory, no single logos ever comes automatically charged with a positive (or negative) valence? In the case of the doctor and her patient, the former may think very differently about an illness than the latter, though according to the human-measure doctrine, the latter's thoughts are no less "true" to her own perception (cf. *Theaetetus* 167a). And if the doctor's view of a case is different than the patient's, how would we know which of the two is most sound, for even doctors are fallible and antilogic is decidedly opposed to argument based on authority alone. In the apologia, Protagoras mentions only the distinction between "better" and "worse" judgment (167b); but slightly later, Protagoras again "pops his head up" (from the grave) to address this important point (*Theaetetus* 171d). His theory (as presented by Socrates) does indeed maintain that no perception is truer than another, just as "no individual or state is wiser than another" (172a); however, when the question becomes not "what is true" but "where does the advantage lie," then Protagorean theory "will admit a difference between two advisors . . . in respect of truth" (172a). The distinction is crucial. In the realm of medicine, the patient's interpretation of her health cannot be "wrong," but it can be limited and might well benefit from exposure to an antilogos (an alternative view) that provides a perspective which the patient doesn't at present possess. As Versenyi puts it, the wise person in this case is the one who can help patients regulate their interaction with the physical world by arguing for or against the appropriateness, or advantage of a certain course of action (1963, 34). That is, the Sophist will make distinctions between opposing logoi based on practical judgments regarding what will turn a particular situation to best advantage. In turn, this judgment must be convincingly presented to the patient for the potential advantage to have its effect.

Put another way, the subject that Protagoras teaches is the ability to discriminate amongst opposing positions, an ability of particular importance because individual perspectives vary so widely. Versenyi points out that if all people and cultures were the same, we would not need education in what Protagoras calls the political art: "(t)he same laws, customs, and way of life would apply to all, and, although it would still need a wise [person] to devise these, once they were found political art would be expendable, and Protagoras would not have to be sent to Thurii to frame a constitution" (1963, 33). After all, why send a wise man to make new laws if one size fits all? Or recall once again Protagoras' analogy between the Sophist and the doctor (*Theaetetus* 166d-e). For the doctor, each patient has a different body, with a different history, and so will manifest a particular condition in a significantly different way. Thus, while the knowledge that the doctor draws upon for aid in treating a case may be based on certain stable and uniform (though not incontrovertible) concepts about the nature and function of the human body, the specific diagnosis and prescription for the particular patient will have to be adjusted to the unique situation. "What agrees with one person, in one situation, may not agree with that person, or with other people, in other situations" (Versenyi 35). Or, as Protagoras puts the case elsewhere, "so diverse and multiform is goodness" that the same treatment which is healthy in one circumstance may be deadly in another (*Protagoras* 334a-b). In sum, the hallmark of any Protagorean inquiry into wisdom, politics, ethics, or pedagogy is that circumstance is all.

What the physician must do in each case is to develop or—as the rhetorician would say—to "invent" a solution, a logos, an argument in response to a very concrete circumstance that is based not on what is good in general but on what is beneficial, advantageous, effective in this particular instance. Like the physician, the Protagorean Sophist, whether in the guise of the teacher, statesman, or moralist, is always a "practitioner," a pragmatist who seeks to provide specific counsel that leads to the health of the community, that nurtures our natural endowment for justice and respect, that enhances the possibility that we can display, on a case by case basis, that potential for *arete* with which we are endowed. Such, at least, is the Protagorean view of moral purpose, a view which displaces the strict hierarchy of ethical absolutism with an emphasis on utility and a faith in the power of argument itself to reveal ideas that are consonant with virtue.[21]

What, then, is the nature of antilogic in this strong ethical climate and how does its practice enact the virtues promoted by Protagorean social and moral theory? Here is my own approximation; though, in this instance, any semblance of historical reconstruction is moot because the Greek record is too thin, Roman developments too likely to intrude, and my own interests too inextricable. What I offer instead is an abbreviated sketch of ideas to come based on the discussion just passed. At this point, it seems appropriate to look both ways.

We can begin with the bedrock of antilogical practice: the effort to counterpoint one opinion, position, or logos with another, contrasting opinion, or antilogos. The motive for this effort lies squarely in the human-measure doctrine—in the relation between knowledge claims and knowing subjects and in the subsequent "partiality" of any single claim. According to the principles of "perspectivism" outlined in Chapter 1, all relevant logoi offered in argument are *part of* what can be said and *partial to* certain aspects of the case. Grant this partiality and it only makes sense to augment any single idea with additional ones, since we are likely to understand little of the case as a whole if we know only our own part. To borrow from Hamlet, there is a good deal more to know than can be imagined from one's own perspective. In ethical terms, this acknowledgement of partiality is an act of honesty, a recognition of human limits, of human fallibility. The antilogican in possession of only partial knowledge must be prepared to discover that another logos is stronger than his/her own initial presumption. As Nathaniel Teich puts it, we are not ready for argument until we can admit that we might be wrong (1987).

From the antilogical perspective, our partiality is also an incentive, a motive for taking the widest possible view of the case, for seeking comprehensive familiarity, for giving every logos its due. This will mean extending the conversation as widely as is feasible, most especially to voices one is least likely to imagine on one's own, the opposition most likely to draw one's attention to an undiscovered side of the question. Invention remains incomplete, writes Thomas Sloane, until all the available possibilities have been imagined (1985). To do this, to seek thoroughness in one's review is an expression of justice, or *dike*, the extension of equal treatment to other participants in controversy. It is also an invocation of *aidos*, of respect for others and what they have that we don't, an acknowledgement that to make progress we need to be in contact with them.

So, to be in argument is to commit oneself to others, to cultivate a relationship through dialogue. In the antilogical dialogue, the standards of honesty, justice, and respect are always in force, along with other considerations. For example, while it is an act of "charity" or respect to credit any argument that seems sound, it is also honorable to suspend judgment while in the process of appraising alternative considerations. As we will see in Part II, a healthy scepticism is only prudent amidst a welter of probable claims. Prudence also dictates a refusal of "certainty" (a resistance to the total acceptance of any single logos) since there is always an antilogos lurking somewhere. Things change, logoi are modified, new voices enter the parlor of discourse, making certainty always a little suspect. And yet, antilogical scepticism and the temporary suspension of judgment are ultimately complemented by pragmatism, the need to find answers that function in the world of practical affairs. When invention has been thorough and the rhetor knows what logoi are available to be reconciled, decision-making is a matter of finding a way to agree, of dialogical cooperation. At this point, authentic dialogue is more than the discovery of good ideas; it is—in its capacity to coordinate differences—the ground for the social contract.

At its core, then, antilogic is oriented toward people and is a matter of ethos, of how we view ourselves and how we interact with others. Because it is uninterested in transcendental appeals, its ethics do not focus on universal principles, which themselves serve as the standard subject of formal ethics. Instead, antilogical ethics are concerned with individual acts of conduct, of being together in controversy without surrendering courtesy; *ad bellum purificandum*. As such, they are an exercise of character, not an act of obedience to principle. There have been bad Sophists, as Plato claimed; and there will always be those like Phidippides who seek personal advantage through Sophistic ingenuity. In the end, the only guarantee of *eupraxis* is the willingness of rhetors to commit themselves to good relations as well as *orthos logos*. Like democracy, the process is voluntary. All of which is small reward perhaps, unless one also believes that good practice makes good things happen.

Let me summarize. When, as will happen, one logos runs counter to another, it is the rhetor's job to recognize one's own initial partialities, to seek out and test all available options, and to defend the strongest conceivable position. The goal of this effort is not to carry the day for one's self or to establish the "truth" beyond doubt; rather, the Protagorean rhetor seeks to identify those features of a case that make a particular position the most advantageous under specific circumstantial conditions. Sometimes, this effort will involve advocating what seems to be a "weaker" position (rather than a "worse" one) because presumption resides on the side of a "stronger," more conventional opinion (cf. Schiappa 109, 113). In any democratic culture, antilogic thus becomes the means by which minority opinion speaks in opposition to the entrenched dominance of conventional wisdom, as well as the means for defending the open forum of ideas upon which democratic culture depends (cf. Dewey 1927). In different circumstances, when the rhetor's own position is ascendant, antilogic becomes a method of diplomatic refutation which nonetheless insists that all sides have a full hearing. For only when the "weakest" case is given full voice will it be possible to realize the full dimensions of a controversy, to build the strongest, most comprehensive logos imaginable, and to prepare the way for mutual acceptance of the most convincing position. In yet other circumstances, when no particular position is dominant, or when the rhetor has yet to make a judgment regarding the utility of a specific claim, antilogic becomes a heuristic for assessing the merit of each alternative, for reconciling conflicting opinions, for cobbling together some middle way, or for creating from disparate clues a logos of one's own. In all such cases, antilogic is a bulwark against myopia, intolerance, and tyranny as well as an agent of convergence in the midst of diversity. Seen thus, the goal of antilogic is *homonoia*, or like-mindedness, a judgment shared by those with dissimilar views. But, of course, any consensus or harmony in controversy will be temporary because antilogic is always ready with another argument, always prepared to take on the dominant logos, always filled with the abiding hope of altering our judgment for the better and achieving an even stronger, more advantageous position from which to reason and act.

Does all this seem a paradox? Can an insistence on the arguments of our opponents (or, if you will, on dissensus) actually be the ultimate function of a *logon techne*?[22] If so, it is a paradox that Heraclitus and Thucydides, along with Cicero, Quintilian, Erasmus, Montaigne, Shakepeare, Donne, Milton, Dryden, Vico, Blake, De Quincey, Dostoyevski, Cardinal Newman, J. S. Mill and others could accommodate.[23] As for Protagoras, he had to endure the claims of critics, like Aristophanes and Aristotle, that antilogic sought simply to deceive us all (see *Phaedrus* 267a). Protagoras' response to these charges, as far as we can extrapolate from slender sources, is suitably antithetical: antilogic places the laurel wreath on the rhetor who demonstrates the ultimate respect for the opinions of others by insisting that all logoi, the unjust and the unconventional as well as the powerful and persuasive, are given voice. It is because of its ethical courage—its willingness to resist privilege and put each position at risk for the sake of learning more—that antilogic belongs, according to Protagoras, at the center of the curriculum as well as at the heart of our public and private efforts to grapple with the experience of controversy. Just what rhetorical principles will structure that curriculum is the topic of the next chapter.

Chapter 4

The Rhetorical Form of Antilogic

"All things produced by our own reason and ability, the true as well as the false, are subject to uncertainty and debate."

Montaigne, "Apology for Raymond Sebond"

"When someone opposes me, he arouses my attention, not my anger. I go to meet a man who contradicts me, who instructs me. The cause of truth should be the common cause of both."

Montaigne, "Of the Art of Discussion"

We can begin this examination of rhetorical form with an artificial but helpful distinction between antilogic as a *techne*, or set of practical skills, and antilogic as *praxis*, a more general capacity for action. We know from David Roochnik that the term *techne* was derived from the Indo-European root for "wood," and that a *tekton* was a woodworker (1990, 18). Over time, the Greek term came to indicate specific skills "uniquely possessed by one member of the community," especially those skills that required an ability for making intellectual calculations, so that the *technai* came to include arithmetic as well as carpentry. Roochnik concludes that all such skills are relatively precise (e.g., medicine), have a determinate subject matter (woodworking, numbers), and are value neutral (can be used for good or ill; 20-21). With this working description, we can posit certain features of antilogic that could be classified as "technical"; but we should be careful in doing so, for an over-scrupulous attempt to calculate the precise form of antilogic runs the risk of diverting attention to the peripheral and hypostatizing what in actual practice is dynamic.

Like *phronesis*, antilogic is basically unscripted and improvisational. By and large, it operates beyond the strict confines of a specific craft or *techne* and abjures the determinate features of, say, the syllogism, or even of dialectic as defined in the *Phaedrus* or the *Topics* (see Ch. 2, sec. 2). After all, if antilogic is oriented toward people rather than propositions, if it is historical, situational, dialogic, and pragmatic, how are we to map those adjustments that develop as a result of the diverse times, places, persons, and purposes involved? And yet, as Eric Havelock notes, the general form and function of Protagorean rhetoric are not "wholly uncalculable" (1957, 200). So, while my goal in this chapter is to clarify the constituent features of antilogical form, I will try to do so with deference to the inherent fluidity of its actual function.

In the first part of the chapter, I focus on examples of early Greek antilogic from Thucydides, Euripides, the *Dissoi Logoi*, and Antiphon. Specifically, I intend to survey significant features and strategies in these historical texts in order to build a preliminary picture of the formal options that qualify antilogic (at least partly) as a *techne*, an art or skill whose form and procedure is stable enough to be identified and taught. It may be that the features we can discern in these early examples are especially pristine because they are nascent, appearing before later developments complicate what is fundamental.

These technical appraisals are brief, though I hope not too cursory. To extend the technical analysis of exemplary texts would, at this point, give the mistaken impression that the goal of antilogic is to create an artifact, something we can assemble according to appropriate technical procedures. To really comprehend antilogical form, we need to articulate these technical features dialectically; i.e., in relation to principles that inform antilogic as *praxis* (cf. V. Kahn 1997, 164). I intend *praxis* here to mean the capacity for action informed by logos, reason, reflection; and, by principles of *praxis,* I mean those guiding or internal ends that set a standard for effective action (see Garver 1994, 206-213). In arguing that antilogic is more than a compilation of technical skills, I attempt in this chapter to identify those guiding principles that predicate effective antilogical practice, distinguish it from other argumentative rivals, and constitute its standard of excellence. In brief, I posit the multivocal, oppositional, and dynamic elements of antilogic as its governing principles, those features that constitute its unique *praxis* and motivate its particular form. To clarify these principles, I appeal again to some early examples of antilogical form. This time, however, I complement historical review with modern theoretical commentary as a way of extending my analysis beyond the original form of antilogical practice. My hope is that by discussing the technical features and guiding principles of antilogical argumentation in direct relation to one another, we can a build a comprehensive base for understanding the ensuing developments in antilogical genealogy.

But before we take up examples of antilogic in its Ur-form, let's summarize what we know about the rhetoric of our subject so far. In brief, antilogic operates within the context of Protagorean relativism, a theory of perception that places all knowledge claims in relation to people, discounts transcendental appeals as unavailable, and so assumes that multiple, conflicting claims will issue from diverse human perspectives. We also know that antilogic is a practical response to this epistemological theory, a discursive process that seeks workable knowledge by assuming the partiality of all single claims, rejecting any assertion of cognitive privilege, and insisting that all claims must be argued. Antilogic organizes the resulting arguments by placing opposing claims in direct contact with one another, calling attention to the pro and contra of their interaction, and attending to the evolution of ideas that follows from juxtaposition. We know, then, that antilogic emphasizes direct dialogue between interested parties as an unparalleled method for testing existing claims and inventing new ones. Of course, openness to dialogue typically signifies a common commitment to the issues at hand, and antilogic seeks to operate both as an agent of commonly accepted solutions and as an instrument of social and political virtue. Finally, we know that if scepticism, inquiry, and dialogue promote questions, antilogic will also seek answers in the common advantage of those involved. Antilogic thus finds its impetus in human "partiality" and endemic disagreement, its goal in the resolution of the common good, and its *raison d'être* in the dialogue of interlocutors whose diverse and multiform logoi are the materials out of which we measure all things.

But, of course, all this is hardly even a silhouette. Our next best step is to turn to the proto-forms, those early experiments in antilogical structure and purpose that appeared in the age of Protagoras. We will see what a few good examples and some hermeneutical investment can do to add a bit of color and depth to the picture.

1. EARLY FORMS OF ANTILOGICAL *TECHNE*

The simplest structure for antilogical discourse is undoubtedly the paired or parallel speeches of the kind we find in Thucydides. The great historian says that his own work has been complicated by "the want of coincidence between accounts of the same occurrences by different eyewitnesses, arising sometimes from imperfect memory, sometimes from undue partiality for one side or another." Under such circumstances, he believes it wise "to give the grounds alleged by either side" as the best indicator of the complexities involved and as the soundest means for arriving at informed historical judgment (1.22-23). Echoes of Protagoras resound in this passage; in particular, the influence of the human-measure's perspectivism and the role of contending arguments in the construction of history. We know that Thucydides, like Euripides, was a pupil of Prodicus, a Sophist slightly younger than Protagoras; and we know that throughout *The History of the Peloponnesian War,* the concepts and strategies of Sophism are at play (cf. de Romilly 1992, 76-78; Rankin 1983, 98-103; Solmsen 1975, 10-46). Specifically, Thucydides is inclined to describe events from a particular human vantage point, or to present different individual reports of the same event, or to provide testimony on the same topic at different historical moments, all of which foreground the specific human observer as the controlling historical perspective. Thucydides also notes that there are typically more than two sides in any political debate, but he selects the most contentious contrasts as indicative (3.36.6), a narrative choice that aligns his history with the two-sided public debates (or contests) made popular by Protagoras in Athens. Finally, there is in both Thucydides and Protagoras a belief that conflict is a natural human tendency (2.61.1), a view both men appear to derive from empirical observation (cf. Ch. 2, sec. 3). Of course, the historian, especially the war historian, is less free to indulge in idealism than the philosopher; so there is also in Thucydides a tragic recognition that the resolution of conflict is too often the work of force, too seldom the result of debate (5.89).

Thucydides' proclivity toward what will become known technically as *in utramque partem* (presenting both sides of a case) is on display immediately in his *History* in two sets of opposing speeches, the first pair addressed to the Athenian assembly (1.31-45), then a matching set addressed to the assembly at Sparta (1.68-78). For our purposes, the well-known debate over the fate of the Lesbian city of Mitylene is a good indicator of Thucydides' ability to adopt antilogical method for historical work (3.36-48). The point at issue is whether or not the Mitylenean population should be severely punished for secretly cultivating a traitorous alliance with Sparta, Athens' enemy. Thucydides presents two speakers on the Mitylene question from the Athenian assembly, one coming directly after the other. Cleon, the most influential statesman in Athens following the death of Pericles, argues that democracy must occasionally act harshly in order to maintain the hegemony of its own "good laws" (3.37-40). As Rankin points out, Cleon echoes Callicles in Plato's *Gorgias*, both of whom maintain that "power is more important than justice or democracy" (1983, 106). The oppressive militarism of this position is countered by Diodotus, who argues that the opposite is the case—that a tolerant response to Mitylene treachery will prove more practical and advantageous to the Athenian cause.

Here are two antithetical speeches on the same topic, of the same relative length, delivered to the same audience, in the same venue, and which take up an almost identical set of concerns in a roughly symmetrical or parallel order. Both sides in this debate share considerable common ground: both agree on the importance of the Mitylene defection with regard to Athenian imperial power, and both are primarily concerned not so much with a forensic judgment regarding the colony's crime as with a deliberative decision on what is best for Athens' future interests (3.44).[1] In addition, both speeches begin with discussion of theoretical points: first, should the Athenian assembly reconsider its previous vote on Mitylene; and second, what is the value of political oratory to the state (the particular topic is the influence of Sophistic argument on public debate; and, interestingly, both speakers prove notably Sophistic in their own discourse; cf. 3.38 and 3.42-43). These parallel exordia are followed by parallel discussions of the issues involved, including Athens' prior treatment of her colony, the political repercussions of the situation, and the financial costs of the various options. Both speakers then provide a concluding recommendation: Cleon closes with an exhortation to protect the interests of empire by acting severely where severity is due; Diodotus argues that severity will only aggravate the situation and that mercy will yield greater benefits, including enhanced status for Athens. In other words, Diodotus reverses Cleon's claim and argues that self-interest in this instance demands that Athens adopt a more tolerant posture: "I consider it far more useful for the preservation of our empire voluntarily to put up with injustice, than to put to death, however justly, those who it is [in] our interest to keep alive" (3.47). Throughout this debate, Diodotus is more patient and deliberative in his analysis, more principled and less passionate in his ethos, so that it is not just the arguments but also the character and manner of the orators that are antithetical in this instance.

There is, of course, much about this rhetorical encounter that is outside the scope of our present interest. As noted, both speakers argue from expedience: Cleon that retribution is the practical way to prevent further rebellion, Diodotus that excessive repression will itself prove impractical. Rankin points out that there is little in the way of epideictic flourish; the fate of thousands hang on this debate, so expedience dominates (110). For our own purposes, the Mitylene debate is a classical presentation of arguing *in utramque partem*, first one side, then the other. It is just this kind of simple but clarifying strategy that we imagine characterized Protagoras' own two volumes of *Antilogies*, now lost. Thucydides allows the two contrasting speeches, the *dissoi logoi*, to stand directly opposite one another. Each speech covers the same set of contingencies in roughly the same formal order, so that "the very elements [of the argument] that might seem gratuitous and artificial when each side was considered in isolation become rigorous means of assessment and comparison once the two arguments were taken together" (de Romilly 1992, 86). Indeed, comparison and contrast is the master *topos* for this form of antilogic; i.e., we consider, in turn, two isomorphic assessments of a single, complex case, and two polarized opinions on how best to act in response.

The result is an approach to inquiry in two columns, pro and contra, in which the facts are set side by side in orderly fashion and judgment is made possible by juxtaposition. This rather simple *techne* is based on the notion that careful discrimination by contrasting alternatives is both a direct route to knowledge in its own right and a rational way to prepare for *euboulia*, or informed judgment. Diodotus remarks that "the good citizen ought to triumph not by frightening his opponents but by beating them fairly in argument" (3.42). And while the triumph of the good citizen is not always the case (though Diodotus' tolerance did carry the day in this instance), antilogic in the form of parallel speeches seems clearly designed to allow participants to decide the merits of the case on balance. Of course, there are other, more subtle forms of antilogical reasoning in Thucydides which go beyond the "A then B" parallelism of this debate (see de Romilly 1988); but this first step, while simple, is meaningful enough: in a world filled with contradictory opinions, we take the measure of what can be known through comparison. Nor are parallel comparisons quite as limited as they might seem, for there is no reason, as a next step, why a *tertium quid* or third possibility should not evolve out of or be cobbled together through the examination of similarly parallel logoi. But paired speeches in direct and symmetrical opposition is a place to start.

The antilogoi in the Thucydidean example present the distinct choice between punishment and lenience. When we turn to Euripides, the issues become considerably more complicated. Of course, to turn to dramatic literature is to acknowledge that the drama in particular had been intuitively alert to the *techne* of *dissoi logoi*, as the works of Sophocles indicate. In the case of Euripides, however, there is a direct link to Protagoras, for not only was the dramatist a student of Prodicus but also, perhaps, of Protagoras himself (de Romilly 1992, viii). According to Solmsen, a number of Euripidean tragedies are text-cases in arguing *in utramque partem*; and, in fact, in the fragment "Antiope," the playwright declares that

> On every subject it would be possible to set up a debate of double arguments, providing one was skilled in speaking. (1.189, in de Romilly 1992, 77)

I have chosen a single Euripidean example for our review. In his *Orestes* (408 BCE) we find parallel antilogical speeches in response to Orestes' murder of his mother Clytemnestra, a case of mythic and monumental complexity (see also Aeschylus' *Oresteia*, presented in 458 BCE, half a century before Euripides' play). Clytemnestra, you may recall, is the wife of Agamemnon, leader of the Greek host in the Trojan War, the leader who is murdered by his wife and her lover when he returns home from war to his palace in Argos. Orestes, in retaliation for the killing of his father, murders his mother and must argue the justice of his case in order to escape being stoned for matricide.

In Aeschylus, Orestes is tried before the Areopagus (in the first convening of this legendary tribunal of judges); but in Euripides, Orestes pleads his case against Tyndarus, the father of Clytemnestra, who, while condemning the crime of his daughter, nonetheless seeks the prosecution of his grandson, Orestes, for her murder. Tyndarus insists that matricide is outside the law, that Orestes could have imposed a "just penalty" on his wayward mother and earned honor for doing so; but in killing her instead, his grandson has placed himself in the same category as his mother: "for while he was justified in thinking her wicked, he acted more wickedly himself in killing her" (ll. 491-541; quote ll. 505-6). Tyndarus' case, then, is built upon a strict conception of the law (*nomos*), which we can consider to carry presumptive force in this instance as the traditional or conventional position, the "strong" case based on conservative appeal. Kerferd remarks that *nomos* is typically conceived of as prescriptive and normative, "never merely descriptive" (1981, 112). Tyndarus is arguing that no one can transgress the *nomoi* and pretend to be acting in accord with justice.

Orestes' arguments follow directly after those of Tyndarus (ll. 544-604), and his defense assumes an entirely different perspective: he admits that in killing his mother he is officially "outside the law," a risky point to concede; but he also argues that, seen from another perspective, his actions are "within" the law and have earned him an "alternative title" as "the avenger of my father" (ll. 546-48). His rhetorical strategy is to set his arguments directly "against" those of Tyndarus, a strategy similar to Diodotus, though Orestes goes further. He argues that his primary allegiance had to be to his father, and that in acting to punish the crime of infidelity and the murder of a king, he is actually "the benefactor of all Greece" (ll. 565-66). What is more, he had been commanded to act by the god Apollo and so was obedient to both human and divine law.

Here is a case in which there is no middle ground, in which the antilogic of paired speeches works to reveal irreconcilable differences in perspective (an irreconcilability reflected in the polarized deliberations of the assembly at Argos; see ll. 884-957). And yet, antilogic continues to operate as a means for the rational assessment of the situation despite dramatically conflicting logoi. Indeed, the two opposing positions promote thinking about complex issues (on human nature, law, justice, and commitment) that is more subtle than either individual argument; so once again antilogic operates as the agent of invention, of new ideas that transcend the limits of any single logos. There are also new developments in the "technical" repertoire on display. Orestes is attempting here to reverse the standard order of convention by placing his individual commitment above "vulgar law"; and because he must make his case to an assembly of authorities, he is clearly advancing what is (in this rhetorical situation) a "weaker" case against a "stronger" one. He does so in typical Sophistic fashion, by claiming that the opposing position is inadequate and that his own conception of/perspective on justice is equally reasonable and more beneficial.[2] What is more, he employs his opponent's claims but reverses them in his own favor, arguing that what seems a transgression of law is actually a defense of higher justice.

In parallel speeches like those of Thucydides, one logos typically lays out its evidence irrespective of the position of the other, and it is left for an arbiter to decide what to do. In the speech of Orestes, however, we see the rhetor building his argument in direct contradiction of his opponent's. This latter strategy is referred to as a *peritrope*, a turning-of-the-tables, in which the ideas or arguments of one logos are reversed in the anti-logos of the respondent. In Euripides, Orestes builds his defense around the notion that while matricide is conventionally "outside the law," the revenge of a slain father supersedes matricide because Orestes' victim was herself a murderer and because Orestes is defending the community by avenging its monarch. In the later terminology of Hermagoras' stasis theory, the point in contention or *summa quaestio* is one of definition; i.e., the issue is not "did he do it?" (*an sit*) but "what did he do?" (*quid sit*) and does it qualify as murder (see *De Inventione* 1.10-19, or Quintilian 3.4). In Aeschylus, the case of Orestes solicits a roughly equal split of opinion on this complex issue; but in Euripides, Orestes speaks last and succeeds in redefining what it means to act "within the law." In the process, he reverses the logos initiated by his opponent.

In this case, antilogic becomes, quite literally, an "argument against," with opposing ideas serving as heuristic and refutation providing an opportunity to clarify one's own case. As a method for addressing both sides, *peripeteia* becomes a standard feature of antilogic's technical inventory. We can examine a new variant of "arguing against" in the treatise known as *Contrasting Arguments*, or the *Dissoi Logoi.* This brief document is unpolished and often simplistic, perhaps the work of a student. It takes the form of a *techne eristike*, a manual of argument, and seems clearly the work of a Protagorean admirer. Protagoras himself did write a book on *The Art of Debate* and, as noted, two books of *Antilogies*, all of which have been lost (*DK* 80 A1). The anonymous *Dissoi Logoi* does, then, provide some idea of the structure of antilogic as it was practiced (and taught?) not long after the death of Protagoras (ca. 420 BCE). The first four chapters, in particular, present distinct antilogies (on what is good and bad, seemly and shameful, etc.) which clearly reflect Protagorean philosophy and which, correspondingly, argue by contraries and reversals. In Chapter 1, we are told that different people hold different views of the good and bad, and that "a thing might be . . . at one time good and at another time bad for the same person," a direct reflection of the human-measure doctrine (1.1). In Chapter 3, the writer introduces the notion that contrasting arguments are often advanced "on the matter of what is just and unjust" and then proceeds to delineate two polarized positions: "Some say that what is just and what is unjust are two different things, others that the same thing is just and unjust" (3.1).[3] The first position is assumed to be dominant, the "stronger" thesis; but by appeal to a litany of antithetical examples, the writer seeks to present the question in varied contexts, to collapse the distinction between binary categories, and in so doing to undermine the logic of the leading thesis. For example, in contradiction to the common-sense position that it is unjust to lie to one's parents, the author argues that deceit would be justifiable if one had to induce one's sick parents to take a life-saving medicine to which they were averse (3.2). To paraphrase comments the author makes at another point, "different occasions take an action that may seem one thing and make it something different" (2.19-20).

The ethics of this position aside, we can see that the motive for the peritrope in the case of deceiving one's parents is both refutative and constructive; it would overturn the logic of the opposition (one must never lie) and use that refutation as support for a new logos (ethical decisions are always contextual). For our study of antilogical *techne*, this dual purpose constitutes an addition to the inventory of useable methods: in the *Dissoi Logoi*, the same rhetor is first prosecution, then defense; or, in other words, the same rhetor articulates both sides of the case. Whereas Diodotus and Orestes lay out opposing positions in sequence, this rhetor, regardless of modest gifts, is him/herself looking in both directions and doing so inside the limits of a single utterance.[4] You will recall from our discussion of antilogical ethics that the Janus-like contemplation of both sides ideally requires a suspension of judgment and is motivated by a desire to obtain a comprehensive view of the subject. The author of the *Dissoi Logoi* is a long way from disinterested (see 1.2, 2.2, 3.2), but he is not dismissive of what "some others may say" and he can offer a relatively balanced examination of the position he would contradict (see 6.2-6). Moreover, in the process of developing his own side of the story, he also cites and develops a contrasting set of arguments.

These exercises in antilogical reasoning have the feel of the lecture hall; no practical action is contemplated and the rhetor is at leisure to simply speculate, sometimes rather wildly. But they also display an inclination, however rudimentary, to contemplate issues *in utramque partem*; that is, to move literally in both directions at once. This particular approach to argumentative inquiry will become the heart of antilogical practice; it will become more rigorous, more technical, more refined. Nonetheless, the primary *techne* of arguing both sides is on display in embryo in the *Dissoi Logoi*. For a final look at Greek developments of antilogical form, we turn to Antiphon, a writer of considerable ingenuity whose *Tetralogies* not only explore new territory but also summarize what we have seen so far.

Antiphon was an Athenian contemporary of Protagoras (c. 480 - 411), a logographer, a possible teacher of Thucydides, and the victim of a failed political conspiracy. There are four separate *Tetralogies*, each including four mock speeches on a single legal case, two each for the prosecution and the defense, who take turns (P, D, P, D). As noted in Chapter 3, one of these tetralogies (#2) appears based on the famous incident of "manslaughter by javelin" which Plutach reports originated in a conversation between Protagoras and Pericles. All of the tetralogies are clearly conceived in the Protagorean manner, with antilogic here taking on new forms that substantiate Antiphon's reputation for cleverness. Each case involves a murder for which no evidence is determinant, so each hypothetical rhetor must build a case on probability only. But because there are four separate turns here, there is a substantial increase in argumentative complexity as each speaker seeks to respond to the thrust and parry of debate with the utmost in dialogical ingenuity. As such, these speeches have the ambience of a performance piece, not unlike Gorgias' *Encomium on Helen*. Nonetheless, as formal experiments in what can be done with this new *techne,* they are entertaining and impressive.

In Tetralogy #3 (*DK* 87 B4), the prosecution begins with a brief but confident appeal to common sense: a young, powerful defendant killed an older man while he (the defendant) was drinking; the defendant obviously deserves to be convicted. Brevity here is a rhetorical strategy, implying that the evidence is *prima facie*. The defendant speaks for himself and asserts, in contrast, that the victim was the aggressor and that he, the defendant, received the first blows and did no more than return them. Moreover, the defendant reveals that the victim did not die on the spot but was taken to a doctor some days later for a treatment that other doctors considered dangerous. It was only after this controversial treatment that the victim died. As a result of this revelation, the defendant claims that the prosecution is guilty of "unjustly plotting my death," the very crime for which the defendant himself stands accused (2.7). This last move is a classic peritrope, an accusation of one's accuser that reverses the evidence and supports a conclusion that stands the opponent's case on its head.

At this point (i.e., after the two opening speeches), the level of rhetorical ingenuity escalates and the arguments become almost too complex for summary. The prosecution maintains that this shocking defense (in which the old man is said to be responsible for his own death) is characteristic of the defendant's outrageous behavior, that had the victim not gone to a doctor his opponent would have said he died through neglect (another reversal), and that even if the victim was "the contriver" of the first blow, the defendant was "the contriver of the murder," a verbal ploy worthy of the linguistic distinctions of Prodicus (3.2-4; see *Protagoras* 337a-c). All these points are not merely answered but transformed in the last speech of the defense which portrays the accused as involuntarily involved by the dead man's aggression, as a victim himself who did not even kill his attacker. In a characteristic paradox, the defense rests with the assertion that to exact the death penalty in this case would be "to kill the innocent in your attempt to punish the murderer" (4.10). Taken together, these speeches display the entire panoply of existent antilogical technique: there is point-by-point contradiction, this time amounting to a total refutation of everything offered by the other side; there are reversals galore; there is new evidence available through a change in perspective; there is the substitution of a seemingly weaker for an originally probable case, etc.

And yet, while the argumentation is dazzling (and Sophistic), we haven't moved that far from the opening arguments: Antiphon's version of the paired speeches is an attempt to burrow deeper and deeper into the complexities of the problem. In sum, the demands of contrary argument promote an ongoing process of refinement in which accusation and challenge by one's opponent require an intensified search for "good reasons," a keener ability to identify weaknesses in one's case, a sharpening of understanding through criticism of plausible alternatives, and a progressively greater grasp of the case in its totality. The charges of sophistry and disregard for the truth that we confronted in the last chapter would clearly be relevant if the motive for

Antiphon's antilogies were to represent actual forensic pleading; obviously, his spokesmen excel in turning any argument to advantage. But if this *techne* is approached as a guide to invention, then antilogic becomes a process for assessing the situation in light of all possible interpretations; for generating more and more ideas; for honing one's articulation of these ideas; and, most importantly, for coming to understand the complex, variable nature of an actual controversy and preparing to act in accordance with that mature, comprehensive understanding. Despite the forensic context of these speeches, Antiphon reminds the attentive reader that antilogic is more than a clever *techne* for defending a choice already made; at its best, it is a method of inquiry into good reasons for reasonable choices.

Antiphon presents an appropriate place to halt these investigations into technical lore. To this point, we have examined what we might call the conventions of antilogic as a rhetorical genre. These conventions of practice are analogous to such dramatic conventions as the reversal of fortune at the outset of Shakespearean comedy or the holy marriage (*hieros gamos*) at its end. We turn now from specific examples of antilogical technique to its general principles of *praxis*, its guiding or constituent ends and internal standards of excellence. In the process, we move from the parts to the whole, from particular conventions to what we might call the *mythos* of our subject—the fundamentals that guide its form and define its purpose. Aristotle refers to the *mythos* of drama, its animating form, as the "soul" of its practice (*Poetics* 1450a; cf. Sloane 1985, 282). By analogy, we turn now from exterior form to the soul of antilogic.

2. THE PRINCIPLES OF ANTILOGIC AS *PRAXIS*

By my account, the chief principles of antilogical argument are these: it is **multivocal**, **oppositional**, and **dynamic**. Given the previous discussion, none of these principles should prove surprising, but their particular contributions to antilogical practice are so crucial that without the presence of any one of them antilogic could hardly be said to exist.

Multivocality. I begin with the most obvious, but also with the most fundamental idea: argument, according to the two-logoi fragment of Protagoras ("there are two opposing logoi present concerning everything"), can never be singular, univocal, one-dimensional; it must always be multiple, dialogical, pluralistic. The inherent multiplicity of argument allows for no single stand or stasis, no claim to cognitive privilege, no absolute position. Put another way, no argument on its own is more than partial; no claim, regardless of how dominant, more than probable. As a result, the single argument standing by itself is a false synecdoche, a part pretending to be the whole. The real argument is found instead among the *multiplex ratio disputandi*, the many ideas in dispute. Without this multiplicity, there is no argument.

To claim multiplicity as one of the constituent principles of argumentation may seem tautological; it is to say that a difference of opinion involves more than one side. But the point needs to be made because the univocality of argument is a standard feature of the "current-traditional" paradigm for rhetoric, a tradition dominant in rhetorical theory and pedagogy at least since Whately's *Elements of Rhetoric* (1828), with deep roots in the "new philosophy" of the 17C and persistent in its advocacy of logic as the defining feature of argumentative practice (see Berlin 1980, Rorty 1979, Toulmin 1995). According to this tradition, the successful rhetor is involved in a solo performance, in constructing a logically consistent, rigorously detailed line of reasoning that can stand on its own merits. This form of argument is best described as demonstration, or *apodeixis*—reasoning that would transcend multiplicity and achieve certainty. And it is argument as demonstration that continues to dominate most contemporary textbooks on argumentation (Mendelson and Lindeman 1999). But while the goal of certainty may be suitable for some forms of scientific discourse, it is a masquerade to present *apodeixis* as the paradigmatic form of practical argument, which is inherently ambivalent in the sense that it oscillates between/among opposing logoi.

Our first principle of antilogic, then, is that no position, stand or logos in any argument is ever complete in and of itself; it always exists in relationship with others. In turn, this principle entails specific implications for the practice, even the form of antilogic. In keeping with prior considerations, we can now say that antilogic is the vehicle for inquiry into the multiplicity of possible positions that develop in response to open questions. Given the irreducible diversity of human interpretations, antilogic seeks, as Sloane puts it, to turn in multiple directions at once, to comprehend the various "modes, manners, means, and opinions" that one encounters within the perimeters of a given controversy (1985, 32). In sum, antilogic aspires to what Erasmus (himself a master of antilogical practice) describes as *copia*, an abundance of ideas, a "Protean variety" of conceptions which taken together render an idea, issue, controversy in all its fullness (*Copia* 1.33). And yet, what in particular does the antilogician do to insure this fecundity, this abundance, this multiplicity of material?

First and foremost, the antilogical rhetor adopts a sceptical independence from any particular side in controversy. Such scepticism is based on the assumption that because every position is in itself partial, no position is beyond challenge. The principal means for mounting such challenge is to take the opposite side, to draw attention to the viability of an opposing position. Alternatively, the antilogical rhetor must also be willing to suspend judgment and to offer respect to all sides in controversy so that every logos has an opportunity to make itself known.[5] Antonius, the great pragmatist of Cicero's *De Oratore*, outlines just such a procedure:

> It is my own practice to take care that every client personally instruct me on his affairs, and that no one else shall be present, so that he may speak the more freely; and to argue his opponent's case to him, so that he may openly declare whatever he has thought of his position. Then, when he has departed, in my own person and with perfect impartiality I play three characters, myself, my opponent, and the arbitrator . . . . When I have thoroughly mastered the circumstances of the case, the issue in doubt immediately comes to mind. (2.102-4)

Antonius is here employing impersonation (*prosopopoeia*) as a method of inquiring into the issue upon which his case turns. In pursuit of this critical insight, he exhibits a paradoxical blend of openness and scepticism, a willingness to grant any position the right to make its case at the same time he submits it to challenge.[6] To pursue this equivocal position, the antilogical rhetor must approach the exchange between partners in debate as an authentic conversation; i.e., as a dialogue in which each party is at liberty not only to take one's turn but also to put one's questions, so that attention flows in both directions and all logoi are in the limelight as well as under scrutiny.[7]

On this point, Socrates remarks with admiration that Protagoras himself not only can deliver both long speeches and short responses but that when he (Protagoras) asks a question, he is also "perfectly capable . . . of waiting and listening to the answer—a rare accomplishment" (*Protagoras* 329b). The implication is that the Protagorean mode of argumentation involves not just the capacity to speak in a manner appropriate to the occasion, but also an ability to ask good questions (questions that offer one's interlocutors a real opportunity to declare themselves) and to listen to the answers with care (since a contribution to understanding may come from any quarter). Contrast this practice with Socratic dialogue, which is directed by and allows questions from a single speaker, which moves in a logical progression toward a goal often preconceived by that speaker, and in which other voices are most often foils for Socrates himself. Alternatively, antilogic attempts to adopt the protocols of authentic dialogue, protocols that encourage all partners to speak and respond in a collective effort to arrive at the fullest possible understanding of the subject (see Havelock 209-10; Epilogue, sec. 2).[8] Just after the passage in the *Protagoras* cited above, Alcibiades refers to "the give-and-take of arguments" at which Socrates excels (Guthrie's translation in *Plato: Collected Dialogues* 336b-c). An alternative translation offered by Eric Havelock refers to the "technique of rendering and receiving discourse" (Havelock 1957, 208). It is just this balance between rendering *and* receiving that captures the spirit at work in antilogic; i.e., the willingness to speak and listen, to question and respond, to challenge and defend. In sum, if there are always two or more sides to every question, then, as a matter of ethical practice, all sides in argument must take turns.

Of course, this emphasis on equity between all participants in debate (on the standards of civil conversation, on the balance between speaking and listening, on the willingness to take turns, on all those features of antilogic that are inherent in its multivocality) does not imply that when it comes to judgment all logoi are the same. Prodicus, who is famous for his lexical distinctions, remarks that

> (t)hose who are present at discussions of this kind [i.e. public debates] must divide their attention between speakers impartially, but not equally. The two things are not the same. They must hear alike but not give equal weight to each. More should be given to the wiser, and less to the other. (*Protagoras* 337a)

That is, while antilogic insists that all participants and perspectives in debate receive fair hearing, not all logoi have the same contribution to make to enhanced understanding. It is the work of practical judgment to discriminate among conflicting options, to test alternatives, weigh probabilities, and to decide on the basis of what is most appropriate. Until that time, however, antilogical practice will assume that (a) every logos has the right to be heard and (b) every logos is open to question.

The motive for what we might call "the open forum of argument" is a simple one: any position may possess the germ of an idea that leads to better understanding, even perhaps to consensus (*homonoia*; cf. Bialostosky 1991, 20). Havelock refers to this democratic conception of argument as part of the "liberal temper" inherent in the Protagorean method of controversy. He argues that as aliens in Athens, unable to speak directly to the assembly, the Sophists had a special interest in expanding the franchise for public speech (216). In this context, all sides must be given a voice in debate because democracy depends upon the consent of the governed and the negotiation of consensus, a process Havelock refers to as "the epistemology of public opinion" (230). It follows from this political position that "the formulation of antithetical positions" is more than a "convenient instrument" for arriving at agreement (ibid. 237). Rather, antilogic is a critical instrument for transforming the multivocality of public discussion into a systematic procedure by which heterogeneous groups agree upon a common purpose. That purpose is never reducible, as far as antilogic is concerned, to the rhetorical equivalent of *e pluribus unum*, because no single logos can ever be conceived of as possessing final authority. Nonetheless, the *homonoia* towards which antilogic moves can be said to manifest the justice and respect for all that Protagoras posits as the foundation of human culture.

Having sketched in the basic features of multivocality and aligned it with both the norms of conversational discourse and with the "liberal temper" of 5C Athens, it remains to briefly update this "guiding principle" by reference to more contemporary theory. There are, as mentioned, branches of 20C thought that have much to contribute to the reconstruction of antilogic. My reference to these sources is brief, in part because this territory is well-known to many readers, in part because I would use only what is needed to develop the topic at hand. That said, the Derridean critique of binaries and the Bakhtinian concept of dialogism are especially helpful in redefining multivocality in contemporary theoretical terms.

Most readers are well aware that post-structuralism, like the two-logoi fragment before it, challenges the notion that signification results from any freestanding, autonomous entity or logos. For Saussure, any single signifier possesses only an arbitrary relation to the thing signified and accrues meaning only through its relation to other signifiers (1959, 67-69). For Derrida, not only is no signifier autonomous, but all bear the mark of their relationships to others, marks "inscribed" in the history of the sign itself (1982, 13-15). Language is thus a system of relations and distinctions, a system in which any single signifier carries the trace of its multiple connections and relationships.

But unlike antilogic, or at least unlike antilogic in its initial Greek phase, post-structuralism insists that the inherent differences that mark the signifier (and, by extension, any attempt to argue meaning) need not be articulated within a field of binary opposites. You will recall that the Greeks consistently presented opposing pairs as diametrical, polarized contrasts (see Ch. 2, sec. 3). Classicist G. E. R. Lloyd explains that such duality provided "abstract clarity and apparent comprehensiveness" and so allowed for the construction of simple theoretical models that accommodated a mass of complicated data (1966, 66). It is just this reductivist, dualistic tendency that has been under attack by late 20C theory. Derrida considers the theoretical status of opposing pairs a misplaced metaphysical assumption, the reflection of what we imagine to be the quintessential nature of Being, which is then wrongly extended to all lower-order categories. Not only do such binaries disguise their dependence on history, they also present themselves as natural hierarchies which insinuate the dominance of one member into cognitive habit and conventional status. At the close of his essay "Signature Event Context," Derrida notes that deconstruction provides a means for "intervening in the field of opposites it criticizes," an intervention intended not simply to complicate the binaries it confronts, but also to "displace" or disperse these foundational pairs through a direct assault on the systematic impulse to think and communicate through contraries (in Bizzell and Herzberg 1184).

The Derridean critique or "displacement" of binaries begins with the scrutiny and rejection of what Derrida calls the "nucleus" of any opposition (1982, 10). In the case of argumentation, it is the nucleus of pro and con that must be called into question. Once identified, we can then transform this illusory binary into a plenitude of relations, relations that deny the implicit tyranny of "for and against", "better and worse" and reveal instead the multiplicity that actually informs controversy. In effect, the post-structural critique of binaries supplies a rationale for extending the scope of the *dissoi logoi* and for reconceiving the context of argumentation as plural rather than double—there are not just two sides to any question, there are as many sides as there are active voices in the open forum of discussion (see Govier 1988). Provided with this critique, we turn next to Bakhtin, who offers further support for the reconception of argument under the heading of *multiplex ratio disputandi.*

Within the lexicon of Bakhtinian terms that bear on our subject, there is a collection of ideas, all allied with dialogism, that enrich and expand our understanding of the processes of antilogic.[9] Among these terms are answerability, alterity, (re)accentuation, the other, and the double-voiced, all concepts by now well known to many rhetoricians (see Ewald 1993, Farmer 1998). Briefly put, dialogism would insist (with Derrida) that any single utterance is unavoidably marked with traces of prior and prospective positions, so that even the opening move in an argument is "but one link in a continuous chain of speech performance" (1963, 72). Moreover, dialogism implies that all language is directly *addressed* (the word has special resonance for Bahktin) to respondents who, in turn, take on a particularly active, participatory role, agreeing, disagreeing, augmenting, applying, etc. This responsiveness "lasts for the speech's duration"; but, more importantly, it also serves to shape the making of the utterance itself. To employ Bakhtin's powerful terminology, there is no "mute perception"; the roles of speaker/respondent are always "interanimated," so that (in the famous phrase) the word is a "two-sided act," a "bridge" with dialogical traffic moving in both directions (1981, 3; 1986, 85-86, resp.).

But dialogism by itself continues to imply a two-part process in which the word looks alternatively toward what has been and what might be said. What sets Bakhtinian dialogics apart is the added notion of heteroglossia, the concept that language use is dialogical in more than a diachronic sense, that our utterances are "over populated" with a polyphony of disparate echoes that reveal an "internal dialogism" between the utterance itself and a host of other voices (1981, 279). Heretofore, we have thought of the pro and con of argument as monads confronting one another within strict discursive boundaries. But in a heteroglossic approach to argument, monads are dissolved since their medium of interaction, the living utterance, is itself "furrowed" with the many voices that already and unavoidably populate the area of controversy (1986, 92-93). When we enter into argument, we consequently encounter an arena already charged with discourse, or, as Bakhtin describes it, a "tension filled environment of alien words, value judgments, and accents," a "dialogically agitated" field where we "weave in and out of complex interrelationships, merge with some, recoil from others, intersect with yet a third group," then a fourth, and a fifth, etc. (1981, 276-77). Instead, then, of formulating argumentative positions in bipolar relation and according to strict oscillations of question and answer, assertion and response, instead, that is, of participating in a platform debate, we engage in argument as participants in what Bakhtin refers to as "the world symposium," a discursive site of "unfolding social heteroglossia," a domain of many, many voices that we must negotiate if we wish to weave a claim of our own into the ongoing conversation (Clark and Holquist 1984, 60).

Exactly how antilogic is to operate in the open forum of argumentation is an issue for consideration in Part II of *Many Sides*. Suffice it to say for now that even the snippets of Derridean and Bakhtinian theory I have appropriated here serve to enhance the profile of multivocality as the first principle of antilogical *praxis*. In sum, we know that pro/contra, prosecution/defense, even stronger/weaker are artificial binaries long assumed, now impeachable; that all utterance, including argumentative claims, are hybrids, jointly constructed by multiple participants; and that, by virtue of the polyglot nature of the open forum, antilogical discourse will be "dynamic, diverse, and vivid," characterized by what Bakhtin calls a "jolly relativity" (1984, 109-110). Of course, there is no guarantee that multivocality itself will lead to practical benefits; but as *praxis*, multivocality is a virtue (*arete*) in itself, an excellence to be pursued for its own sake.

Nonetheless, there is a certain recoil in opening up the argumentative process to the bewildering multiplicity that constitutes the universe of possible opinion. A few years ago, I read in a contemporary journal that "[w]hen fallen man is made the measure of all things, as the Left would have it, strange and horrific consequences inevitably follow" (*Chronicles: A Magazine of American Culture*, June 1995, 45). I feel reasonably certain that this writer is unaware of the precedent for his sentiment in Plato's critique of Protagoras and the human-measure fragment. The fact remains, however, that the target here is an unrestricted form of democracy that provides an open platform for multiple voices. And yet we know what happens when argument is kidnapped and held in thrall by one side, whatever side that may be. James Kasteley, in a brilliant article on Euripides' *Hecuba*, describes a situation in which those who hold power and make decisions are uninterested in open discussion and peremptorily insist on a logos compatible with their own interests (1993; cf. Thucydides 5.89). In such a situation, when multiplicity is abandoned (or even simply reduced to a set of contraries), the space within which rhetoric operates is drastically restricted. Antilogic would reverse these demographic as well as political restrictions and insist that unless all voices are allowed to "render and receive" in the open forum, the quest for understanding through discursive interaction is doomed to partiality.

You may recall that when antilogic functions as an invention procedure, the rhetor must postpone judgment in controversy until after she has prepared herself *in utramque partem* (for either side in debate). The *praxis* of multiplicity, founded on Protagorean principles and informed by dialogism, would solicit just this postponement, a hesitation in the face of varied possibilities that is motivated by both tolerance and prudence. Through this process of listening, exploring, and juxtaposition, the rhetor would work to extend the letter of the Aristotelian desideratum and seek to discover *all* the available means of persuasion, *all* the voices that call out for attention in any given case (cf. *Rhetoric* 1355b 27-28). Once filled with this plenitude, we should be prepared to argue with less prejudice, as well as on any side.

Oppositionality. Our first principle, then, is that no stand in any argument is ever complete in and of itself. All stands, regardless of how compelling they may seem, require a complement, a second, a third, or a fourth logos to respond to the inherent partiality of the first. These additional logoi, and their connection to one another, bring us to antilogic's second guiding principle: the relationship between/among the two or more logoi of any argument tends to be *oppositional*, the two logoi in practice are anti-logoi, related to one another by difference and tension. Whereas multivocality indicates simply a pluralism of potential voices, the diverse logoi of argument are actually in antithetical relation to one another—contrast in argument becomes contradiction, the existence of the "other" implies an opponent.

Technically, there are several forms of oppositional relationship. For example, "opposites" themselves are simply "any pair of terms between which we apprehend, or imagine a contrast or antithesis" (such as night and day); whereas "contradictories" must be mutually exclusive (as in odd and even numbers), and "contraries" are opposites that admit some contextual variability (as in tall and short, since Jean may be tall in relation to John but not when compared to Jane) (Lloyd 1966, 86-90).[10] For our purposes, it is enough to recognize that opposition in argument is typically antagonistic, with one side putting itself forward as demonstrative, as the foreground that ignores, obscures, and would efface the background. This totalizing tendency is itself a logos that antilogic would oppose; for as Sloane says: "the one question on which there are not two sides is whether there are two sides to every question" (1985, 124).

This second principle may also seem tautological since any argument automatically assumes opposing views in contention. But the point to make at present is not that different positions are always available but rather that all logoi stand in a complementary position vis-à-vis one another; they are, in fact, not simply in proximity but also and more importantly in relationship. That is, even though opposing positions enter into contact from opposite sides of the field, as it were, their combined appearance opens up opportunities for mutual benefit unavailable to any member individually. We can begin to get a sense of these opportunities by appeal to Bakhtin's theory of alterity, which focuses on the perceptual relation of opposing perspectives. The theory as a whole is replete with implications, but the crux of the matter is summarized in this passage by Clark and Holquist, describing what they call alterity's Law of Placement:

> You can see things behind my back, such as a painting or passing clouds, that are closed to my vision, while I can see things that your placement denies to your vision, such as a different painting on the other wall or other clouds moving behind your head. This difference determines that although we are in the same event, that event is different for both of us. (1984, 70).

The kinship of this concept with Protagorean perspectivism should be obvious—the Law of Placement assumes that we see "the same event" differently based on our location. And with regard to the present argument, Bakhtin's alterity neatly clarifies the potential contribution of oppositionality to argument: by virtue of our encounters with others, we can begin to glimpse what is unavailable to our own perspective but within the ken of relationship. Argument is thus supplied with a motive for oppositional *praxis* in the escape it offers from the limits of our own specific location and the partiality of single views in general. In effect, the principle of oppositionality encourages rhetors to reconceive the potentially tense relationship with opposites, to replace the notion of argument-as-war with argument-as-opportunity, to transform confrontation into collaboration, even perhaps to let go of enmity in an act of gratitude for what alterity can do. If this *peripeteia* seems itself an act of sophistry, there is, nonetheless, a strong case that such a move is compatible with Protagorean theory.

Once again, the Protagorean anthropology of the Great Myth is helpful. In it, human beings are first supplied by Prometheus with "skill at the arts" sufficient to keep them alive; but because they lack "political wisdom," they are unable to congregate and protect themselves through numbers against stronger creatures (*Protagoras* 321d-22c). Worried that humans might be destroyed, the gods give them a second legacy: respect for others and a sense of justice, gifts that allow them "to bring order to our cities and create a bond of friendship and union" (332c). Since antilogic is the Protagorean means by which we inquire into the unavoidable disagreements that routinely disturb the harmony of human community, it is at moments of dispute (or *agon*, which means not only a "struggle" but also a "contest" or "debate") that a sense of justice and respect for others is most in need. De Romilly refers to the *aidos* of the Great Myth as "the seed of the social contract" (166); and Liddell and Scott translate the term not just as "respect," but also as "honor," "regard for others," and even "reverence."[11] It is not hard to see how the notion of reverence for one's opponent in debate could operate as the binding force for community in general or for any group involved in debate. Such reverence for opposition is the position taken by Prodicus, a disciple of Protagoras and a temperate voice in the often-heated debate between Protagoras and Socrates. At a point in the *Protagoras* when Socrates is inclined to quit the discussion, Prodicus says:

> I add my pleas, Protagoras and Socrates, that you should be reconciled. Let your conversation be a discussion, not a dispute. A discussion is carried on among friends with good will, but a dispute is between rivals and enemies. In this way our meeting will be best conducted.[12] (337a-b)

It is the invocation of *aidos*, of respect for others—especially one's opponents—that allows for the transformation of argument from a debate with one's rivals into the kind of discussion that solicits the goodwill for which Prodicus calls. Just how this attitude or atmosphere can actually be created and employed, especially in our classrooms, is a subject for later discussion (see Ch. 5, sec. 2 and the Epilogue, sec. 2). For now, our concern is with the paradoxical principle itself, with amity-in-opposition.

According to the Protagorean approach to "discussion" and "debate," significant disagreement presupposes a mutual concern for the problem at hand. As authentic conversation (i.e., dialogue informed by *aidos)*, antilogic becomes the shared method by which private, diverse, even partisan interests can be dealt with in a manner that relies upon negotiation and that aims at both justice and understanding, even in those instances when reconciliation and consensus are out of reach. Michael Billig, a social psychologist, remarks that "(w)here everyone agrees with one another, it is not possible for discussion to flourish" because the "spirit of contradiction" is a primary catalyst for the production of new ideas (83). On the other hand, adds Billig, "when two people are said to be having an argument—that is, they are ill-disposed to each other—they might, in point of fact, be refusing to communicate by argumentation" (84). Antilogic insists on difference in order to get things done through discourse, to solve problems; but the problem exists for both participants, so one's opponent in debate becomes an ally in the search for understanding. Heraclitus, who stands as a predecessor for the paradox of oppositional reasoning, notes that "all things come to pass through conflict" and that "the counter thrust brings together" (*DK* 22 B8), to which Charles Kahn adds that "from the perspective of wisdom" the act of negation always possesses some "positive value" (1979, 193). Such is the view of antilogic: each anti-logos in debate has a necessary clarification to offer, a contribution to the common objective which only its particular partiality can offer. When Protagoras enters into debate, he can wait, listen, and calculate the nature of that contribution. He is, in effect, enacting Wayne Booth's golden rule of rhetorical exchange: "[P]ay as much attention to your opponents' reasons as you expect [them] to pay to yours" (1974, 149).

Not all of us can wait so well and be so tolerant; often, we need an arbiter, like Critias in the *Protagoras*, who will act with authority to prohibit the degeneration of debate into "open faction" (Havelock 218). In a classroom that operates antilogically (like Quintilian's), the instructor will often be called upon to perform the function of Critias. In any venue, the goal of antilogic (as informed by oppositionality) is not to overthrow one's opponent in some dazzling solo performance or unanticipated peritrope, but to extend the scope of the discussion by adding a different perspective, thereby revealing what is masked by one-sided thinking. Opposition, then, may be instinctively greeted as a personal affront; but if it is accompanied by respect, that recoil can be replaced by a recognition of its salutary effects, of its ability to generate new possibilities unimagined on one's own, possibilities that can lead not only to better understanding but also to better relationships. The first step toward such benefits, however, is to liberate the concept of disagreement from its false association with discord and to reconceive opposition as an indispensable ally to argumentation, a *praxis* whose actions will prove their own reward.

This effort to reconfigure the role of opposition in argument has been advanced by several schools of contemporary theory, mostly in rhetoric and communication studies, all of whom offer interesting continuities with the antilogical tradition. There is, for example, the field of Rogerian rhetoric and the more expansive tradition of dialogical communication into which it has been incorporated. Begun in the 1960s and early 1970s as an attempt to employ the indirect, non-combative therapy of Carl Rogers, and influenced by both the thought of Martin Buber and the emerging area of conflict management, Rogerian rhetoric emphasizes empathetic relations between individuals in dialogue as a productive medium for problem-solving (see Brent 1991, 452-57).[13] Richard Johannesen, in his 1971 article, "The Emerging Concept of Communication as Dialogue," states the basic tenet of this approach: "The essential movement of dialogue is turning toward, outgoing to, reaching for the other" (375). Carl Rogers himself fills out this idea with its necessary counterpart: "Once you have been able to see the other's point of view, your own comments will have to be drastically revised" (1961, 333). As an aid to problem-solving, Rogers emphasizes conciliatory dialogue with ideas at variance from one's own, a process that involves encouraging the interlocutor to speak, restating the position of the other side, and refraining from evaluation.

Rogerian ideas were introduced to rhetoric and composition studies by Young, Becker, and Pike through their popular textbook, *Rhetoric: Discovery and Change* (1971), a text that situated composition against the background of technological modernization and promoted Rogerian principles as a means to achieve "greater cooperation" among disparate groups and amidst changing historical circumstances (see Young et al. 223; and Brent 1991, 454). But scholarly controversy arose over the adaptation of Rogers' model to the composition situation. Feminist critics, in particular, asserted that Rogerian rhetoric elicited a certain self-effacement, a tendency to avoid conflict altogether in the attempt to diminish the harmful effects of emotional and evaluative response (see Ede, 1984). As the passage from Prodicus above indicates, antilogic agrees with Rogerian therapy that when respect is lacking, debate can easily degenerate into harsh partisanship and eristic. But antilogic also assumes that the confrontation of antitheses is the catalyst not only for the individual's accommodation of difference and the mutual clarification of views but also (and more importantly) for the genesis of ideas hitherto unknown, perhaps even unknowable to either side in isolation. Ultimately, it is the clash of difference itself, mediated of course by norms of dialogical civility, that is the agent for "discovery and change."

Additional developments of the Rogerian tradition point out ways it can contribute to the psychology of argumentation and, in particular, to the amelioration of the anxiety that naturally attends argumentative exchange. Doug Brent, in a perceptive and generally supportive review of Rogerian rhetoric, envisions the development of an "epistemic dialogue" that has much in common with antilogic (1991, 462; see also 1996); and Nathaniel Teich outlines a Rogerian approach to argument that includes this very antilogical instruction to the student writer: "Do not choose a topic or take a position on it unless you are prepared to be persuaded by cogent reasoning or empathetic understanding to change your mind" (58).[14] In addition, Angela Lunsford (1978), Maxine Hairston (1976), and Paul Bator (1980) help to clarify the important contextual and operational differences between the "adversarial" rhetoric of the courtroom (cf. Aristotle's *Rhetoric*) and the dialogical principles that animate the Rogerian and antilogical approaches. With regard to the present argument, Teich makes the essential point: "The agon of the rhetor can be a struggle *for* something (truth, good reasoning) not necessarily *against* another as hostile opponent" (56). Indeed, the paradox of antilogic is that the opponent, or opposition in general, is the necessary partner in our joint struggle for something that is beyond our individual grasp.

A related emphasis on struggle and conflict in the composition process can be found in the fertile research area of collaborative learning, most notably in John Trimbur's well-known "Consensus and Difference in Collaborative Learning" (1989). Trimbur's primary interest is in unpacking the political implications of social constructionist theory and, more specifically, in deconstructing the concept of consensus as an appropriate ideal for the collaborative process. But the specific context of his interest is the composition classroom, and his critique of collaboration can be read as an exposition of the oppositional principle that informs the practice of antilogic.[15] According to this critique, the "politics of consensus" works to induce an acceptance of normative standards and a certain passivity in the face of unequal power relations that determine who and what count as meaningful contributors to the discussion at hand. Trimbur's response to the canonical view (that consensus is an appropriate ideal for the collaborative process) is distinctly antilogical; he explores consensus in terms of "conflict rather than agreement" because "consensus cannot be known without its opposite" (608). When seen from this opposing perspective, the collaborative process reveals itself as a multivocal exchange which will invariably "give voice to the conflicts inherent in an unequal social order and in the asymmetrical relations of power in everyday life" (609). Like argument, then, collaboration always takes place within "an agonistic framework of conflict and difference," an environment that calls for "a rhetoric of dissensus" (609).

In order to develop this alternative rhetoric—a rhetoric that doesn't exile difference to the margins and stigmatize struggle as uncooperative—Trimbur argues that we must distinguish between consensus as "an aculturative practice that reproduces business as usual" and consensus as an "oppositional" practice that challenges prevailing conditions through the operation of dissensus (612). We must, in other words, open up opportunities for antilogoi to assert themselves in response to the dominant (and often domineering) norms that would control argument. What we don't learn from Trimbur is a *techne* of dissensus that will allow us to realize the pedagogical desire for "heterogeneity without hierarchies" (615). But there are clues elsewhere that collaborative-learning research is at work on the problem. Rebecca Burnett, for example, argues that collaborators drafting a business document are measurably more productive if they engage in substantive conflict; i.e., if they adopt such strategies as deferring consensus, posing alternatives to one another, and voicing "explicit disagreements about both content and rhetorical elements" (1993, 144). She distinguishes substantive conflict from its affective (interpersonal) and procedural (methodological or strategic) counterparts, and contends that when collaborators are alert to the fact that opposing the logoi of one's collaborative partners can be a cooperative rather than a contentious gesture (a struggle for rather than against), then conflict can itself produce calculable benefits. Her studies of this process in action identify a number of ways that collaborators can manage conflict, and she correlates these with the quality of the collaborative product. In so doing, she offers provocative empirical support for an ancient rhetorical practice that has strong empirical roots (see Ch. 2, sec. 3).

One final tradition of scholarship that addresses the role of conflict in argument deserves mention, though the importance and nuance of its evolving contribution to a refigured conception of argument outstrips any pretense of summary here (see Berrill 1996, Fulkerson 1996). I refer to feminist approaches to argument. As noted above, there are feminists who have complained that the passivity of Rogerian argument deprives the female rhetor of the very assertiveness and sense of agency that convention has perennially kept from her (see Lassner 1990). Alternatively, there is also a sustained body of feminist criticism (see Tompkins 1988, Frey 1990, Lamb

1991 and 1996) which maintains that argument is not simply masculine in origin and aggressive by nature, but that it is also, as I note elsewhere, "blatantly adversarial"—the rhetorical equivalent of the Vince Lombardi approach to competition ("winning is not the most important thing; it is the *only* thing"). The result of these powerful, antilogical critiques is a distinctly feminist response to argument that recognizes the inevitable hierarchies at play in controversy and at the same time seeks an approach to opposition that is, as Catherine Lamb writes, "fair to both sides" (1991, 11). I would like to close this appraisal of oppositional *praxis* by referring to one such response, Susan Jarratt's "Feminism and Composition: The Case for Conflict" (1991).

The author of a vanguard study in the revival of Sophistic rhetoric (*Rereading the Sophists: Classcial Rhetoric Refigured*, 1991), Professor Jarratt argues that the feminist "case against argument," based on resistance to conflict, is pedagogically unproductive. Instead, she would promote "overtly confrontational feminist pedagogies" inspired by the practice of "the first Sophists" (1991, 106). She is particularly concerned that conflict-averse classrooms will not prepare students to negotiate the oppressive discourses of racism, sexism, and classism and will leave dominant discourse intact (108). Our educational institutions, she writes, reproduce and enforce "the power and control of the existing social order"; but they also "allow students and teachers to challenge, oppose, and resist those forces" (118). In response, she calls for a pedagogy of "productive conflict" that will address the "social complexities of our classrooms and the political exigencies of our country in this historical moment" (111). Her model for productive rhetorical conflict is the *dissoi logoi*, which assumes that "public action can be guided by informed debate among members of a democratic community" and that conflict (synonymous here with oppositionality) "can serve as a measure of the political effect of a discursive practice" (114). More specifically, she argues that

> [t]he discursive method driving both feminist and Sophistic ways of negotiating change through discourse is argument. In both the prephilosophical fifth century, BC, and in the current postmodern antifoundational context, rhetorical positions stand temporarily as grounds for action in the absence of universally verifiable truth. (121)

Though Jarratt doesn't refer to antilogic by name (other than to use the term *dissoi logoi*) or develop its protocols in detail, the discursive method which grounds this politically-oriented pedagogy is clearly Protagorean (cf. Jarratt 1991, 49-53). Nonetheless, this essay offers a powerful argument in behalf of "contending with words" and the role of conflict in contemporary classrooms. Moreover, in support of her argument, Jarratt surveys an impressive group of rhetoricians and educational theorists (including bell hocks, Ira Shor, Kathleen Weiler, and Patricia Bizzell) who are also calling for a new way of talking about contention, controversy, and conflict. Protagorean antilogic, with its compound emphasis on respect and opposition, stands poised to contribute to this discussion.

Dynamism. So, according to Protagorean principles, no question open to argument can avoid the pluralism of multiple perspectives, and, once articulated, these multiple positions, stands, claims will constellate themselves as logoi in opposition. The last member of the trio of guiding principles is entailed in the concept of opposition. For now, we will assume, as did the ancients, that such opposition presents itself as a contrasting pair, or, at least, as logoi collected within the opposing hemispheres of a controversy. The important point is that, at any given time, one logos will always be seen as stronger. That is, there is typically a presumption of superiority in favor of one argument over its opposite, so that the weaker argument must carry the burden of overturning its dominant counterpart. We explored this notion of dominance in relation to the stronger/weaker fragment which argues that one logos (usually less prominent) can be substituted for another when the new idea or formulation proves to be stronger than its contrary (i.e., more conducive to the general good of those involved) (see Ch. 2, sec. 3; also, *Theaetetus* 167c-172a). Jacqueline de Romilly remarks that this rhetorical doctrine ("substituting the weaker for the stronger logos") depends upon the possibility of convincing others and getting them to change their minds (1992, 190). This emphasis on change is implicit in all antilogic and as such forms the core of our third principle of antilogical *praxis*; i.e., the structure of antilogic is inherently dynamic, it assumes continuous movement on the part of those involved in controversy as they respond to the progressive development of the argumentative process. Moreover, the dynamic momentum of antilogic is perpetual and will forestall any attempts at closure: though we must, on occasion, take stock of where we are and make decisions based on our present understanding, antilogic will always look beyond the present and dominant logos to an antilogical "other" which will respond to and modify existing perceptions. I will close out this section on rhetorical form with a brief examination of this dynamic resistance to closure.

In a general sense, dynamism is implicit in the very form of antilogic, a form that bypasses any single argument as inherently partial and becomes itself only in the interaction, the biplay, the ongoing give-and-take between successive logoi. Nor can antilogic be identified with the simple presence of two opposing positions, since it is possible to hypostatize any set of contraries into a sterile and static binary (like good/evil) which leads to entrenchment and deadlock, the antithesis of dynamism. Instead, antilogic invokes opposing logoi for the purpose of moving the discussion along. The concept of motion is crucial and links antilogic with the most basic conception of dialogue, a conversation in which partners continually trade responses. Hegel hints at a similar conception when he notes that "the reason why dialectic first seizes upon motion as its object lies in the fact that dialectic is itself this motion" (in Gadamer's *Hegel's Dialectic* 13). If anything, antilogic is more responsive than dialectic to subtle shifts and steady motion, because the former is "thrown" (in the Heideggerian sense) into the world of contingency, into a Heraclitean space where ideas are always in flux, whereas dialectic (as a strategy for discovering an impersonal and objective truth) would lift argument outside the time and flux of its particular setting. As such, each new turn in the progress of antilogical exchange marks a step into a slightly different spot in the stream as new articulations, perspectives, circumstances add to the flow of argument and modify the course of its development.

The emphasis of antilogic, then, is on movement, on the interaction of logoi, and not on any set pattern of reasoning, or even on the ultimate result (some final, triumphant argument). To appropriate Jim Corder's term, truth is always "emergent" in antilogic, always in the process of coming into being through the progressive juxtaposition of multiple, contradictory logoi (1994). In a sense, then, the principle of antilogical dynamism dictates the goal of the argumentative process as progress itself, or momentum, an action rather than a resolution, the ongoing action of reason in response to controversy. Moreover, with apologies to William Blake, we can fill out his famous aphorism, "without contraries, there is no progression," by noting that, given *multiplex ratio*, progress is potentially limitless (see Gross 1986). But in order for the movement of antilogic to be movement forward, there must be change on the part of the participants.

Consequently, the dynamism of antilogic would subvert any tendency to persist in an attachment to a single, immutable logos. As Nathanel Teich tells his students, you can not begin an argument in good faith unless it is possible that you might change your mind in the process (58). Similarly, Kathleen Welch writes of dialectic that it "breaks down the monolithic tendencies of many students" and requires them to "shift stance" (1993, 137). Such is the shift that Cicero's Antonius makes as he impersonates the different players in debate; it is the shift that Quintilian requires of his students as they prepare *in utramque partem*, by taking up each side in a controversy. G. B. Kerferd conceives of the process in a more restricted sense when he writes that antilogic proceeds by establishing "a contrary or contradictory logos in such a way that the opponent must either accept both *logoi*, or at least abandon his first position" (63). It seems to me that the range of potential responses to a preceding logos is somewhat greater than the two options that Kerferd allows; we might, for example, augment, reduce, revise, refute, ignore, agree, applaud, restate, or refine a prior argument. But on the essential point Kerferd is undoubtedly correct: antilogic requires that we listen to the other side and respond, and this response will typically involve the modification of prior commitments. In sum, we move the argument forward by introducing opposition, the dynamics of which will require from rhetors and audience alike a change, however slight, in the nature of their affiliations. From the antilogical point of view, to resist change (whether from passivity or dogmatism) is to abjure argument.

The dynamism of antilogic, therefore, dictates that the opening position in any argument is always provisional, that the progressive nature of the argument will transcend one's initial presumptions, regardless of how dominant those beliefs may seem. But the progress of antilogic doesn't stop with a single reversal, a surprising peritrope, a simple turning-of-the-tables which leaves the once-weaker logos in permanent command. If "truth" in antilogic is emergent, always in the process of becoming, then its progressive movement is a little like Zeno's well-known paradox; i.e., no matter how far we have come there is always a ways to go, always another argument to hear. To a large extent, the resistance to closure that characterizes the antilogical process is an extension of the two previous principles of antilogical form. The multivocality of the antilogical process will ensure that there is yet another voice to add to the dialogue; and even if the number of players who might contribute to an argument is limited, the progressive nature of the discourse allows (requires?) them to repeatedly re-enter and respond, to "render and receive" again and again, as the evolving discussion enters new ground. As Bakhtin would have it, every utterance entails an answer, so there is no true way to finish off the conversation. With regard to opposition, it too is never foreclosed. Thomas Sloane cites the 16C Italian writer Francesco Guicciardini on this point:

> In all human decisions and actions there is always a reason for doing the opposite of what we do, for nothing is so perfect that it does not contain a defect. Nothing is so evil that it does not contain some good, just as nothing is so good that it does not contain some evil. (1985, 65)

The essential scepticism on which antilogic is based dictates that there is always some correction, some adjustment, some critique to offer to the logos of the other side and, in turn, that our own critique must also be answered, in an act of infinite progression. In this sense, the dynamism of antilogic is experimental; i.e., it advances by critique, but it also operates as a projection, as an effort to calculate what is not yet known, to approximate a "truth" still emergent. The focus of opposition and critique may seem to be on the past, on precedent, on the logos that is already advanced; but the salient feature of antilogic in its dynamic form is its connection with the future, so that each new logos in the progress of argument is a gesture in the direction of discovery (see Dewey's "Recovery" 1970, 23). And while that progress is never complete and the future never arrives, there is something of an imperative for antilogic to keep the conversation going (cf. Rorty 1979, 389-94).

Among contemporary rhetoricians, it has been William Covino and Victor Vitanza who have most vigorously attacked the tendency of discourse to seek closure. In *The Art of Wondering*, Covino works to resurrect a repressed tradition in which rhetoric is "a mode of *avoiding* rather than *intending* closure" (1988, 9). From Covino's own antilogical perspective, Plato's *Phaedrus* with its "interplay of ambiguities," Aristotle's *Rhetoric* with its attention to "the discovery of multiple perspectives" and Cicero's *De Oratore* with its own "drama of perspectives," all represent this "forgotten" tradition of rhetorical open-endedness, which his own argument would advance in opposition to the dominant stress on logic and closure in the history of rhetorical theory (19, 28, and 44, resp.). And in the critical, stylistic practice of Vitanza, we find a similar pleasure in antilogic and an antipathy to closure. Vitanza writes that his methodological purpose is to "initiate the construction of a continuous *dissoi logoi* . . . to pick up an argument, to examine it for a while, and then to drop it midway in order once again only to pick it up from still another vantage point; . . . in other words, *to sustain . . . without any resolution . . . the motion of the whole*" (1987, 69). The approach to argument on display in Covino and Vitanza and inherent in the dynamism of antilogic would, in Kathleen Welch's apt phrase, "transcend stasis" (1993, 136). It would substitute as the appropriate argumentative attitude a tolerance for perpetual incompleteness in place of an insistence on conclusive resolution; it would open up the conversation still further to voices that have yet to be heard rather than narrow the options in the search for ultimate "truth." As noted above, decisions must be made despite the fact that decision-making will interrupt the progressive development of any argument. But the practical necessity of coming to *euboulia* or good decisions in matters of controversy does not justify the conclusion that a particular decision is the last, best position, the one argument left standing at the end of debate. As Michael Billig maintains at the end of his own book on *Arguing and Thinking,* there is no such thing in argument, or thinking, as "the last word" (1987, 250-56). Instead, the last word must be answered because what is at issue in argument is not some finite resolution but the process of discovery itself.

In "Socrates and the Confusion of the Humanities," William Bouwsma argues that it is Protagoras, not Socrates, who should be accorded the title of "founding father of education in the humanities," and that "if we are to understand the humanities as the expression of some coherent tradition, we ought to think very seriously about what Protagoras stood for" (1982, 13-15). Part I of this book has been an attempt to respond to this injunction and to think seriously about what Protagoras stood for: about the basic principles of his worldview (Chapter 1), the discursive methods by which he put his theories to practice (Chapter 2), and the practical, ethical, and rhetorical dimensions of that unique practice (Chapters 3 and 4). In this effort at rehabilitation, it is essential to recall that Protagoras thought of himself principally as an educator and that his theoretical and methodological innovations are all put to service in his pedagogical program (*Protagoras* 317b). It is rhetoric that is at the heart of the Protagorean educational enterprise, and it is argument that is at the heart of Protagorean rhetoric. Seen thus, the art of antilogical argument is the Archimedian point from which the world itself can be moved.[16] It remains now to be seen what happens to the theory, practice, and pedagogy of antilogic when the leverage it would exercise is placed in Roman hands. Happily, those hands belong to one of the most skilful orators (Cicero) and one of the most accomplished teachers (Quintilian) of their own or any age.

Part II

Roman Developments in Practice and Pedagogy

Chapter 5

De Oratore and the Development of *Controversia*

"Since a wise man can be mistaken, and a hundred men, and many nations, yes, and human nature according to us is mistaken for many centuries about this or that, what assurance have we that sometimes it stops being mistaken, and that in this century it is not making a mistake?"

Montaigne, "Apology for Raymond Sebond"

"As for Cicero himself . . . [he] was without obligation to any party, following what seemed probable to him now in one sect, now in another, keeping himself always in Academic doubt."

Montaigne, "Apology for Raymond Sebond"

When teaching Cicero's *De Oratore*, I begin by asking my students to cite a favorite passage in the dialogue. There are usually some standard choices: the outline of the three duties or offices of rhetoric (2.29.128-30), Antonius' method of invention by impersonation (2.24.102-3), and invariably Crassus' denunciation of "the absurd and unprofitable and reprehensible severance between the tongue and the brain, leading to our having one set of professors to teach us to think, and another to teach us to speak" (3.16.61).[1] These are all weighty moments, and they usually lead to engaging class discussion.[2] When my turn comes to cite a favorite passage, however, my choice routinely meets with stares and silence. For I choose an inconspicuous moment at the outset of the dialogue when Scaevola, a relatively minor character, contradicts the impassioned opening statement of Licinius Crassus, the man whom Cicero calls the most illustrious orator of his day (*Brutus* 38.143).

The circumstance is this: Crassus has just delivered a stirring epideictic on the power of discourse to gather scattered humanity into one place and "to lead it out of its brutish existence in the wilderness up to our present condition of civilization" (1.8.33). Crassus here is the voice of emergent humanism, the Isocratic patriot at the head of an evolutionary ascent from ignorance to eloquence, the mouthpiece of Cicero himself. Nonetheless, as soon as Crassus has finished his encomium on eloquence, Scaevola responds with a courteous but serious challenge to each claim in Crassus' high-minded thesis. For all your eloquence, says Scaevola, you carry your argument too far. My own response to this fleeting moment is that it is not only a stunning *peripeteia*, or reversal of expectation, but that it is also a dramatic announcement at the outset of the dialogue that—in the realm of rhetoric and for the practice of argument—no position is sacrosanct, everything must be argued, for there are always two sides, or more, to every question and we should always be prepared *in utramque partem*, to examine all sides of the case.

In response to this initial *peripeteia* (or peritrope), Crassus himself proceeds to turn the tables on Scaevola, only to have his own eloquent arguments, in turn, repeatedly questioned and routinely rebutted by others throughout the dialogue. Such reversals are an engaging part of the drama of *De Oratore*; they are also, as readers of Part I will recognize, standard elements in the arsenal of antilogic. But like so many other parts of the Greek *paideia*, antilogic has now changed names. In Rome, argument by contraries is called *controversia*; and in the three centuries from Antiphon to Crassus, a variety of modifications in its theory and practice have naturally taken place. This is the Hellenistic period, traditionally set from the death of Alexander in 323 BCE to the defeat of Anthony by Octavian and the advent of the Roman Empire in 31 BCE. Beyond the momentous political events of the age, this is also an active time for philosophy, with the development of Stoicism, Epicureanism, and scepticism and a dominant role for philosophy in advanced education. Despite changing times and practices, however, argument in the *De Oratore* is clearly consistent with the antilogical principles and patterns outlined in Part I; i.e., we can readily identify the invocation of the *dissoi logoi* (divided or opposing claims); the incorporation of multivocality, oppositionality, and dynamism; the dialogical patterns of give-and-take; the suspension of judgment; the concern for practical ethics and pragmatic results; and, above all, the framework of probable knowledge and prudential judgment that structures the entire process. Indeed antilogical *techne* and *praxis* are in play not only in the *De Oratore* but throughout Cicero's mature philosophical corpus. In these works (especially the main group of 44-43 BCE), Ciceronian inquiry invariably proceeds in what we can call the "controversial" manner; i.e., by critical juxtaposition, often in dialogue, of the multiple views that fill out the landscape of opinion on the topic at hand.[3]

Kenneth Burke writes that his own primary purpose in *The Grammar of Motives* is to "express an *attitude* toward language embodied in a *method*" (1969, 441). In Cicero, the "attitude" toward language and argument that we studied in Part I is intact, but its methods are refined. Heretofore, we have dealt only with excerpts of argumentative practice, limited illustrations of this or that antilogical strategy. With the *De Oratore*, we are presented with a full-dress performance by one of history's great rhetors. Consequently, we have an unparalleled opportunity to take stock of our subject in detail. In this chapter, then, we move not only from Greece to Rome but also from general overviews to focused critical scrutiny, to an account of ancient *controversia* that I hope captures the subtlety of the art as practiced by a master. In particular, we will concentrate on Book I of the De *Oratore* because that is the site of the most consistent dialogical interaction among characters. There are additional, interesting developments of *controversia* in Book III that we will also examine, though in Books II and III the major speakers (Antonius and Crassus, respectively) hold forth in more-or-less uninterrupted fashion (see 2.4.16). But in Book I, with its regular shifts between speakers and reversals of position, we can focus on the discursive relations among characters, on patterns of assertion and response, defense and revision, on the accommodation of one speech (or *logos*) by another, and on the *praxis* of *controversia* in concrete detail. In sum, it is in the episodes of argumentative exchange, when rhetors must not only "render" their own logos but also "receive" and respond to that of their interlocutors, that the Ciceronian attitude and method are most notably on display.

However, my purpose in this chapter is not simply to analyze a distinguished rhetorical model. My interest is also and fundamentally pedagogical: I will argue that Cicero's pedagogical stance, as represented by the dialogue's leading figures, is uniquely compatible with his rhetorical theory and particularly instructive for contemporary teachers of argument. This argument (along with that of the next chapter on Quintilian) marks a definite shift toward the pedagogy of our subject. The Protagorean program is clearly informed by pedagogical concerns (see Ch. 2, sec. 1), while the dissemination of the Greek *paideia* is, in large part, a pedagogical event, with Greek teachers of rhetoric and philosophy transporting the new learning throughout the Mediterranean. But in Rome, pedagogy becomes increasingly formal, and rhetoric itself increasingly identified with its pedagogical manifestations (for reasons I will note). This growing attention to pedagogy, its historical significance, and the role of *controversia* in the process will become a central theme of Part II. At present, however, we must prepare for a change in the protagonist of *Many Sides*, for antilogic is about to re-enter with a new name and costume. And, as is characteristic of intellectual drama, the set is arranged and the prologue best spoken by history itself. So, before we explore the attitudes, methods, and pedagogy of *controversia* as these are developed in the *De Oratore*, it will help to look briefly at the developing history of the controversial method. For behind the practice of argumentation at work in the *De Oratore* stands Protagoras and the Sophistic tradition of arguing both sides, and between Sophism and Cicero comes the complex history of Hellenistic philosophy. Taken together, these influences provide a backdrop for the rhetorical restoration enacted by Cicero.

1. FROM ANTILOGIC TO *CONTROVERSIA*

According to M. R. Wright, the young Cicero translated Plato's eponymous dialogue on Protagoras, though there is little sustained discussion of the great Sophist in Cicero's mature canon (1991, 1).[4] Nonetheless, the line of continuity is there, both in terms of the theory of knowledge that underwrites speculation about discourse and the discourse method that follows from this theory. The Protagorean influence is, of course, mediated by almost four centuries of intellectual history and muted by the absence of Protagorean texts, even in the age of Cicero. So, before we survey the relevant events of these Hellenistic centuries, it may help to remind ourselves of the main tenets of Protagoreanism so that we can trace the often obscured lines of its influence through various post- Sophistic permutations.

Diogenes Laertius lists some fourteen books by Protagoras (ca. 490-420) on subjects ranging from philosophy to government, theology to mathematics (*DK* 80 A1).[5] None of these works has survived, though Protagoras himself remained well-known as the originator of the *homo mensura* doctrine, the anti-foundational theory that concentrates on the relation of individual perception to the nature of knowledge (*DK* 80 B1; see Guthrie 1969, 3.187). In the absence of *ipsissima verba* (Protagoras' own words), I have interpreted the human-measure doctrine to mean that knowledge is relative to the source or perspective from which it is derived and that different perspectives on the same experience will yield different (re)constructions of that experience. And, according to the Protagoras of Plato's dialogues, this variability in our perceptions does not imply the naïve belief that all views are equal. Rather, while human perceptions of the real and true are always contingent and while one perspective may not be "truer" than others by universal, invariant standards, one logos can, nevertheless, be distinguished from among the alternatives as more useful or advantageous given particular circumstantial conditions (*Protagoras* 334b, *Theaetetus* 167b). When the epistemological pragmatism of this doctrine is translated into discursive practice, the result is an approach to argument based on the recognition that if knowledge is local and partial, knowing subjects will naturally produce opposing claims (antilogoi) and that some of these oppositions can be equally well defended. As Diogenes Laertius puts it, Protagoras was "the first to say that on every issue there are two arguments opposed to each other" (*DK* 80 A1). The practice that follows from this famous maxim is a form of argument in which comparative reasoning determines the greater or lesser efficacy of competing claims, arguments, perspectives, logoi by examining them in relation to one another. Moreover, antilogical practice maintains that by purposefully placing opposing claims in juxtaposition rhetors can not only minimize the unfair advantage of a conventionally stronger position but also generate a consensually supported proposition that both adjudicates conflict and leads to prudent action.

Between the ages of the first Sophists and Cicero (106-43 BCE), argumentative practice is routinely adapted to changing perspectives and conditions (see Buckley 1951; Hankinson 1995; Long 1955, 1974; McKeon 1950). And while the historical record is thin, there is enough evidence from Cicero and others to argue that antilogic, along with dialectic (its methodological "counterpart") are the original models for emerging forms of disputation that develop in concert with the philosophical controversies of the Hellenistic age. The range of these controversies extends to virtually every field of knowledge (ethics, nature, politics, religion, epistemology, etc.), and competing voices include numerous, minor schools of thought, along with the Stoics, Epicureans, and Academics. The theoretical complexities that issued from these debates required sophisticated procedures of reasoning in order to weigh the lines of thought and locate the position that elicited the greatest confidence. In Cicero's mature view, such reasoning was best governed by "considerations of probability and practical significance" (McKeon 1950, 55).[6] This emphasis on probable reasoning and practical standards echoes the Protagorean perspective, but only indirectly because (as noted) the works of Protagoras are no longer available in the first century BCE. For direct support in his theoretical and discursive practices, Cicero turns most consistently to what Michael Buckley calls the "operational procedures" of the Hellenistic Academy, whose dominant figures following Plato and Aristotle are Arcesilaus in the third-century BCE and Carneades in the second (1971, 148).[7]

The centerpiece of the Academy's procedures was the practice of *controversia*, of speaking for and against each side (*in utramque partem*), a practice that Cicero refers to as the Academy's "custom" and which he learned as a student of Philo, the Academy's representative in Rome (*Tusculan Disputations* 2.3.9; cf. *De Oratore* 3.36.145, 2. 21.80; *De Finibus* 5.4.11). Cicero approaches *controversia* as a philosophical concept embodied in an operational method; in turn, the method itself is distinguished by both its heuristic and critical capacities (McKeon 1950, 7; see *Academica* 2.3.7, *De Oratore* 2.38.157-61). By most accounts, Cicero inherits this theory-based method from the Hellenistic Academy; but by my reckoning, the *fons et origo* of *controversia*, its base and origin, is in Protagoras.

The Academy had been founded by Plato in 369; and, at his death in 347, its leadership passes to his nephew, Speusippus. At this point, Aristotle, who had been associated with the school for two decades, departs, though he remains a member, even after establishing the Peripatos (at the Lyceum), his own school in Athens, in 335. Circulating the work of these two masters (whose doctrines were regarded as basically consistent) was a major task of the post-Platonic Academy (see Long 1974; Powell 1995). Cicero is clearly familiar with both. He is one of the first to adapt the Platonic dialogue to Latin, he experiments with and elaborates on its form throughout his career, and he adopts strains of ethical idealism and metaphysical interest that are fundamentally Platonic. Regarding Aristotle, Cicero appears to know the *Rhetoric* and the *Topics* well. With these precedents as a guide, he traces the practice of *controversia* back to Socrates (*Tusc. Disputations* 1.4.8) and Aristotle (*De Finibus* 5.4.11; *De Oratore* 3.21.80). This genealogy is simplified but reasonable given the loss of Protagorean works, the profound influence of the Socratic method, and the passages in Aristotle that outline argument by contraries (see *Topics* 101a; *Rhetoric* 1355a, 1557a, 1402a). Moreover, this history is reinforced in the years following the fourth century BCE (Aristotle dies in 322) when the Academy moves away from its initial philosophical ground and comes, through its association with scepticism, to embrace a form of antithetical reasoning and disputation centered on thinking through contraries.

As a distinct philosophical doctrine, scepticism is traditionally connected with Pyrrho of Elis (c. 365-c. 270 BCE), and Pyrrhonism is replete with Protagorean traces. As defined by Sextus Empiricus in his *Outline of Pyrrhonism*, scepticism centers on the ability to set up antitheses that account for "the equal weight which characterizes opposing states of affairs and argument," a direct echo of the Protagorean *dissoi logoi* (1.8). Pyrrho himself wrote nothing, but his views are recorded by Timon of Phlius, a follower, and much later by Sextus and Diogenes Laertius (second and third-century CE, respectively). In brief, Pyrrhonism starts from the premise that all perception is both sense-based and individual and, therefore, reflects only what appears to be the case to the knowing subject, not what an object actually is in itself. Given this theoretical context, Pyrrhonists concentrate (as had their Protagorean precursors) on the relative nature of perceptual responses, the natural contradictions in personal views, and (as noted) on the "equal weight" that can be found in contrary positions (cf. Diogenes Laertius, 4.101). In response specifically to the prevalence of contradiction in human affairs, Pyrrho promotes not just the suspension of judgment (*epoche*) but a certain freedom from dogmatism that comes from maintaining a "sceptical" distrust of any belief because the supporting evidence is only probable (see Hankinson 1995; Schmitt 1972, 5-17). According to A. A. Long, such ideas, like much else in Hellenistic philosophy, are the continuation of concepts "inherited" from earlier thinkers (1974, 3). In particular:

> [c]ertain problems to which Pyrrho drew attention had already been recognized by earlier philosophers, who put forward different kinds of answers for resolving them. Pyrrho's scepticism has its closest conceptual connection with Protagoras among his predecessors. (79)

And, as Long also points out, Pyrrho's emphasis on equally weighted opposing positions looks forward as well as back. While he founded no school himself and appears to have had few direct followers, Pyrrho's ideas do find a home in the post-Platonic Academy, where they are taken up by Arcesilaus (315-240 BCE), who assumes the leadership of Plato's Academy in 265 and who turns the "New" Academy in a decidedly new direction from that imagined by its founder.

In the *De Oratore*, Cicero notes that Arcesilaus believed that "neither the senses nor the mind can perceive anything certain," a position not exactly in line with the Platonic drive for absolutes (3.67). The operational method that Arcesilaus invokes in support of his basic scepticism is to insist that everything must be argued and that no conclusion is free from contradiction. According to Cicero, the method of Arcesilaus can be traced to the conversational practice of Socrates, a practice that challenges all dogmatic assertions as self-contradictory and frequently ends in *aporia* (see *Tusc. Disputations* 1.4.8). Cicero adds elsewhere that by adopting the Socratic method and arguing against the opinions of all men, Arcesilaus "led the majority to withhold any consent from either position more easily" (*Academica* 1.12.44-45; cf. *De Finibus* 2.1-2). The expressed purpose of this method was to debunk dogmatic, insupportable claims and to discover what probabilities existed that could sustain challenge and support action. But in actual practice, Arcesilaus' predominantly critical approach is less concerned with assent than it is with refuting claims to certainty, achieving "freedom from error," and arriving at the suspension of judgment that is the sceptic's mark of wisdom (Von Arnim in Long 1974, 92). In Arcesilaus, therefore, the antilogical strain that survives in Pyrrho is revived and institutionalized, but it is also truncated, appearing now as a purely negative dialectic—oppositionality overwhelms assent, argument is dominated by rebuttal, and the "equal weight" of opposing ideas yields only reason to doubt. A more comprehensive approach to the antithetical method is reclaimed in the next century by Carneades (219-129 BCE), whom Cicero credits not only with the restoration of *controversia* to the Academy but also with the introduction of *in utramque partem* to Rome itself (*De Oratore* 3.18.67-68; *Tusc. Disputations* 5.4.10-11).

Like Pyrrho or Socrates, Carneades wrote nothing; but, as the head of the Third Academy and as both a compelling speaker and a subtle dialectician, his influence was considerable (Diogenes Laertius 4.62). Like Arcesilaus, and in line with the sceptical tradition, Carneades was prepared to challenge any dogmatically held belief, though his principal attacks were reserved for the Stoics. Hankinson refers to him as a "sceptical controversialist"; and controversiality—in the sense of "preparedness for disputation"—is the hallmark of his sharp, antithetical criticism (1995, 96-108). But Carneades did not confine himself to negative critique only. He posits a theory of plausibility (*to pithanon*) as a positive criterion for practical decisions (see *Academica* 2.104). Such a criterion may at first seem paradoxical, a commitment on Carneades' part to an objective standard that he denies to others; and yet to assert plausible rather than necessary claims and criteria is to avoid dogma by allowing for error. To illustrate the process he offers this example: a dark and uncertain form lying coiled in the corner could be a snake; but, given all the evidence, it is more likely to be a rope (from Carneades, in Hankinson 112).[8]

By virtue, then, of placing opposing claims in relation to one another, we can distinguish the plausible from the implausible to the extent that a reasonable, if fallible decision can be made regarding the efficacy of a claim under existing circumstances. In contrast with Arcesilaus, Carneades' approach to antithetical reasoning is what Buckley calls "two-voiced" in its insistence that both "positive and negative positions be represented" (1971, 92). In the process, the radical scepticism of Arcesilaus' unrelenting antitheses is replaced by a more balanced method which seeks, through pro *and* contra assessment, to calculate the potential plausibility among competing logoi and in so doing to provide sufficient assurance for decisions regarding action (cf. Hunt 1954, 42).

Theoretical originality is not, however, the only basis for Carneades' influence. In 155 BCE, he and two other Athenian philosophers (a Stoic and a Peripatetic) visited Rome as ambassadors of their schools. During the visit Carneades gave his famous lectures on justice, which Cicero describes this way:

> When Carneades had been sent by the Athenians as an ambassador to Rome, he discoursed at length on justice . . . . On the next day, he overturned his own discourse with a speech putting the opposite position, and undermined justice which he had praised the previous day . . . [all] in the style of a rhetorical exercise arguing on both sides. (*De Republica* 3.9)

This intellectual *coup de théâtre* and the interest it generated serves to announce the arrival of philosophy in Rome (Powell 1995, 13-14). It also marks a highly public appearance for controversial reasoning, a gala performance of the Academic merger of doctrine and method, an event not lost on Cicero, who refers to the "inspired intellectual acumen and rhetorical fluency of Carneades" (*De Oratore* 3.18.67-68). This merger of acumen and fluency, attitude and method provides a base for Cicero's own intellectual operations. Or, as Buckley puts the case, "the antithetical method begot *controversia*" (1971, 92); to which I would only add that the antithetical scepticism of the Academy that begets the *controversia* of Cicero had itself been begotten under the auspices of the antilogical tradition.

Cicero himself inherits the Carneadean position of modified scepticism and argument *in utramque partem* from Philo of Larissa, who led the Fourth Academy from 110 to 83 BCE and who, like his Academic predecessors, was an ambassador to Rome, arriving in 88, when Cicero was 18 and studying law. Because he was Cicero's teacher and because we know the Academy principally through Cicero, it is difficult to distinguish Philo's ideas from more general Academic positions (see Hankinson 1995, 116ff). However, like Carneades, he seems to have taught that sense-impressions are undependable and that knowledge claims are subsequently indeterminate; but despite this uncertainty, we can pursue the probable by avoiding dogmatism and examining ideas on either side of a question. For Cicero, a particular benefit of this teaching is that it allows for various positive positions (in ethics, theology, politics) as long as one claims only probability for them or, more precisely, greater probability for one position than its rivals. Another benefit for Cicero was the inclination towards rhetoric and debate that follows from Academic scepticism; and indeed, Philo himself was a teacher of rhetoric as well as a philosopher (cf. Powell 1995, 20-23). But it is Cicero's own synthesis of rhetoric and philosophy that gives Academic method enduring form. It is hardly likely that the "academic" debates of ancient ideologies will stir much interest for most contemporary readers (even granting a certain commonality of intellectual interests). Rather, ancient *controversia* comes to life for us in the *De Oratore*, a literary adaptation of a philosophical method in which rhetorical procedures are proposed for testing probable claims sufficient for practical action. In *Many Sides*, the New Academy, its leaders and their methods serve as prologue to this premier performance of controversial reasoning.

Cicero announces his own methodological preferences in statements such as the following: "the sole object of our discussions is by arguing both sides (*in utramque partem*) to draw out and give shape to some result that may be either true or the nearest possible approximation of the truth" (*Academica* 2.7-8; cf. *De Officiis* 2.2.8; *Tusc. Disputations* 2.3.9; *De Oratore* 3.36.145). But he does more than simply state the case for controversial reasoning. Throughout his philosophical canon, he attempts to "draw out and give shape to" the topic at hand by presenting *multiplex ratio* in dynamic interplay; i.e., by placing multiple characters in dialogical settings where they defend their positions with vigor and competence. The goal of these "deliberately polyvalent exercises" is not so much to resolve the questions at issue but to model the procedures by which "the nearest approximation of the truth" can be discovered (Swearingen 1991, 153). Specifically, in the *De Oratore*, instead of discussing *controversia* formally (i.e., by abstracting its general nature and detailing its logical parts), Cicero chooses to instantiate within the form of the discourse the ideas on rhetorical method that this very discourse would address. In a word, Cicero chooses to *perform* his topic and, in so doing, to allow his readers direct access to *controversia* as a process or operation rather than simply a doctrine. In response, if we approach the text as "the imitation of an action," we may be able to identify what Crassus and Antonius insist cannot be apprehended apart from the particular form in which it appears. We may also learn something about controversial pedagogy, which (by exposing the student directly to the interaction of competing logoi) seeks to stimulate an attitude towards and understanding of the topic that is independent of professorial dicta.

2. THE DRAMA OF BOOK I

I suggest that we approach the *De Oratore* as a drama in which the plot is informed at every turn by what we might call the spirit of friendly contradiction (cf. Billig 1987, 237). The dramatological approach to the dialogue, however, requires that we attend not just to the plot, but also (as Aristotle points out) to character, setting, language, and theme (*Poetics* 1450a-b). So by way of preparing the stage, let us begin with character and setting, both of which significantly condition the way we view the unfolding plot.

According to Cicero's *Brutus*, his history of Roman oratory, Crassus and Antonius were both "consummate orators and the first among Romans whose diffuse elegance rivaled the glory of the Greeks" (36.138; in Watson 1970, 301). Marcus Antonius (143-87 BCE) was especially distinguished for his skill in invention, for the strategic manner with which he marshalled his arguments, and for his prodigious memory (*Brutus* 37.139; *Tusc. Disputations* 5.19.55; Wilkins 1979, 17). Crassus (140-91 BCE) was famous not only for the unrivaled beauty, dignity, and wit of his language (*Brutus* 59.215; *De Oratore* 2.28.121; 3.9.33; 3.43.171) but also for his wide culture, vast learning, and powerful appeals to the passions (*De Oratore* 3.9.33; *Brutus* 43.158). In addition, he was particularly noted for his ability at *altercatio*, the quick trading of opposing ideas in debate, a talent on display throughout his career in the Forum (Wilkins 12; cf. Watson 1970, 307). The other regular participants in the dialogue, Cotta and Sulpicius, were not only the "two most approved orators" of the generation following Antonius and Crassus, they were also both devoted followers of the older men; Cotta of Antonius and Sulpicius of Crassus (*Brutus* 55.201, in Watson 1970, 321). As A. S. Wilkins puts it, both younger men were "bound by the closest ties of respect and affection to those whom they looked upon as their common masters" (1979, 17; cf. *De Oratore* 1.21.97). Given the fact that the dialogue itself proceeds according to the prompting of the younger men who would solicit their mentors' views on the practice of oratory, the text can be described as a master class in which the teachers exhibit the accumulated expertise of their long and distinguished careers. Taken together, these four characters, along with several others (including Scaevola) who appear more briefly, make up the "burgeoning pluralism of views" that the *De Oratore* presents on the topic of oratory (Swearingen 1991, 134).

But there is more that we should know about the cast than simply their proficiency in oratory. It is interesting to note that Antonius was not only the grandfather of Marc Antony (the triumvir and Cicero's fatal enemy) but also the father of Cicero's partner as Consul of Rome and an early victim himself of the political violence that swept Rome under Marius (Wilkins 1979, 1). Sulpicius also met death by proscription (as did Cicero), while Cotta was banished from Rome the year after the fictional date of the dialogue for his part in one of the many political intrigues of the day (*De Oratore* 3.3.11). Even more telling than such lethal politics, however, is the fact that Cicero chose to set the dialogue several days before the death of Crassus (September, 91 BCE) who, after delivering a passionate speech to the Senate against the sitting Consul of Rome, immediately fell ill and died (3.2.6). There is, then, a certain irony in the passionate adherence of these four great speakers to the virtues of public debate in the face of a political fanaticism which routinely erupted into violence and, in the process, obviated the power of oratory (Wilkins 1979, 9-26; Wood 1988, 31-32). But if the dialogue is politically ironic, it is also a panegyric, an homage to the skills of mentors whom Cicero himself knew and admired. Cicero and his brother studied as young men with both Crassus and Scaevola; Cicero later visited Scaevola regularly at his home to discuss the law; and Antonius was a friend of Cicero's uncle (*De Oratore* 2.1.2; *Brutus* 89; Wood 1988, 43). This intimacy of relationship between the author and the great orators of a departed age may well account for the utopian tone of the dialogue, as the speakers gather at the beautiful Tusculan villa of Crassus, during the early fall, away from the hubbub of the Roman Games, and on the very eve of the passing of the host. Seen in this way, the text becomes a hymn to the methods by which eloquence can employ disagreement in the service of mutual understanding. Moreover, the setting of the meeting is an idealized site for discourse, a privileged space where the guiding principles of controversial *praxis* are invoked as sources of strength and productivity, where rhetors in argument "struggle for" a common understanding rather than against one another. *Ad bellum purificandum*. It is what we might wish our own classrooms could be like.

At the time of the dialogue itself, the major figures introduced above were allied in a battle for control of the court system, an effort led by the tribune Livius Drusus (who would be murdered shortly after) and supported in the Senate by Crassus (Wilkins 1979, 5-6). In early September (91 BCE), however, all political action was suspended in order to celebrate the *Ludi Romani*, or Roman Games. During this interlude, many of the senators and nobles retreated to their villas, and Wilkins refers to it as a "plausible fiction" that the characters in the dialogue should congregate at the estate of Crassus (1979, 6). The first day of the sojourn is spent discussing the political crisis of the moment and "the state of politics generally" (1.7.26). On the second day, however, during a morning walk, Scaevola sights a plane tree in Crassus' garden which reminds him of the tree under which Socrates and Phaedrus conduct their own famous dialogue on rhetoric (cf. *Phaedrus* 229a-230e); and so Scaevola proposes that the party divert itself from their political anxieties by "imitating" the Platonic precedent (1.7.28).

Whereupon, Crassus introduces "a conversation on the pursuit of oratory, with a view of relieving all our minds from the discourse of the day before" (1.7.29-30). The party thus commences what we might call "games" of an alternative kind.

In the prologue to the ensuing drama, Cicero invokes the *controversia* that distinguishes the rest of the text by claiming that there has always been a scarcity of great orators (1.2.8-4.16). Later in Book I, however, Antonius will refer to the "vast supply of talent which I see existent among our fellow citizens" (1.21.95); and Cicero himself, in the *Brutus*, will catalog the many exemplars of Roman rhetorical excellence. Obviously, there are two perspectives on this question, and the single, monological voice, even when it is that of the author, is subject to challenge. Moreover, Cicero's espoused motive for presenting the "old story" of Crassus and his colleagues is to respond to the wishes of his brother, Quintus, and to supplant the "crude" effort of his early treatise on rhetoric (*De Inventione*, c. 86 BCE) with a more mature appraisal of the topic (1.2.4-5; *De Oratore*, 55 BCE). But his return to the subject is already conditioned by controversy since we are told that Cicero and his brother have for some time disagreed on what leads to eloquence (1.2.5). The *De Oratore*, then, emerges as itself a response to controversy, an answer to an unsettled question, an investment in the ongoing dynamism of argument. As Bakhtin points out, we never enter discourse "from the sidelines" because the discussion is always already in progress (1981, 277).

The oppositionality of Cicero's prologue is further delineated in the differences that Cicero notes between the various methods of oratorical instruction. Is this "incredibly vast and difficult" subject to be best understood by systematic training or by observation of "those to whom the highest honors of eloquence have been awarded" (1.5.17; 1.6.23)? Is emphasis to be laid on "the refinements of learning" or "on a sort of natural talent and practice" (1.2.5)? At this early point, such questions are matters for dispute, not subjects for resolution. But Cicero is insistent on one issue in particular: let us prompt the students of oratory, he says, to surrender the notion that

> they can gain their coveted object [oratorical expertise] by reliance on rules or teachers or methods of practice employed by everybody, but to rest assured that they can do this by the help of other means. (1.5.19)

This passage might well serve as another of my favorite moments in the *De Oratore*; it certainly articulates one of the most persistent themes in Cicero's *opera rhetorica* (*De Oratore* 1.31.145; 2.11.44-45; 3.21.125;

Orator 3.15; *Brutus* 76.263; see also Leff 1989, 123-24; Swearingen 1991, 157-160). That is, rules, teachers, and methods inevitably seek to determine the nature of practice. But in the actual experience of rhetorical exchange, preconceived rules are particularly inadequate because rhetoric is—above all—conditioned by the existential circumstances of the specific case. The concept is pedagogically daunting: if we are teaching practice instead of doctrine, and if practice can best be known "in process," then what the teacher of rhetoric professes is the art of making good decisions in particular rhetorical circumstances (cf. *Theaetetus* 166a-167d). More on this later; for now, what about those "other means" for achieving rhetorical prowess? At this point, the narrator appears more interested in suspending judgment than providing solutions. So we leave the prologue and its narrator in *aporia* and turn our attention to speakers who are "the most eloquent of our nation, and of the highest rank in distinction of every kind" (1.6.23). Readers will have to discern for themselves by what particular means these gifted orators have achieved the "coveted object" that is the subject of the ensuing dialogue.

As I noted at the opening of the chapter, Crassus begins the dialogue with a panegyric on the powers of oratory which not only echoes Cicero's own preface but also recalls the commonplaces of Isocrates, who originates the notion that it is by the refinements of speech that we rise from "brutish creation" and exercise "wise control" of the state at-large. Given the political turmoil of the Roman setting, however, it is not surprising that this opening protreptic (in which Crassus encourages his young friends to "go forward" and cultivate oratory) should meet with opposition (1.8.32-35). That is, when approached as a dramatic action taken within a particular context rather than as a platform oration on the glories of civic humanism, Crassus' pathos-laden exordium on the powers of oratory calls out for what members of the New Academy might call an antithetical corrective. And Scaevola, his respondent, is the very model of "courteous" correction (1.9.35; see *Brutus* 58.212).

The elderly lawyer (and an official state augur) begins his rejoinder thus: "On his other points I am in agreement with Crassus . . . but the two following [ideas] I cannot grant you . . . " (1.9.35). Rhetorically, this response starts with an effort at identification followed quickly by contradiction; at the level of syntax, it enacts the basic pattern of "yes, but," a locution that operates as something of a master-figure (or schema) for the controversial method.[9] The initial emphasis is upon the connection/ relationship between the interlocutors. And this sense of partnership sets a standard for the ensuing debate in which the willingness to speak against (or contra-dict) is sustained and indeed enhanced by a full measure of mutual respect (a quality that follows from the *praxis* of multivocality; see Ch. 4, sec. 2). In this case, Scaevola is polite but decisive in his opposition to the twin notions advanced by Crassus: that oratory is the foundation of civil order and that the gifted speaker must be broadly educated. Through appeals to example and authority, he argues that the case for oratory is considerably more complex, more problematic than Crassus suggests. Crassus had begun with a philosophical, even Platonic approach to the issue (if these terms are construed to mean an assessment of the topic in its ideal form), whereas Scaevola responds by historicizing Crassus' encomiastic remarks (1.10.44). The older man's speech is characterized by concession and rebuttal, by a reassertion of the *multiplex ratio* that Crassus had ignored, and by a notable effort not to taint his interlocutor in the process of speaking against the argument itself. There are two positions now (*dissoi logoi*), and the existence of division and conflict will lead to the invention of ever-more-refined logoi. "Without contraries, there is no progression."

The pattern or structure of Crassus' response to Scaevola is basically similar to the "yes, but" of Scaevola's own retort, though the logoi to be accepted/rejected are, of course, reversed. After an exordium on philosophy's dismissal of oratory ("I too have heard that argument"), Crassus begins to position his own revised claim for oratory in contradiction to, but not in direct rebuttal of Scaevola. Note the parry-and-thrust pattern of the following constructions: even if one accepts the narrow view of the philosophers, "nonetheless . . . [the] critic must grant" that oratory does possess certain positive elements (1.11.48); or, if we take the philosophers' position to the limit (i.e., that rhetoric is conducted totally without reference to knowledge), how then can we explain the orderliness, grace, and fullness of expression (copiousness) that so many orators plainly manifest (1.11.48)? With the aid of these locutions, Crassus is inventing a progressively more complex case by arguing *in utramque partem*, on both sides of the case. In essence, he asserts that "I will accept this, but not that," as he does when he says

> *granted that* the topics of [the orator's] discourse may be found in certain other fields of research, *yet* their actual style is the peculiar product of this pursuit which we are now discussing and investigating, and of no other. (1.12.49; italics mine)

Shortly after, he adopts the following syntax: "when he has granted. . . still he will assert" (1.12.54); "while I acknowledge . . . yet consider" (1.13.55); and "when I have allowed . . . it is nonetheless true that . . ." (1.13.56). The effect of these circumlocutions is to internalize *controversia*, to incorporate the arguments of the other into one's own reconstruction of the case. The goal here is not to develop a single line of reasoning that will yield a final, totalizing claim; rather, Crassus has begun to circle around his topic, to supplant the notion of a direct endorsement of any one proposition or claim with a tendency toward reversals, inversions, ironies, and oppositions. He accomplishes this by adapting a form of response that places agreement and revision in a state of syntactic tension, all in an effort to both answer one's interlocutor and to render the issue in its complex ambivalence.

The exchange with Scaevola had begun with a relatively simple assessment of the efficacy of oratory. But in Crassus' hands, and in response to the stimulation of Scaevola's critique, the purview of the discussion has expanded; oppositionality has led to new terrain. At issue now is the relation of rhetoric to both knowledge and craft, or *techne*. For the present, Crassus seems to indicate that both knowledge about the subject and knowledge about one's craft are necessary and that neither is independently sufficient for true eloquence. In Book III, he will develop this position into a comprehensive vision in which eloquence harmonizes the seeming antagonism between knowledge and style, *res* and *verba*, *ratio* and *oratio*. At this very early point in the drama, however, we have already reached a new perspective on the question of oratory, one neither so laudatory, nor so critical as the opening salvos by Crassus and Scaevola, a perspective which accommodates opposition and in so doing begins to acknowledge the unavoidable *multiplex ratio* so obviously available on this complex topic.

Crassus rounds out his speech by contradicting Socrates (1.14.65), by lauding the intellectual versatility of the "finished orator" (1.15.65-66), and by asserting that he has himself endeavored to accomplish the ideal he describes (1.16.71). In response, Scaevola smiles and remarks that

> in the very speech you [Crassus] have made against me, you have by some trick so managed matters as both to grant me what I said did not belong to the orator, and then somehow or another to wrest away those things again and hand them over to the orator as his absolute property. (1.17.74)

Scaevola is clearly jesting here; for while Crassus has indeed granted some of his claims, it is not by trickery that he has at the same time contradicted the basic argument of his interlocutor. Kerferd writes that it is "the essential feature" of antilogic that both pro and contra arguments "could be expressed by a single speaker as it were *within* a single complex argument" (1981, 84). Such expression is "multivocality" in action; a concrete instance of discourse as a "two-sided act" (Bakhtin) or the "double voice" employed by Carneades. Crassus has adapted the practice by compounding agreement and disagreement within a single utterance. He has also avoided any hint of dogmatism that pretends to resolve the issues by offering an intimation of certainty. Nonetheless, Scaevola sums up Crassus' rejoinder accurately: you agree and disagree with what I said, but on balance you work to substitute an opposing logos, though you do so with enough graciousness so as not to offend me. In other words, the rhetoric of his son-in-law places a high premium on *conciliare*, on winning the favor of one's listeners (2.27.115). As we are told later, this effort at conciliation requires the rhetor to display not only dignity and achievement in one's own bearing, but also "the tokens of good nature, kindness, calmness, loyalty, and a disposition that is pleasing and not grasping or covetous" (2.43.182). These are all qualities that will placate the mood (*affectus*) of the interlocutor, especially when that interlocutor is being contradicted (see May 1986, 3-5). Once again, *controversia* appears as a matter of ethos, of how one acts in response to controversy (see Ch. 3, sec. 2). And yet, despite the persuasiveness of Crassus' oratorical persona, there remains room for Scaevola's own rejoinder: "let us see whether you may or may not be attributing to [the orator] more than the real facts of the case allow" (1.17.77).

At this point, however, there is a cast change, as Antonius steps forward, for the first time, to take up Scaevola's side in the debate. For those who recognize the eminence of the two orators now at center stage, it is an important moment, the elevation of the discussion to the very highest level. Given Antonius' renown as an attorney and the premium that Cicero places on *conciliare*, we should hardly be surprised that Antonius too will begin with an act of identification. As we are now beginning to realize, such respect is typically the prelude to an antilogical reversal. So Antonius begins, "Crassus, to my mind you establish your case" if, that is, we grant you your assumption that it is possible to master the "vast and difficult" subject area that you claim for the rhetor (1.18.80). Since Antonius will prove particularly sceptical about the possibility or need for such learning, his opening response amounts logically to a "yes, but"; rhetorically, it is an act of courtesy which establishes his own persona as a cooperative rather than competitive partner in controversy. And indeed, Antonius will end this first speech by paying the very highest compliment to Crassus: that he is the one truly eloquent speaker that Antonius has heard (1.21.95).

The most interesting aspect of Antonius' response, however, is the method by which he chooses to develop his alternative position: i.e., by narrating an Athenian debate over the very issue that these Roman friends are discussing. Like Crassus, Antonius has spent time with the learned men of the Academy: "pretty much the same as those whom you [Crassus] have lately mentioned" (1.18.82; cf. 1.11.45). Both men share, then, a very similar experience (which further serves the goal of identification); but as Protagoras' human-measure doctrine would have it, their assessment of that experience is very different. In Crassus' version of the Athenian debate, all of the philosophers maintain "with one voice as it were" that the orator is "shut out from all learning," and Antonius' account confirms this "reading" (1.11.45). First Mnesarchus, then Charmadas (one Stoic, one Academician) insist, according to Antonius, that eloquence requires the study of philosophy. But "(c)ertain Athenians [mainly politicians and attorneys] argued on the other side," a response which "roused up" the pique of Charmadas, who replies hotly that rhetorical texts are filled with "trumpery" and that rhetors themselves are ignorant "even of the true principles and methods of eloquence" (1.19.85-87). His opponents, in turn, cite Demosthenes and other orators as examples of a "consummate wisdom" that Charmadas would claim exclusively for philosophy; to which he adds that rhetorical prowess is a natural talent, not an organized skill over which the rhetoricians can assume pedagogical authority (1.20.84-93). We have here the give-and-take of opposing sides, though the scales certainly seem to lean in favor of the philosophers. So much so that Antonius tells us that he was "won over by these same views" and actually writes a small pamphlet on the disparity between present-day oratory and true eloquence (1.21.94). And yet the whole narrative is fraught with irony: by inference, Charmadas (as a present-day figure) is not "truly eloquent," nor does the head of the Academy operate according to his own institution's principle of always taking the opposite side in debate (see 1.18.84). Instead, Charmadas and company dogmatically (i.e., monologically) insist on their own claim to philosophy's preeminence in a scene that is the ancient equivalent of an academic turf-battle.

What is especially curious about all this is that Antonius seems for the moment to be persuaded by a position that is actually antithetical to his own, by an argument that would have us believe that "no one could speak [eloquently] unless he has mastered the philosophical teachings of the most learned men" (1.20.93). Such a proposition more nearly approximates the stand of Crassus and certainly contradicts the emphasis on practical experience that Antonius will eventually maintain. But in this instance, Antonius recalls that Charmadas "looked like [he was] persuading me," which at very least causes us to wonder if Antonius' "little pamphlet" is not the "crude" product of his own unfinished youth in much the same way that the *De Inventione* was eventually viewed by Cicero (cf. 1.21.94 and 1.2.4). The important point is that Antonius in this scene appears to model the skills that *controversia* would promote: that is, he listens to the claim of his Greek interlocutors and, in response, is willing to rethink his own presumptions.

According to Michael Billig, the whole goal of antilogic is not to sustain one's own position, but "to draw one's attention to another side of the matter" (11). Antonius' willingness to scrutinize and modify his own initial position resides at the core of the antilogical process, a process that is nothing if it is not dynamic, if the collision of opposites does not operate to prompt a shift in stance by the interlocutors. As I have noted before, those who will not accept the possibility that their minds might be changed in response to debate are not ready for argument in the first place (see Teich 1987). Antonius responds to the Athenian debate with a readiness to change, however slightly and for whatever duration, his own affiliations. In doing so, he enacts for the benefit of the younger participants in the dialogue the fundamental notion that to resist change is to abjure argument.

In the context of Antonius' soon-to-be articulated predilection for pragmatism and common sense, this early inclination toward the philosophical perspective is an instance of dramatic irony (cf. 1.54.233). His employment of autobiography and narrative also serve to concentrate attention on "human life and conduct" as opposed to "well-worn maxims" as the appropriate vehicle for instruction in discourse (1.15.68; 1.31.137). In the process, the concrete personal experience of this distinguished orator becomes part of "the other means," the controversial alternative to reliance on rules as appropriate guides to achievement in oratory (1.4.19). Antonius' story of the Athenian debate, then, is not simply a dramatization of *multiplex ratio* or of his own willingness to contemplate a topic *in utramque partem*; it also presents a pedagogical counterpoint to the monological approach of the standard classroom lecture, an antilogical argument in favor of specific, practical cases rather than normative precepts as the preferable mode of instruction in argument. However, despite Antonius' best efforts to display the controversial attitude, the student-characters (Cotta and Sulpicius) in this Roman revival of the "illustrious Academy and Lyceum" (1.21.98) persist in soliciting just the kind of static theorizing about argument that Crassus, in particular, would make every effort to avoid (1.21.98-99; 1.26.119).

Shortly after Antonius' narrative and directly following Crassus' proviso that he is willing to discuss only his own, personal approach to oratory (1.22.102), Sulpicius asks Crassus "if you hold that there is any such thing as the 'art' of oratory?" (1.22.102). This is just the sort of general, theoretical question that Crassus disdains. Nevertheless, after some dismissive references to the idle erudition of "talkative Greekling(s)" (1.22.102; cf. 1.11.47 and 1.51.221), he does proceed to address the subject, though he initiates this discussion with an exercise in antilogical reasoning that is bound to frustrate any hope on the part of his audience for reducing the issue to systematic clarity. He begins: "I think there is either no art of speaking at all or a very thin one"; i.e., if approached from the perspective of "exact knowledge," of *episteme/cognitio*, of the thing as it is in its essence, "there seems to be no such thing as an art of oratory" (1.23.108; cf. 1.14.60, 1.51.219 and *De Officiis* 1.43.153-155). If, however, we approach "the practice and conduct of speaking" as these actions have been observed in rhetoric's most exemplary practitioners, then "I do not understand why this should not be regarded as an art" (1.23.109). Again, we have a double-voiced construction, the pro and contra response within a single speech, the internalized antilogoi which balance opposing possibilities in an effort to formulate the most probable position. But in this case, the subtle juxtaposition of "yes" and "no" is itself immediately undercut as Crassus adds that in either case (art or not art) "certain other qualifications are of greater consequence for the attainment of eloquence" (1.23.109).

Antonius immediately recognizes that this question (what is "more profitable to oratory than . . . Art herself?" 1.24.110) is the crux of the matter, a well-defined stasis question upon which both Crassus and his eager students can agree. The question at issue, then, becomes not one of definition (what is it? or *quid sit*) to be resolved by appeal to abstracts; rather, we are faced with the issue of how argument operates, or better yet, what conditions or qualities best prepare one for the practice of oratory (*quale sit*).[10] What follows from this sharpened focus on the matter at issue (*summa questio*), however, constitutes yet another dramatic reversal or *peripeteia* of expectation. Crassus first proceeds to deliver his "own opinion" that natural talent is crucial (1.24.113-25.115), that diffidence is useful to an orator (1.26.119-20), that the standard for oratorical evaluation is especially demanding (1.28.129-30), and that the *sine qua non* of the art is good taste, the rules of which escape easy formulation (1.29.132). But Crassus seems discontent with this method, perhaps because it is too declamatory, too monological, and he proposes to "shift subjects" and to "chat at last in our own fashion" (i.e., dialogically; 1.29.133). But Cotta refuses to hear of such a change, refuses to respond to this request on the part of his own teacher to recast the manner of discussion and instead insists that the lecture continue. When Crassus relents and agrees to explain (i.e., to lecture about) his own "habitual method" of oratory, Sulpicius exclaims with great satisfaction, "Cotta, behold our longed-for day . . . we are now about to learn from his own lips everything that we have long been desiring" (1.30.136).

My own students are also thrilled at this moment in the dialogue. They too typically feel that Crassus has been obscuring his actual opinions and hiding behind a veil of argumentative evasions, perhaps for dramatic effect. Like Cotta and Sulpicius, they want to be told what is really important here, what will be on the test. But instead of carefully delineated *praecepta* (or rules), what the Roman students get in this instance is more irony, more dramatic *peripeteia*, another performance by the practiced actor/orator who gives voice to opposing logoi and so refuses to resolve the controversy or to have "the last word" on the subject. Crassus begins his response with a *tour de force* rehearsal of the Roman rhetorical curriculum, including the five canons of rhetoric, the various stasis categories, the parts of an oration, the modes of proof, the genres of oratory, the commonplaces, the *topoi*, and the basic elements of style, all those matters that "employ nearly all the learning of your professors"—all compressed by Crassus into two paragraphs, approximately 350 words, all so concise as to be hardly comprehensible, much less persuasive (1.31.138-45). Nonetheless, he adds that "if I were to call this learning useless, I should be lying"; and yet

> to my thinking the virtue in all the rules is not that orators by following them have won a reputation for eloquence, but that certain persons have noted and collected the doings of men who were naturally eloquent. (1.32.146)

The structure here again is "granted, yet," the peritrope of antilogic: I will respond to/identify with my interlocutor in so far as is possible, but I will also distinguish my own logos through courteous contradiction. This subtle display of controversial reasoning is capped off by an elegant chiasmus, a figure emblematic of the antithetical nature of Ciceronian argument (see Quintilian 9.3.85): "eloquence is not the offspring of the art, but art [the offspring] of eloquence" (1.32.146).

Matters continue in this vein for a bit, with Crassus on writing (1.33.150-51), on declamation, on transcription, on real-world practice, and on arguing *omni re in contrarias partes* (everything in opposing ways; cf. 1.34.158 and 2.53.215), until he comes to a breathless halt, noting that "I have poured out all my ideas" in the manner of a "chance patriarch" (1.34.159). But of course, Crassus is no routine lecturer whose learning is to be condensed into formal rules and confined within an uninterrupted declamation. He is, in fact, "the most eloquent of our nation" (1.6.23), a man used to and disposed toward the verbal sparring of the Senate. To ask him to play the pedagogue is to cast the protagonist in a supporting role. Nonetheless, his own performance, glutted as it is with ideas, provides a masterful deconstruction of the classroom lecture as a serviceable medium for dealing with the particular ideas at issue. The lecture format, which promotes unchallenged monologue rather than dialogical exchange, is particularly *ineptus* or silly under the circumstances (see 1.24.112). It is no surprise, then, that Cotta responds by saying that "so great was the speed of his words, and so swiftly winged his discourse that, while realizing its rushing energy, I could hardly follow the traces of its advance" (1.35.161).

From the point of view of *controversia*, there is much that is problematic about this or any extended monologue. Especially when delivered by a speaker of such overwhelming authority, the monologue tends to reify or give substance to its ideas, to posit a determinate nature for its logos even though, as in this case, the topic is supremely indeterminate, always conditioned by the kairotic nature of the particular discourse situation (see 3.55.210-12). As we know, antilogic/*controversia* always looks upon the single, dominant logos as "partial" in at least two ways: as comprising only part of the entire argument and as disposed toward one side or another in any matter of controversy (cf. Ehninger 1979, 103). Crassus has attempted to call attention to the partiality of his own claims, to circumvent the possibility that his remarks will be "taken out of context" and reified as general maxims, by including antilogical elements within his own speech: i.e., by routine appeals to irony and by syntactic reversals. But he cannot fully control the way in which his argument will be received. Stripped of the biplay which conditions actual conversation (the "chat after our own fashion," 1.29.133), there is a powerful inclination to ignore the necessity of responding to every logos, to embrace the notion that truth can have a single voice, and to accept that voice as bringing the conversation to closure.

It is little wonder, then, that Crassus' monologue should, at least temporarily, bring the discussion at Tusculun to a halt: "When Crassus finished these observations, a general silence ensued," in response to which Scaevola inquires, "why are you two [Cotta and Sulpicius] so silent? Does nothing come to mind on which you would like to question Crassus further?" (1.35.160). The circumstance is, unfortunately, a classroom commonplace: the lecturer carries on for so long and is so filled with information that when he or she finally stops and turns to the class for questions, the students are in a stupor. Cotta is able to summon up enough presence of mind to say that listening to Crassus' lecture has been like visiting a "richly stored mansion" with its special treasures stored away: "I discerned the wealth and magnificence of [Crassus'] talent as through some wrappings and coverings" (1.35.161). As Heidegger and Derrida have pointed out, discourse obscures as much as it reveals. It is the duty of the other side, the antilogos, to speak up, to resist silence, and to address what is inevitably left out or obscured by even the most authoritative logos.

Scaevola, at this point, continues to act as mentor to the younger men by prompting them to ask questions and press "the master of the house" on that which remains uncertain (1.35.162). But the very abundance of the monologue has stymied Cotta, and the *auctoritas* of his master's ethos is simply too much for the younger man to challenge (see *De Officiis* 2.9.34 and 3.30.109). So despite Cotta's own maturity and stature (he is 33 at the time of the dialogue and a well-known attorney; Wilkins 1979, 20), he remains inhibited by a concern that any issues he might raise "may well seem the elementary concerns of schoolboys" (1.35.163). In other words, it is very hard to talk back to one's teacher. So the students appeal to Scaevola to generate a rejoinder to his son-in-law's eloquence, which the distinguished lawyer does in the standard antilogical way, by praise and blame: "I heartily approve" of your efforts here, Crassus, but "these everyday and hackneyed maxims hardly [deserve] the attention of a man of my years" (1.36.165). Scaevola then (re)turns to the discussion of a topic that he himself finds more profitable; namely, what is it that the rhetor should know and, in particular, how much knowledge of the law is necessary for the practicing orator.

Prompted by his interlocutor with this clue to a discussion much more congenial to his own inclinations, Crassus immediately launches into an extended digression on common law in his longest speech of Book I (36.165-46.203). Antilogic once again prompts a significant advance in the speaker's own line of reasoning. The digression itself will pave the way for other extended detours in the *De Oratore* (e.g. on humor 2.54.217-2.71.290; on philosophy 3.15.56-3.22.81). Like almost everything else in this complex drama, these digressions have certain thematic implications. In the first place, the entire dialogue can be seen as a digression from the actual problems of state that occupy these influential Romans on the first day of their retreat (1.7.26). But more to the present point, the digressions are an organizational counterpoint to the linear progression of the discourse as a whole (which moves generally from invention, arrangement, and memory to style and delivery—the five canons of rhetoric). These interruptions serve as formal antilogoi, deviations from the orderly, logical, linear pattern that would make this text a treatise on rhetoric, and pull it back in the direction of conversation, with all the unplanned detours of spontaneous discourse, detours in which something interesting might happen (see Leff 1989, 122). This improvisational quality, in which antithetical positions continually respond, review, revise, and restate themselves, is exactly what gives antilogic its heuristic, generative potential; it also serves to remind teachers that the most productive part of a discussion is often not in one's lesson plan.[11]

As a scene within the pedagogical drama of the *De Oratore*, Crassus' digression on the law operates not only as support for Crassus' more general claim—the importance of knowledge to oratory—but also as an exhibition of all those elements of the art that the students would reduce, against their teacher's wishes, to guidelines (cf. 1.5.19). Crassus' oration on the law is itself replete with legal knowledge, distinguished by its copious approach to the subject, and substantiated with specific examples. Moreover, it is marked with its own internal digression designed to appeal to its listeners (a digression within a digression on the nature of *techne* or art; 1.41.186-1.42.188), and it is capped with a stirring display of eloquence in a peroration on the preeminence of the *lex romana* and the power of the eloquent lawyer as "the oracular seat of the whole community" (1.44.197-46.203). With this powerful oratorical performance, Crassus has avoided the "thin and bloodless" discussion of *techne* (talk about talk; 1.13.56) and concentrated on modeling for his students the kind of oratorical *praxis* that they might follow. In response, Scaevola at least seems to feel that Crassus has "done enough and to spare" for his students by showing them the "Door" and the "Way" to eloquence (1.47.204). But as all teachers must come to realize, however begrudgingly, we can't walk through the door on behalf of our students, we can only point the way. When Sulpicius, like Cotta before him, asks for more particulars on "the methods and the theory of these studies" (i.e., "please give me a map of the way"), Crassus demurs and calls on Antonius, whom he knows will meet this last speech with a powerful antilogos (1.47.205). In the face of repeated demands to "clarify the assignment," the best pedagogical option seems to be to provide an alternative point of view.

Sulpicius responds to the switch in speakers by noting that "from the lips of Antonius we shall be learning your (Crassus') own views too" (1.47.206). But while the level of oratorical wisdom may remain the same, Antonius represents the antilogical other, an oratorical opponent motivated specifically by the intention to systematically counter all the main points in Crassus' own position. Antonius does, however, follow Crassus in method if not in matter. He too promises to concentrate specifically on his own practice (his own "habitual method"), an act of agreement over the stasis at issue that confirms the relationship between the two opponents (cf. 1.30.135 and 1.48.208). Antonius also begins with a prolepsis, a disclaimer that he does not know the subject very well and that his own practice is not based on traditional teaching (1.48.208). The disclaimer serves to ennoble the ethos of this well-known orator by adding humility to his other obvious accomplishments; but prolepsis also serves as an antilogical effort to anticipate and offset potential opposition (Quintilian 9.2.16-18; Billig 1987, 239). In this case, Antonius immediately acts to propitiate Cotta and Sulpicius, to forestall their predictable objection that his response to Crassus will share with Crassus a lack of theoretical clarity.

The substance of Antonius' argument is introduced with a direct comparison between the philosopher and the orator. The former is one who strives to know, to master, to "follow out as a whole the theory of right living" (1.49.212). This definition, with its emphasis on conduct rather than on knowledge for its own sake, is very Roman, a Ciceronian variant of the Aristotelian *phronemos*, the person of practical wisdom (*Nic. Ethics* 1143b-44a). "But the orator," adds Antonius, "I myself do not picture as Crassus did"; i.e., instead of possessing "omniscience in every topic and every art," Antonius' more pragmatic portrait is of someone who can simply speak well and convince (1.49.213). The definition looks forward to Quintilian's orator as *dicendi peritus* or "skilled at speaking" (12.1.1, cf. *De Oratore* 1.61.260); but the crucial point is that, in Antonius' conception "of oratory," there is a "vast difference" between the gifts of knowledge and of speech (1.49.215). Like Socrates, whom Crassus will condemn in Book III, Antonius seeks to clarify the distinction between "the science of wise thinking" and that of "elegant speaking" (3.16.60-61; cf. 3.19.72). Indeed, in approaching the distinction between philosophy and oratory, Antonius is attacking the very crux of Crassus' argument—the latter's passionately held belief that eloquence involves the synthesis of thinking and speaking, or better yet, that it folds wisdom and craft together in an action that transcends both. But while Antonius will admit that there have been great public figures (like Pericles or Crassus) in whom wide learning and oratorical ability are allied, he will also insist that oratorical skill is not necessarily dependent on advanced learning (1.50.216; cf. 1.10.44). Rather, Antonius' orator need only "taste" what belongs to others (1.50.218). Such familiarity is sufficient because philosophy and oratory are "two distinct things" and "consummate eloquence can exist quite apart from philosophy" (1.54.233). The breach between Antonius and Crassus on this critical point is about as wide as it can be, so that the *controversia* that the students see dramatized offers a clearly defined opposition, two logoi that are decidedly antithetical. And as Antonius proceeds with this, the longest speech in Book I (1.48.209-61.262), he apparently intends to press this antithesis on every point which his opponent has raised.

In particular, Antonius argues against the requirement that the orator possess detailed knowledge of the law by showing that when distinguished attorneys are at odds, it is eloquence, not legal wisdom that carries a case (1.55.234-57.245); he argues that in any individual debate a reasonable understanding of the particular points at issue is enough since the orator is always able to consult authorities (1.58.246-59.253); and finally, he argues that an active life devoted to public discourse cannot be expected to allow adequate leisure for the demanding requirements of philosophy (1.60.254-

61.262; cf. 1.18.81 and 1.61.256). Along the way, he routinely adopts the syntactic antilogic of "yes, but" (see 1.55.235), he balances praise and mild censure of his opponent (see 1.52.226-27, 55.230, 57.245), and he proves himself fully capable of dramatic irony by arguing against the need for knowledge of the law while at the same time displaying significant grasp of the *lex romana* (1.55.235-1.59.251). Unlike Plato, who—according to Cicero—employs his considerable eloquence to attack oratory as inferior to philosophy, Antonius employs his own eloquence to liberate oratory from the unnecessary constraints of philosophy (1.11.47).

At the end of this systematic counter-statement, Sulpicius and Cotta "appeared to be in grave doubt as to which of the two speakers' discourses bore a closer resemblance to the truth" (1.61.262). Perhaps that doubt is a good thing, perhaps these two are altogether too ardent in pursuit of certainty about a subject that will always be conditioned by competing probabilities. But to me, and I hope to my students, the more interesting topic of speculation is the sincerity of Antonius' persona as the *advocatus diaboli* on almost every claim his host and colleague makes. Certainly Crassus has his own doubts about Antonius' sincerity: "you are making our orator something of a mechanic [*operarium*, or workman]," he says, and

> I rather suspect you are really of a different opinion, and are gratifying that singular liking of yours for contradiction [*mirifica ad refellendum*: wonderful power of refutation], in which no one has ever outdone you. (1.62.263; cf. Wilkins 1979, 227)

If such is the case, then Antonius is exercising the argumentative methodology he learned on his visit to the Greek Academy (see 1.18.82ff and 2.38.161); and in fact, Scaevola describes this second day's discussion as "reasoning . . . in the Greek mode," by which he means arguing an opposing position for the purpose of more readily discovering the probable truth (2.3.13; cf. *Tusc. Disputations* 1.4.7-8). On the third and final day of discussion, then, Crassus calls attention to Antonius' altered perspective (2.10.40), to which Antonius jokes that "yesterday it was my design, if I should have succeeded in refuting your argument, to steal these pupils from you" (2.10.40). Nor is such a disguise out of keeping with Antonius' own rhetorical habits. In the famous passage we have reviewed before, Antonius outlines the method of role-playing that is his routine preparation for a complex legal case. During a private interview with his clients, he begins by presenting the arguments of the opponent, so that the client "may argue his own [case] and openly declare whatever he has thought of his position." Then,

> when he has departed, in my own person and with perfect impartiality, I play three characters, myself, my opponent and the arbitrator. Whatever consideration [*locus*, or line of argument] is likely to prove more helpful than embarrassing I decide to discuss, whatever I find more harmful than good I entirely reject and discard the topic concerned. (2.24.102)

Such is Antonius' own method of *in utramque partem*, of exploring the various sides of the case; and the typical result, he adds, is that he quickly comes to understand the crux of the dispute (2.24.104). I would claim somewhat more for Antonius and for the method of dramatic impersonation (*ethopoeia*) that he employs throughout Book I of the *De Oratore*. That is, by acting out an opposing logos, one which confronts his interlocutor but which is not strictly congruent with his own views on the topic, he is able to clarify to the fullest extent that which would remain hidden in an ordinary conversation, a conversation that had not accentuated the guiding principles of controversial *praxis*, not (as in this case) placed the operational (or mechanical) perspective in direct opposition to the philosophical one.

Antonius will suggest later that the "lines of defense" that one might appraise in any given argument may well be infinite in their variety (2.31.136). But *multiplex ratio* aside, without the effort to oppose the dominant logos in a particular argument, we can never be sure that we have made a sufficient effort to discover what we don't already know, to uncover that which lies beyond the horizon of our own initial presumptions. Giambattista Vico, writing in the 18C, indicates that "we cannot be certain we have dealt with something in all its essential respects if not all the possible pertinent questions have been asked" (in Mooney 1994, 134). According to antilogic/*controversia*, it is only one's opponents, with their different frames of reference, who can prompt us to recognize what had been obscure to our own view or who can conceive of (and share) those "pertinent" matters about which we are unaware. Thomas Sloane, a distinguished expositor of *controversia*, puts the matter succinctly: "one must debate both sides . . . or one's *inventio* will remain not fully invented" (1989, 462).

Throughout Book I, oppositionality has been the key to expanding the scope of the discussion, to coaxing from the speakers ever more effective articulations, to allowing the argument to be comprehensively invented or conceived. According to Cicero, this progressive series of antithetical inventions leads ultimately to the closest possible approximation of the truth. But since "truth" itself is always dynamic, always in the process of being further defined and more fully apprehended, we can only hope to approach it through the continued juxtaposition of conflicting opinions rather than to confine it in a summary assertion of wisdom (see *Academica* 2.20.60). Michael Leff remarks that the *De Oratore* "instructs by being what it cannot explain" (1989, 124). In this context, Antonius and Crassus are professors who prefer to conduct their seminar in argumentation by acting out the operations of antilogic; and in the process, they argue the proposition that only through controversy can we proceed towards knowledge.

3. *CONTROVERSIA* AND THE SINGLE SPEAKER IN BOOK III

The dramatic interaction between illustrious orators in Book I gives way predominantly to monologue in Books II and III, first as Antonius develops the topics of invention and arrangement from his distinctly utilitarian perspective, then as Crassus returns to discuss philosophy, style, and eloquence. Of particular interest to us at this point is the way the single orator invokes the controversial method without benefit of an argumentative partner. I have already noted Crassus' ability to incorporate both pro and contra positions "*within* a single complex argument" (Kerferd 1981, 84). In Book III, he continues to model for his students the Academic method of inquiry and argument *in utramque partem*; this time, however, he is largely on his own and, as a result, he offers some original variations on controversial reasoning that help fill out the *De Oratore's* compendium of argumentative practices.

The ostensible subject of Book III is *elocutio,* or style (see Douglas 1957, 20). In Book II, Antonius had already covered the first two canons of oratory and left it for Crassus, in Book III, to develop "the proper method of embellishing" the topic (3.5.19). But, in fact, the theme of *elocutio* is the focus of only intermittent attention for much of the first two-thirds of the book, as Crassus routinely digresses from technical considerations to take up instead the profound relationship between rhetoric and philosophy. These digressions serve to displace the topic of style from a purely functionalist context and to entangle it with matters of a much deeper nature. Between 3.19 and 3.143, then, Crassus oscillates between the functional and the philosophical, between micro and macro considerations connected with style in what amounts to an invocation of *controversia* at the level of form or structure. Cotta acknowledges this exercise in *multiplex ratio* when he notes that while Crassus was indeed "cast for the part of speaking about stylistic embellishment," he was also regularly "caught by the flood-tide of [his] genius and carried away from land out to [the] deep sea" of philosophy (3.36.144-45). Nonetheless, Crassus' expansive, controversial approach to the topic has persuaded Cotta "to come over entirely to the side of the Academy"; i.e., to the school's "two-fold method" of arguing antithetical positions on every subject (3.145). So Crassus, through this method of inquiry, has managed not only to complicate such a seemingly routine subject as style, but he has done so in a way that inclines his students to weigh the relative merits of alternative positions, to recognize their own initial partialities and prejudices, and ultimately to build for themselves a more comprehensive understanding of style, an understanding that takes a step closer to the topic's fullness.

In sum, Crassus folds theory into practice during Book III and in so doing prompts his students to look for more in the subject of style than the technical advice of a training manual (cf. Leff 1989, 118). This blending or merging of theory and practice, content and form is the hallmark of Ciceronian eloquence; but before we take up that complex topic, we should dwell for a moment longer on the unique way in which Cicero approaches the stylistic category of *ornatus* or ornament. At 3.37, Crassus introduces the four elements or "virtues" of style as outlined in Theophrastean doctrine (correctness, clarity, ornament, and decorum), and his discussion of these elements does indeed plot the basic development of Book III (see esp. 3.37-90, 3.148-212), despite the digressions into philosophical concerns. But even within the technical discussions, Crassus the teacher employs *controversia* as his operational method. At 3.96, when it is time for Crassus to discuss "the embellishment of oratory," he makes a contrast between what Elaine Fantham calls "intrinsic *ornatus*" and "applied ornament" (1988, 276). The latter is, of course, the standard conception of ornament. Cicero describes it as "the flowers of language and the gems of thought" that may be distributed like "brilliant" jewels throughout the discourse as "a source of decoration" (3.25.96). This decorative approach to ornament, however, has distinct limitations, and Fantham argues persuasively that Cicero posits the notions of *varietas* and *satietas* (variety and satiety) as inherent checks on the use of stylistic embellishment (1988, 276-82). Even the concept of simple decoration, therefore, carries within itself a counter-argument; so that in order to grasp the concept effectively, we must approach it in a controversial, antithetical frame of mind.

More germane to our topic, however, is the contrasting conception of intrinsic *ornatus*, a form of embellishment that is distributed throughout the entire body of a discourse. According to Fantham, Crassus conceives of discourse "as a body whose attraction derives primarily from . . . health evenly diffused throughout its limbs" and from qualities such as dignity, charm, refinement, learning, and emotion that are "inherent in the whole work" (1988, 276). Obviously, this conception of inherent ornament is not traditional, not found in Aristotle, Theophrastus, or Demetrius (see Douglas 1957; Fantham 1988, 286-90); and yet in *De Oratore* 3.96-103 Cicero develops the notion that ornament is the verbal extension of rich and various material, the linguistic development of that which is inherent in the subject itself (cf. 3.103). To a considerable extent, this line of reasoning is a continuation of Cicero's constant effort throughout Book III to reconcile philosophy and rhetoric, the thing and the word; and indeed at 3.178, Crassus directly aligns ornament with content in the notion (later connected with Keats) that dignity and utility (in the subject) naturally express themselves as beauty. This alignment of the functional (the duties of rhetoric) and the beautiful (the levels of style) is original in Cicero and a subject of compelling interest in its own right (see *Orator* 69-74, 122-25; also Douglas 1957; Fantham 1984; and Leff 1990). But the point at present is that Crassus, in an effort to develop the seemingly technical subject of "the ornate" for his students, appeals yet again to the controversial method and invents an alternative conception of *ornatus* which allows us to contemplate the topic as a dynamic interaction between contrasting alternatives, between intrinsic and decorative ornament, the "truth" of the subject residing ultimately in the interaction between these alternative conceptions or antilogoi.

While Ciceronian *ornatus* both decorates the surface and partakes of the substance of the matter, the actual duty of effecting a merger of style and matter, form and content falls to the offices of decorum, style's fourth and ultimate virtue. And, in the scope of its operation, Ciceronian decorum offers a significant advance over both the Aristotelian theory of *to prepon* (what is fitting) and the traditional Roman notion of decorum (propriety) (see Aristotle's *Rhetoric* 3.7 and Quintilian 11.1; see also Leff 1990). As Walter Beale points out, it is not unusual for the concept of decorum to take on philosophical dimensions (168-70; see *De Officiis* 1.93). But as an element of style, decorum is most often considered in a fairly routine context as the linguistic adaptation of the material in question to concrete situations and particular audiences. For Cicero, however, decorum becomes not only the means by which language nudges ideas into accommodation with the surrounding rhetorical environment, but also decorum directs the treatment of a specific subject (or *res*) in accordance with the "inherent" qualities of the subject itself.[12] To paraphrase Leff's elegant judgment on the two-fold process of Ciceronian decorum, an appropriate (decorous or apt) treatment of the subject involves not simply an adjustment of language to the constantly shifting circumstances towards which the discourse is directed, it also involves the discovery of language appropriate to the nature and attributes of the topic (1990, 120-22).[13] This new vector of interest for decorum (the attunement of one's style to the "quality" of the subject) will naturally require the orator to possess not only what Cicero calls a "full supply of matter" on the topic at hand (3.125) but also a stock of general knowledge that extends—as Cicero has maintained from the outset of the *De Oratore*—to just about every facet of human behavior (3.54, 3.76). As a result, eloquence (which we can now describe as the successful invocation of Ciceronian decorum) will transcend abstract, theoretical knowledge at just this point: i.e., its ability to merge content (the matter or argument of discourse) with a form of expression (*elocutio*) appropriate to topic and circumstance (3.142-45). What is more, the merger of argument and style, *ratio* and *oratio* represented in eloquence, allows discourse to take on body or substance in such a way that ideas become an active force capable of motivating changes in the world at-large.

Readers may recall that among my students' favorite passages from the *De Oratore* is the critique of Socrates, whom Crassus claims had separated the "tongue and the brain, leading us to have one set of professors to teach us to think and another set to teach us to speak" (3.61). For Cicero, the operations of eloquence, with its integration of *res* and *verba,* specifically address "this absurd and reprehensible schism" (3.61). To put the matter in yet another way, eloquence is the embodiment of wisdom in a form that can be held in mind by the listener and passed between discursive partners. But eloquence is also an action, the process of blending content and form; it is something that people like Crassus and Antonius do, something that others, like Cotta and Sulpicius can, under the best conditions, learn to do (Gage 1984, 156). Consequently, Book III never reduces its subject (*elocutio*) to "thin and bloodless" form as abstract reasoning, never atomizes its topic as a set of pedagogical precepts or "hackneyed maxims." In fact, as an independent subject of scrutiny, decorum (like *controversia*) receives remarkably abbreviated treatment by Crassus (3.210-21). In the place of discursive analysis and the podium lecture, Cicero chooses to illustrate decorum in the person of Crassus, the "consummate orator" whose own discursive behavior models the offices of eloquence.

In the same way, Crassus exhibits rather than discusses the operations of *controversia.* There is in fact more than a passing analogy in the operations of these distinctly Ciceronian practices: both decorum and *controversia* ask rhetors to turn in two directions at once, to coordinate argument and persuasion, to concentrate on particular people and places, and to resolve questions with practical answers. In addition, as forms of action, both are best viewed in motion; i.e., through a concrete performance which exhibits the oscillation characteristic of the act. Any theoretical claim or statement about such practices will necessarily remain abstract and inchoate until the claim is provided with concrete embodiment that puts practice itself on display under specific rhetorical circumstances. It is not surprising, then, that both these concepts receive scant discursive attention from Cicero. Instead, both are featured in and embodied by the actions of the main characters of this oratorical drama. I take this process of embodiment to be a guiding principle of Cicero's approach to rhetorical education, an approach which refuses to separate rhetorical action from its motivating environment and which asks students to "draw out and give shape" to Cicero's rhetorical principles on their own, without the mediation of *pervulgata praecepta* (hackneyed maxims; 2.18.75). In the last section of this chapter, we will begin to take stock of this distinctive pedagogy. If we can come to grips with the instructional interests and methods at work in the *De Oratore*—a work I earlier referred to as "a master class" in controversial argumentation—I think we will be well on our way towards conceiving a pedagogy consistent with the particular attitude and methods of our subject.

4. THE QUESTION OF CICERO'S PEDAGOGY

On the one hand, the pedagogy of the *De Oratore* is predicated on the theory of *imitatio*—the emulation of working models that embody the practice in question in all its nuance and specificity. On the other hand, this same pedagogy opposes abstract standards and general formulae which reduce the richness of contextual reference by separating discursive practice from concrete situations. This Ciceronian position, however, creates a challenge for scholars (like me) who see in *controversia* an approach to rhetoric and argumentation that is particularly relevant to our own cultural, political, and educational circumstances.[14] And yet, any effort to abstract *controversia* from the text for our own use meets with obvious objections. In the first place, there is the routine historiographic problem of extrapolating ideas from such a radically different cultural context and, in the process, distorting their original intention and ignoring overwhelming obstacles to effective adaptation. Such problems are perennial in historical research and have been addressed fully elsewhere (see Kimball 1995; Nussbaum 1997; Oakley 1992). Suffice it to say, with Quintilian, that "of that which is good by nature we may surely make a good use" (2.10.3). I would argue that controversial argumentation shows enough promise as an alternative tradition to the current reigning paradigm for argumentation to at least merit speculation on its potential "good use" in the contemporary classroom. A second objection deals with self-contradiction. For in abstracting a set of pedagogical highlights from the text, am I not attempting to do for my readers just what Crassus himself refused to do for his students; namely, to translate a specific form of rhetorical practice into a general theory that ignores the very insistence on unique discursive moments that a rhetoric of embodiment is intent on promoting?

On this point, I indeed run the risk of appropriating my source in a manner which contradicts its spirit, of hypostatizing in theory what is fluid, dynamic, and always contingent in practice. On the other hand, I take some comfort in the recognition that—as noted at the outset—the primary function of Cicero's published text is didactic, that Crassus is addressing an audience whom he would teach and, as such, the *De Oratore* is firmly conditioned by "instructional" intentions. Consequently, in the notes that close out this chapter, I will begin to extrapolate from ancient practice ideas that may be of value in revising our own pedagogy along controversial lines. These notes, however, are only by way of introduction, an early version of themes on teaching to which we will return later. In Chapter 6, on Quintilian, I will reprise many of the same ideas announced here, though in a somewhat different context. And, in the Epilogue on contemporary pedagogy, the finale of this discussion, these same themes are refashioned in contemporary dress. For now, I take my clue from Antonius and approach the subject of an appropriate pedagogy as a respondent who sees the issues from a slightly different angle than his interlocutor and who would rehearse the impressive, if austere pedagogy of the *De Oratore* in a more pragmatic, if less eloquent key.

In the *De Officiis*, Cicero argues that "a clear view" of any complex topic is available only through a comparative estimate of alternative perspectives on either side of the case (2.2.8). In the *De Oratore*, to provide his readers with "a clear view" of the rhetorical process in particular, he embodies alternative perspectives in the persons of Crassus and Antonius, and he dramatizes the protocols of controversial argument in their exchange. To understand oratory, to comprehend the connection between rhetoric and philosophy, to fathom the operations of eloquence or anything else of substance connected with rhetoric, we are asked to attend to the concrete practice of Rome's most eloquent rhetors (1.6.23). The first principle of Ciceronian pedagogy, then, is *imitatio* , the study of rhetorical knowledge at work in particular instances of rhetorical practice (2.22.89-90). As a cardinal principle of ancient pedagogy, imitation has a rich and noble history, much too vast and diffused to detain us here (see Bonner 1977; Clark 1957; Leff

1997; Russell 1981; Sullivan 1989). Suffice it to say that our neo-Romantic tendency to equate imitation with the surrender of identity and the slavish reproduction of models runs directly counter to the classical tradition. Mimesis, says Aristotle, is a natural part of the learning process; but it instructs only to the extent that students add something of their own to their model (*Poetics* 1447a-b; Russell 1981,108). Two pedagogical points can be made about the model established by the speakers of the *De Oratore*. First, the example set by Crassus and Antonius places a premium on oratorical practice rather than abstract principles, on the decorum of good choices in particular circumstances. For Cotta or Sulpicius to simply reproduce what they see in the model would be inadequate because the value of the example lies in the way that Crassus adjusts rhetorical protocol to the unique demands of the moment. It follows that imitation is successful only if invention intervenes to reframe the original according to the demands of a new time and place. Second, the pedagogy of imitation highlights the role of interpretation in the process of rhetorical production. To imitate a model effectively assumes that the student possesses the interpretive resources to recognize what distinguishes the source. Imitation pedagogy, therefore, recognizes a deep connection between analysis and production: to emulate Crassus, one must be able to "read" his rhetorical actions before reenacting its protocols. These and other "imitation" issues will, of course, require a good deal more attention.

For now, and as a potential subject for *imitatio* , I would call attention to the way in which the *De Oratore* works to recast the concept of argument from an eristic contest in which opponents "struggle against" one another to a civil conversation in which multiple voices "struggle toward" an ever-more-comprehensive understanding of the nature and practice of oratory. This distinction between conflict and *controversia* is captured in the *Protagoras*, when Prodicus remarks that "conversation [should] be a discussion, not a dispute. A discussion is carried on among friends with goodwill, but dispute is between rivals and enemies" (337a-b). Walter Ong makes a similar distinction between agonism, which is a struggle in pursuit of some goal, and antagonism, the struggle against an opponent or interlocutor (in Teich 1987, 56). What is at issue here is "the *attitude* embodied in a *method*" (Burke 1969, 441). The attitude of Crassus and Antonius is one of friendly contradiction, a readiness to transform the natural antagonism that attends dispute into an opportunity for insight and invention that neither party is capable of on his own. This attitude reflects the Burkean purification of argument and is, of course, ideal; but the concrete ideal of the *De Oratore* does lead us to interrogate the dominance of the forensic model as the standard for instruction in argument. The aim of argument as conducted by Crassus is not a victory such as one would seek in court; rather, argument in this context is an inquiry, a process that solicits *multiplex ratio* in the examination of the best, most prudent response to the conflict at hand. In the *De Oratore*, the appropriate model for effective instruction is clearly argument as inquiry instead of argument as war (see Crosswhite 223-68). Prominent among the Ciceronian methods for realizing this new attitude is the process of identification, that initial act of narrowing the gap between speakers for the purpose of *conciliatio* (3.53.205). As we have seen, identification can proceed in any number of ways, including the syntactic variations of "yes, but" and "granted, yet" that we observed throughout Book I (e.g., 1.61.260; see also Billig 1987, 234-35). But the protocols of identification are diverse and subtle, clearly a subject for inquiry and *imitatio* in their own right.

In addition to identification, we have also surveyed an array of other Ciceronian methods worthy of pedagogical exploration, including variants of prolepsis (or anticipating objections), the value of role-playing, peritropes (or turning-the-tables), the importance of give-and-take, internalizing dissent within one's own utterance. We have also noted the Ciceronian emphasis on dialogue over lecture, on strategic digressions from the lesson plan, on *multiplex ratio*, on modeling the willingness to change, and on avoiding the last word. All these Ciceronian practices can be explored as protocols of interest for imitation and pedagogies suitable for extrapolation. But before I leave this sampler of pedagogical options, I would add some additional notes on a related set of topics central to Cicero and his pedagogy.

Michael Buckley writes that the Ciceronian conception of discourse was a "composite of two moments: invention and judgment" (1970, 143). We have already seen that the antithetical method of Arcesilaus tended to ignore the latter, while Carneades reasserts a balance between the two. Ciceronian argument as embodied in the practice of the *De Oratore* extends the Carneadean model by reimagining the protocols of invention in ways that go well beyond the scope of this chapter. Consider, for example, the potential of controversial interaction from the rhetor's point of view. At very least, exposure to opposing positions serves to acquaint us with limitations in our own perspective, limitations that compel reformulation and response, as was the case with Crassus after Sulpicius' initial rejoinder to his idealism. Moreover, the antilogical responses of other participants serve to fill out the rhetor's vision of the controversy and so to stimulate an expansion in one's position as we seek to accommodate new understandings. If Perelman and Olbrechts-Tyteca are right that arguments always develop "in terms of an audience" (1969, 5), then the response of the controversial interlocutor will mandate, in turn, an adjustment in the rhetor's own views. And yet, exciting as the heuristic potential of controversial invention may be as an alternative to the topics-based procedures (invention based on the Aristotelan *topoi*), the classroom exchange of opposing views in a contemporary setting will have to be managed/mediated by the instructor with special care since our cultural conditioning to argument more often suggests combat than the amicable agonism depicted by Cicero. In addition, the teacher herself may choose (as Antonius does) to promote invention through constructive yet oppositional questions (e.g., "what would you say to the claim that your argument is . . .?" or "tell me why the following is not the case?"), questions that can lead to new paths of inquiry as well as require reflexive rigor in scrutinizing one's own initial formulations. However the controversial exchange is managed, the Ciceronian precedent seems to indicate that an intellectual/rhetorical energy is stimulated in rhetors as the mind shuttles between/among alternative logoi and that this dialogic energy can be harnessed for the purposes of improved rhetorical invention (see Sloane 1985, 84).

Related to controversial invention is the ability to suspend judgment while alternative claims are appraised. Teachers of rhetoric, composition, and speech will realize that one of the most basic challenges to instruction in argument is convincing students that their own first instincts are not necessarily the best source for a claim, and that all matters of controversy over human action will present rhetors with an array of probabilities rather than a set of certainties for their consideration (see Cicero's *Academica* 2.7-8; Aristotle's *Rhetoric* 1357a). Under such circumstances, it is only prudent to suspend judgment as we calculate "the degrees of probability" among competing claims in an effort to achieve an understanding of the issues sufficient for judgment and action (Hunt 1954, 42). Cicero illustrates the actions of *epoche* (suspending judgment) in the person of Crassus who, despite his strong commitment early in the dialogue to the rhetorical importance of comprehensive knowledge, desists from delivering any final judgment on key questions until he has listened to a host of alternative responses.

For us, the issue is not a simple one. Teachers must learn not only how to encourage a suspension of judgment in their students but also how to cultivate both a tolerance for uncertainty and a willingness to comprehend competing claims. Alternatively, in requiring students to experiment with multiple roles, we may be raising issues of identity that are potentially threatening. For example, if the neutrality required to negotiate conflicting claims demands an unreasonable surrender of the self, or if the need to understand the claims of others means relinquishing hard-won values or taking on offensive ones, do the lessons of role-playing come at too great a cost? Martha Nussbaum has argued that both the suspension of judgment and the willingness to grapple with alternatives are parts of the process of becoming what the Stoics called a "citizen of the world" (1997, 50-84). I would add that a controversial pedagogy can complement this effort; but the process is a complex one, and the immense cultural differences between the aristocratic, male-only discussion of the *De Oratore* and our own highly diverse classrooms remind us to proceed with caution.

Finally, there is the question of judgment itself, the assessment of ideas, evidence, and arguments discovered during the creative period of invention (Buckley 1971, 89-90). Despite the *praxis* of dynamism, which seeks to keep the conversation going, and despite the tendency of scepticism to continue to ask questions, there is in *controversia* a complementary pragmatism that requires answers in order to act. The operation of *controversia* in matters of judgment is a complex topic since the comparison of probable options will always be less definite than the appeal to universal standards because these options are embodied in novel particulars. This is not the time to do more than mention this subject (for a more detailed discussion of practical judgment, see Epilogue, sec. 3). But the criterion of practical benefit initiated by Protagoras, revisioned by Carneades, and employed by Crassus is a place to start, a useful principle for consideration by participants in argument who will have to learn to negotiate agreements as well as define their differences.

Shortly after completing the *De Oratore*, Cicero wrote to a friend indicating that he was "especially please(d)" with the dialogue and that it "differ(s) from the usual rules that embrace the oratorical method."[15] If we conceive of the traditional rules of oratory and argument as founded on monological patterns of formal reasoning and set in a distinctly adversarial context, then the controversial method of the *De Oratore* does indeed "differ" from usual practice. And yet, Ciceronian *controversia* has its own, alternative tradition. I have argued that this tradition is Protagorean in origin and Academic in development; it takes form in the dialogical interaction of *multiplex ratio disputandi*; and, as McKeon points out, it assumes particular resonance during periods of intellectual upheaval, periods in which the "clash of theories" requires a method for dealing with probabilities and contradictions, periods like Periclean Athens, Ciceronean Rome, the Humanistic Renaissance, and our own time, all those ages that have been loosely labeled Sophistic (1970, 59). At present, the controversial method is of particular interest to teachers who would reject doctrinal certainty as a basis for instruction and would pursue a practice of negotiating the conflicts that inevitably condition the acquisition of knowledge. For Cicero, the controversial method was a way of merging the intellectual virtues of comprehensive knowledge and good speech, but also of aligning both with the practical virtue of right conduct (3.57-59). Indeed, it is because eloquence is so fully engaged with "the life of [humanity]" that it is among the *summae virtutes* (supreme virtues, 3.55). This engagement, however, unavoidably exposes us to the endemic disagreements that characterize any rhetorical culture. As Crassus notes, "our ordinary public life" is replete with "contention and criticism" (3.92). In this argumentative environment, the great advantage of a *controversia*-based education is that it prepares the student to understand and address the competing claims inherent in "the concrete life of the community" (Schaeffer 1990, 9). Which is to say that, for Cicero, pedagogical method is of value to the extent that it serves a larger purpose; and that purpose, as rendered in this text, is to prepare students to play their part in the argumentative exchanges that constitute "the life of humanity."

The young Cicero describes justice as "a habit of mind which gives every[one] their just dessert while reserving the common advantage" (*De Inventione* 2.160). The practice and pedagogy of the *De Oratore* partakes of this same habit of mind; it works to balance the existential fact of individual difference with the possibility of common understanding. In a phrase, *controversia* seeks to maintain justice in argument by engaging opposition in the search for knowledge. We have now a detailed idea of the operational protocols of this habit of mind. We are, however, only just beginning to understand the pedagogical process by which it can be encouraged in others. At this point, we know that the process will involve the pedagogy of *imitatio,* with its corresponding emphases on the concrete embodiment of rhetorical principles and the emulation of those principles in some form of specific practice. Just what form that will be, what "other means" are required to help the student learn to do what Crassus has done, will have to wait until the next chapter. For in Quintilian, Cicero's enthusiastic advancement of controversial method receives comprehensive endorsement, while the hints of an appropriate pedagogy are filled out in lavish, yet practical detail.

Chapter 6

Quintilian and the Pedagogy of *Controversia*

"I would rather make an able man than a learned man, I would also urge that care be taken to choose a guide with a well-made rather than a well-filled head; that both these qualities should be required of him, but more particularly character and understanding than learning; and that he should go about his job in a novel way."

"The bees plunder the flowers here and there, but afterward they make of them honey, which is all theirs; it is no longer thyme or marjoram. Even so with the pieces borrowed from others; he will transform and blend them to make a work that is all his own, to wit, his judgment. His education, work, and study aim only at forming this."

Montaigne, "Of the Education of Children"

Some time ago, I attended a conference on "Isocrates' Rhetorical Education" occasioned by a fine book on the same topic.[1] It was a stimulating event that served to reassert the significance of Isocrates' political and social theory. The conference also sought to explore Isocrates' pedagogy and, in this effort, it was less successful. That is, when the topic of pedagogy was raised at all, it was typically cast in the context of other, presumably larger issues, such as the relation of the *paedeia* to the Athenian *polis*. Indeed, the conference discussion always remained at a distance from what we might call the "practicalities of instruction."

During the conference, one of the participants observed that it was difficult to persevere for long with discussions of pedagogy because pedagogy was inherently "thin discourse." There were no real objections to this claim; but since then, I have found myself returning to the notion that the discourse of rhetorical pedagogy, in particular, is somehow "thin" or insubstantial. What does this mean? Is pedagogical discourse somehow without the deep resources, internal coherence, and intellectual rewards that we require of the best rhetorical theory? More concretely, what works might actually contend as classics of rhetorical pedagogy, as enduring documents worth repeated readings, as thick descriptions of rhetorical instruction?

For example, the *Phaedrus* is about teaching as well as about rhetoric, but does its drama make it a special case? And what about Aristotle's *Rhetoric*, which most readers find valuable but "dry"? Would anyone nominate the *Ad Herrenium* or the *De Inventione* as thick discourse? Or how about the dialectical precision of Boethius, the invective of Ramus, or the amiable humanism of Thomas Wilson? And what happens to the wit of *The Praise of Folly* when Erasmus turns to *De Copia* or the *De Conscribendis Epistolis*? For that matter, do Campbell, Blair, and Whateley, not to mention Bain and Genung, merit Aristotle's comment that writers of *technai* seem to possess only "a small portion" of the art they profess (*Rhetoric* 1354a)?

There are, of course, omissions in my catalog, strategic ones to be sure. But, in general, I would not quarrel with the claim that the bulk of the genre is "dry" and ephemeral, though this need not be the case. What, then, makes the difference? When does the discourse of rhetorical pedagogy take on resonance, becoming something worthy of contemplation and discussion? The previous chapter provides some hints; for while Cicero's *De Oratore* is not a textbook (indeed, Crassus refers to textbooks as "utterly ludicrous," 2.77), it is about teaching oratory and argument and what is needed for their successful practice. And, as I read it, the pedagogy of the dialogue is tied to a single, dominant principle: i.e., the exhibition of oratorical skill as a model for emulation takes precedence over professorial exposition. Put another way, pedagogy becomes substantial by calling attention to the operations of specific linguistic resources in particular rhetorical conditions, to rhetoric *re atque usu* (in fact and use).

On the other hand, we also saw in the last chapter that there are limits to how far we can go in appropriating pedagogical principles from Cicero's remarkable but decidedly anti-theoretical dialogue. So it makes sense to seek amplification in Quintilian, who is not simply a Ciceronian enthusiast, but in whom the Ciceronian legacy is alive and well. Moreover, Quintilian advances Cicero's rhetorical legacy through the development of a pedagogy that is considerably more detailed and functional than what he inherits. But the core of the pedagogy remains the same: Quintilian is unwavering in his commitment to teaching strategies that emphasize learning through practical application. Like Cicero, he promotes rhetorical method rather than theory, he invokes *imitatio* and paradigmatic examples to illustrate concrete rhetorical practice, and he aims ultimately to prepare his students for active involvement in public life. To what extent this pragmatic pedagogy can inform our own approach to rhetorical instruction is a question I reserve for the end of the chapter. For now, I would only argue that Cicero and Quintilian, taken together, offer a detailed argument in support of a thick pedagogy, a pedagogy practical in orientation, yet grounded in theory and worthy of scholarly discussion. We have already sampled something of this pedagogy in the last chapter; but there is a good deal more of the argument to come, especially as regards the "practicalities of instruction."

Naturally, differences distinguish the two Romans. Whereas Cicero (106-43 BCE) was a statesman with an interest in philosophy, history, politics, and just about everything else, Quintilian (c. 40-c. 96 CE) was a successful public orator who dedicated his mature career to teaching. Indeed, he is acclaimed by general consensus as the outstanding rhetorician and teacher of Imperial Rome (M. L. Clarke 1963, 14; Colson 1924, xiii; Murphy 1990, 30), and his great work, the *Institutio Oratoria* (on "The Education of the Orator," 95 CE), has been called "the best guide to the classical rhetorical tradition as it existed in the schools," "a professorial book . . . eminently clear and sensible, authoritative and definitive," "a masterly treatise . . . influential for centuries . . . which tells us more than any other single work about Roman rhetorical education" (Russell 1981, 117; Grube 1965, 285; Bonner 1977, 161, resp.; see Saintsbury 1908, 290). There are also significant cultural differences between the crumbling Republic of Cicero's day and the imperial despotism of Quintilian's, differences marked in their writing and to which we will return. But with regard to rhetoric—its theory and pedagogy—continuity predominates. Consider, for example, their shared commitment to the ideal orator.

Cicero announces his interest in the ideal orator at the outset of the *De Oratore* (1.4, 1.118, 1.128). And, towards the end of the dialogue, he specifically links the "true and perfect" orator with the ability to address "both sides of every subject" (*in utramque sententiam*; 3.78-80). Quintilian raises the same topic early and returns to it often: "My aim," he writes in his preface, "is the education of the perfect orator" (1. pr. 9; cf. 1. pr. 18, 10.1.35-36); and, towards his own close, he frames his conception of the ideal as "that same perfect orator that Cicero also sought to discover" (12.1.19).[2] The analogy, however, is incomplete; what we need is proof that Quintilian is similarly committed to *controversia* as an inherent characteristic of the ideal orator, proof that this chapter intends to supply. At which point, we could claim that classroom practice in *controversia* is Quintilian's favored means of fostering the oratorical ideal. For now, we need to know a bit more about this ideal because the concept provides a *telos* or goal towards which this entire pedagogical tradition is directed.

For both Cicero and Quintilian, consummate oratory is characterized by the convergence of prudence and eloquence in pursuit of the public good. In the *De Officiis*, Cicero describes prudence as the "safeguarding" of human interests (1.43.153); while throughout the *De Oratore*, the ideal orator is described as committed to right conduct as well as good speech, both of which are employed in service to "the life of mankind" (1.68, 3.54-57, 3.76). Similarly, Quintilian describes the "Roman wise man" as avoiding the cloistered study and exercising his virtues of conduct and speech in the "actual practice and experience of oratory in the broad daylight of the forum" (12.2.7-8). Such emphasis on good judgment as expressed in public action has profound effects for education, especially when such pragmatism is seen in opposition to the theoretical preoccupations that characterized philosophical instruction of the day (see Colson 1924, Marrou 1956, Bonner 1977). That is, for Quintilian in particular, philosophy's interest in theoretical topics, formal reasoning, and determinate standards for judgment is simply not compatible with the realities of argument regarding human affairs, which call for neither logic nor theory but rather practical intelligence (cf. V. Kahn 1985, 30). The kind of education contemplated by Cicero in the *De Oratore* and developed by Quintilian in the *Institutio* is focused directly on cultivating this practical intelligence for use in matters of practical controversy, when rhetors are typically called upon to compare alternative options in order to determine what is appropriate under present conditions.

In turn, "determining the appropriate" is the duty of decorum; and you may recall from the last chapter that this duty involves a "two-fold wisdom," a consideration not only of what will serve in the particular circumstance, but also what is best in accord with the subject itself (see Cicero's *Orator*, 1.70-72; cf. *De Oratore* 3.59). Decorum invokes this wisdom by coordinating the choice of rhetorical resources with practical judgment about the topic. As such, decorum is the rhetorical counterpart of prudence, the ability to weigh alternative positions and arrive at judgments that are ethically sound and socially useful. If rhetorical education can foster this ability to respond to the practical demands of daily life with good judgment, then rhetorical training is contributing directly to the oratorical ideal by nurturing the *good* person skilled in speaking (*virum bonum decendi peritum*; 12.1.44). For Quintilian, like Cicero before him, prudence, goodness, eloquence, and decorum stand in metonymic relation to one another in the sense that they are all grounded in a concern for actual practice, or *utilitas* (usefulness; see 4.2.122; see also Greer 1925, 31). Taken together, they represent the *summa bonum* of the oratorical life and the ultimate goal of rhetorical education.

In sum, the ideal orator must not only speak well and be good but must also act wisely in the world by remaining alert to the contingencies of particular rhetorical circumstances. Of course, no one speaker can actually fulfill this ideal, not Cicero, not Crassus, and certainly not Quintilian's students. Nonetheless, the ideal provides a goal for pedagogy; i.e., how best can a pedagogy encourage the prudential capacity to manage the conflicting concerns that dominate human affairs in a complex society? Throughout this chapter, as we examine the concrete pedagogical strategies that Quintilian deploys in response to this question, it may help to keep in mind that specific instructional practices are themselves embedded in a complex ecology of ideas within which they develop and have meaning. The ideal orator is one index of this ideological ecology, a theoretical construct that informs and lends substance to particular practices by virtue of its connections with decorum, prudence, and public commitment, as well as more general associations with Sophistic pragmatism, Academic scepticism and "antithetical" reasoning. Too often we approach the "practicalities of instruction" as a set of discrete classroom practices attentive only to present need and unrelated to history or theory, a tendency that may contribute to the reputation that pedagogy has with many readers as inherently insubstantial. However, with the aid of earlier studies in *Many Sides*, we should be able to re-vision Quintilian's pedagogy as of a piece with the theoretical/historical tradition out of which it develops. When this particular history and pedagogy are taken together, the resulting gestalt is, for this reader at least, anything but "thin."

In past chapters, we have explored in some detail the intellectual background of controversial pedagogy, first in Protagoras, then in Cicero and the heritage of the New Academy. In this chapter, we will concentrate on the specific practices that give concrete form to the pedagogy of rhetorical pragmatism. I will refer briefly to earlier developments in pragmatic instruction, but my overriding interest in this chapter is the pedagogy of the *Institutio Oratoria.* I begin with Quintilian's own use of controversial method in his general speculations about education and rhetoric; i.e., with his own rhetorical practice, a subject that has not been adequately explored but which reveals the extent of his commitment to controversial methods. Next, I review Quintilian's approach to the *progymnasmata*, the progressive series of exercises that constitutes the basic Roman curriculum. We are often told that the *Institutio* offers advice on rhetorical training from cradle to grave; I will argue along similar lines that controversial training informs the *progymnasmata* from beginning to end.

The apogee of this curriculum is the exercise of declamation. My examination of Quintilian's pedagogy culminates in a detailed review of declamation, the capstone exercise that focuses directly on the practice of controversial reasoning. Finally, at the end of the chapter, we return to some earlier topics—dialogue, invention, imitation—topics we first explored in connection with Cicero but which receive a new level of pedagogical definition in Quintilian. It is at this point that we take up the question of *utilitas*: i.e., what is there in Quintilian that may still be of use to us?

As a whole, I approach Quintilian as a Janiform figure, a substantial historical presence who looks back a century to Cicero and past him to the Sophistic origins of pragmatic pedagogy, but who also looks forward, to the Renaissance, when he was much admired, and further still, perhaps to our own time. Once we know more about the pedagogy he develops in the *Institutio*, we should be prepared to extend the question of relevance and address the principal concern of the Epilogue: i.e., to what extent can the tradition of antilogic/*controversia* still serve, *mutatis mutandis*, to enrich our own rhetorical and pedagogical practices? With the close of this chapter, then, we take a giant step forward in time; but we do so with the aid of rhetoric's master teacher, who knew that looking both ways was a process that could be taught, and whose methods for doing so qualify as a thick description of rhetorical education, whether or not we choose to revive them.

1. AN HISTORICAL PRELUDE

It is not enough to say that the Sophists were the first teachers of oratory; more importantly, they made preparation for public oratory a principle feature of a comprehensive educational program (see Marrou 1956; Colson 1924, xxi; see *IO* 1. Pr. 10ff). W. J. Greer writes that

> The Sophists had, from the beginning, been advocates of a practical education, an education which would prepare the [student] to take an independent and commanding position in public life. Its product was the [political person] who . . . became able *to speak successfully on either side of a proposition*. Such a [person] stood forth as an orator qualified to discuss persuasively matters coming before the popular assembly, to win his case before judge and jury, to pronounce a eulogy, or to deliver an invective with telling effect. (1925, 27, italics mine)

The most influential representative of Sophistic pragmatism is, of course, Protagoras, who declared himself with pride to be "a Sophist and educator" and whose subject, as he tells Socrates, is "the proper care of [the student's] personal affairs, so that he may best manage his own household, and also of the state's affairs, so as to become a real power in the city, both as speaker and man of action" (*Protagoras* 317b-318e). Among Protagoras' works (all lost), one book, the *Antilogiae*, appears to have been a textbook, or *technai*, which we assume outlined the antilogical process of arguing either side (Gomperz 1.465, cited in Smith 1918, 211; cf. *DK* 80 A1). Kerferd refers to the practice of antilogic as "the most characteristic feature of thought of the whole Sophistic period" (1981, 85). This being the case, we can also assume that antilogic was well represented in Sophistic teaching. Marrou writes that Protagoras' own system of teaching was firmly based on the methods of antilogic, that it sought to emulate "as near as possible the actual condition of public oratory," and that the system as a whole was "astonishing in its practical effectiveness" (1954, 51-59). This emphasis on pedagogical verisimilitude is also at work in the early exercises in Greek declamation, which insist not only that students argue both sides of a case but that rhetorical training be patterned after the practice of argument in public forums (see Russell 1983; cf. *De Oratore* 1.138). Bromley Smith sums up the early Sophistic educational legacy when he refers to it as "practical preparation . . . in an age where [every citizen] was his own lawyer and lawmaker" (1918, 211).

In Isocrates, this emphasis on practical method is complemented with a strong sense of civic humanism that becomes enormously influential in Rome and beyond. Isocrates does, of course, fault the Sophists for inflated claims about their teaching, for too much attention to display and eristic, and for not enough concern for practical exercises that provide students with the concrete experience of oratory (*Against the Sophists* 10-17; cf. 204). Central to his own educational program is the relationship between rhetorical training and political participation, a relationship which Protagoras had also explored (see *Protagoras* 322c -328d). For Isocrates, engagement in politics implies a broad commitment to the public good, a commitment one exercises through the power of speech, a power that is the central feature of the social contract (*Nicocles* 5-9, *Antidosis* 253-57). In this context, training in oratory is a practical "apprenticeship" to active engagement in public affairs, a necessary prelude, as Takis Poulakos puts it, to "deliberating the destiny of the *polis*" (1997, 4). Moreover, like the conception of oratory promoted by Cicero's Crassus, rhetorical training in Isocrates becomes an organizing framework for all other studies that relate to political culture (Poulakos 8). But Isocrates is not a Sophist, and while he promotes practical exercises and the direct application of one's education to the public good, he does not include in his curriculum the practice of arguing both sides. For the revival of that method, we have to wait for the antithetical operations of Academic scepticism and for the theoretical recognition that in matters of action regarding public affairs, practical decisions involve multiple perspectives and thus require the operations of *controversia* as a guide to prudence (see Ch. 5, sec. 1). And yet, there seems to me adequate justification for Jacqueline de Romilly to declare that

> Protagoras' new teaching truly leads straight to Isocrates, from Isocrates to Cicero, and from Cicero straight to us. (1992, 56)

Perhaps the road is not so straight, and the path to the present is still to be charted; but the line of influence is there, though we have missed a major stop. For if the Sophistic legacy passes through Cicero, which I believe it does, then the Ciceronian tradition, especially in education, passes next to Quintilian. Of course, a century intervenes; and even in this thumbnail history, I should acknowledge that there are, in fact, important changes in the social, political climate during this century, changes that dictate adjustments in the legacy of rhetorical education, adjustments that need to be noted.

Cicero was at work in a period of intense political violence, during which time the Senate was unable to control the huge armies required to maintain Rome's expanding territories. The result was an extremely chaotic political environment and eventually civil war (Starr 1965, 19). Cicero assumes the Consulship of the Republic in 63 BCE; only four years later Caesar and the First Triumvirate come to power, and by 58 BCE Cicero is in exile and would have very little influence on political activity. The effects of these and later events on public oratory were naturally dramatic. Chester Starr sums up the situation this way: "[w]hen one man became sovereign in Rome . . . the significance of political debate waned swiftly" (23). Nonetheless, while Cicero refused to capitulate to Caesar, he was not a participant in the aristocratic plot of 44 BCE to murder him (despite Shakespeare's revision of history). Cicero himself was assassinated the following year by Antony's henchmen. With the demise of the Republic and, in particular, with the death of Augustus in 14 CE, a tranquillity enforced by Imperial despotism dominated Roman life, and along with it came further curbs on the freedom of speech. Books were burned, criticism of the Emperor was stifled, and oratory was given much less room to address public controversy (Grube 1965, 257). Important political decisions were no longer made in the Senate, and court cases were now presented to a single, omnipotent judge rather than tried before a jury of citizens. The concomitant decline in oratory became a favorite subject of Quintilian's first-century contemporaries.

Seneca the Elder attributes the reduction in rhetorical talent to a taste for luxury that came with Imperial peace, but he also notes that there are no longer public rewards for oratorical prowess, an oblique hint at the lost opportunities for public expression under the Emperor (*Controversia* 1, Pr. 7; Grube 1965, 260, n.1). Petronius begins his *Satyricon* with a tirade against fatuous oratory and the claim that "all the arts of speech were fed the same diet and produced nothing capable of lasting" (263). And in his outspoken and ironic "Dialogue on Orators" Tacitus acknowledges that orators receive less honor and repute when "citizens have no choice but to obey their ruler" (40). Under such conditions, it is only natural, as George Grube points out, that "rhetoric took refuge in schools" (1965, 257). But in schools, too, rhetoric began to display ornate, ostentatious tendencies that Seneca the Elder and Quintilian deplore (see *IO* 10.2.27 and 12.10.45). In particular, declamation, which had originally been intended as preparation for practical persuasion, had, according to Quintilian "departed from the true semblance of pleading" and was composed "merely to please" (5.12.17-23; see also 8.6.76; 6. Pr. 3; 8.3.58; and Greer 1925, 29). In such an environment, Quintilian and his pedagogy are distinctly neo-classical in the sense that they look back to the pragmatism of the Sophistic heritage and to the legacy of *controversia* in which the ideal orator was not simply able to, but actually obligated to appraise the *dissoi logoi* available on any topic of controversy.

Quintilian himself was the recipient of a good deal of favor from the Flavian emperors and is virtually silent about political conditions per se in the *Institutio*, which was written in his retirement (Watson 1970, viii-xix; see 3.7.9; 4. Pr. 3; 10.1.91). But within the confines of his own pedagogy and practice, he remains very much wedded to the openness and multiplicity that are central to the Sophistic legacy of antilogic and the Ciceronian practice of *controversia*. As we shall see, his constant concern for his students is that they come to see the contrary possibilities inherent in all humanistic study and that they be well practiced in assessing such complexity *in utramque partem*, by juxtaposing opposite sides (7.3.30). This commitment to the multiplicity inherent in argument and to the need for comparative inquiry is on full display in his own critical method.

2. QUINTILIAN'S USE OF CONTROVERSIAL METHOD

In fact, Quintilian wastes no time in putting his own critical practice on display. He opens the *Institutio Oratoria* this way:

> I was asked by certain of my friends to write something on the art of speaking . . . [because] they urged that previous writers on the

> subject had expressed different and at times contradictory opinions, between which it was very difficult to choose. (1. Pr. 1-2)

Later, while introducing his discussion of rhetorical invention, he notes that his first task is to canvas the infinite diversity of opinions among writers on this subject (*infinita dissensio auctorum*; 3.1.7; cf. 3.1.1). Quintilian's effort to catalogue and respond to this "authorial dissent" is the hallmark of his critical practice. But he also insists that throughout his compendia of opposing antilogoi he will neither adhere superstitiously "to any particular school" nor will he "shrink from expressing his own opinion on certain points." Moreover, he will regularly "leave it to my readers to select what they will" (3.1.22). The first order of business for Quintilian, then, is to survey the *multiplex ratio disputandi* that make-up the landscape of opinion on the point at issue; and, during this effort, he will proceed as equitably as possible in his critique of alternative positions before he settles upon a stand of his own (Murphy 1990, 34; Kennedy 1969, 55).

Colson describes this critical practice as non-*katholika*; i.e., Quintilian begins an inquiry by abjuring universal statements or summary judgments (1924, xxxix). In terms employed by Gregory Clark, Quintilian avoids the natural tendency "to claim [his] own correctness and completeness" (1990, 71). An alternative description may be that Quintilian is distinctly non-dogmatic in two senses of this term. First, he seeks to remain independent of the various schools and ideologies that dominated the educational theory of his day (e.g., the Theodoreans and the Apollodoreans).[3] Second, he would (as noted) accord his readers some latitude to exercise their own judgment in determining a controversy for themselves.[4] On both counts, Quintilian's liberal approach to critical reasoning will have significant pedagogical implications. And yet, while Quintilian insists on his own independence from various "schools of thought," his methods of inquiry have an undeniable kinship with the critical practices of the New Academy.

In Book XII, Quintilian notes that the practices of the New Academy are particularly "useful" because their "habit of disputing both sides of the question [*in utramque partem*] approaches most nearly the actual practice of the courts" (12.2.25). He inherits this position most directly, of course, from Cicero, who summarizes the Academic method in a passage worth citing again:

> . . . . the only object of the Academics' discussions is by arguing both sides of a question to draw out and fashion something which is either true or which comes as close as possible to the truth. (*Academica* 2.8)

Founded on the scepticism of Pyrrho of Elias (4C BCE), the Academic practice of argument by antithesis is as much a philosophical stance (which rejects any dogmatic assertion) as it is a critical method. Given institutional status by Arcesilaus in the 3C BCE and further revised by Carneades in the 2C, the principles of the New Academy pass into the Roman tradition through Carneades' successor, Philo of Larissa, who visits Rome in 88 BCE and who tutors Cicero in rhetoric and philosophy. Cicero, in turn, gives argument by antithesis a decidedly rhetorical turn, transforming the dialectical clash of ideas into a dialogical method of exchange between people, a method he refers to as *controversia*. So while Quintilian may claim that it is unnecessary to "swear allegiance to any philosophical code" (12.2.26) and while his own philosophical interests tend in the direction of moral philosophy rather than epistemological speculation, his practice as a critic and educational theorist clearly reflects the Ciceronian refinements in Academic method. In particular, he ascribes to certain argumentative principles: that all claims must be argued because more than one probability exits, that reasonable statements pro and contra exist for every claim, that judgment is best deferred as alternative logoi are weighed, and that criteria for judgment will be based on a comparative assessment of the advantages/disadvantages of the claims at play. These principles are, in turn, fostered by instrumental teaching strategies, strategies designed to guide the production of specific arguments. But it is worth repeating that these principles are ultimately tethered to theories of knowledge, discourse, and virtue. Given this connection, the practicalities of his teaching should be seen as instruments in support of an organizing theory of discourse.

Examples of Quintilian's critical method are available at almost every turn in the *Institutio Oratoria*. Indeed, these exercises in internal or authorial *controversia* are among the most distinctive features of this text; it seems clear that Qunitilian took seriously his own advice that teachers should exercise their own argumentative skills "without stint" (2.2.8). After the reference to the "contradictory opinions" that opens the *Institutio*, and after a few lines on pre-school education, Quintilian immediately engages the question of whether or not it is better to educate a child at home or at public school (see 1.2.2-17). "Contradictory opinions" fully condition this topic and must be acknowledged, for while "eminent authorities" favor the public schools, "(i)t would . . . be folly to shut our eyes to the fact that there are some who disagree" (2.2.2). These critics, he goes on, are "guided in the main by two principles," and he lays out each of these *contra* arguments in significant detail. What is particularly interesting about this first exercise in argumentation is that Quintilian avoids a simple claim/rebuttal structure and oscillates back and forth between the two contesting positions. One argument is over the claim that "schools corrupt morals," and we begin with an accusation against the schools (2.2.4), then shift to a defense of the respectable teacher (2.2.5), before we turn back to the initial claim in a rebuttal that parents too can be a bad moral influence (2.2.6). The movement back and forth between sides is more fully developed in the second "objection" (that a tutor at home can give more attention to a single pupil than can a large school; 2.2.9), which in this case goes through multiple claims and counter-claims (2.2.9-16) before the *pro* position takes the floor (2.2.17-29). The procedure as a whole operates, as Colson notes, more like a "discussion" than a treatise, and as such is reminiscent of the *De Oratore*, Book III, in which Crassus puts forward opposing positions on style as if speaking in dialogue with himself (1924, xxxix). Careful readers will recognize that this multi-vocal, heteroglossic approach to argumentation is one of Quintilian's most distinctive rhetorical habits.

Similar examples of *controversia* appear seemingly at every turn in the *Institutio*. In Book II, for instance, Quintilian takes up such issues as the choice of an appropriate teacher (2.3.1-9), memorizing commonplaces (2.4.27-32), the controversy over declamation (to which we will return; 2.6.1-2.10.15), and the role of rules in oratorical training (2.15.1-17). The analytical method in all cases is *controversia*: the author first surveys the diversity of opinion on the topic to make sure all reasonable positions have been considered and to weigh the pros and cons of each side. In the discussion of rules (*praecepta*), for example, even though Quintilian would generally support the contingency of the rhetorical process and so avoid "tying myself down to universals in general," he also appraises the advantages of pedagogical prescriptions and acknowledges that "the last word on the subject will never be written" (2.13.14-17). And, in the discussion of declamation, he writes that "I now come to another point in which the practice of teachers has differed. Some have not been content with . . . . Others have merely suggested [that] . . . . Both practices have their advantages . . . . But if we must choose one . . . " (2.6.1-2). The dominant transitional phrases or conjunctive schemes of these supremely non-dogmatic inquiries are *at contra* (on the contrary) and *verum* (on the other hand), as the rhetor works his way through the various nuances of an argument and models for his readers (and students) the actual practice of controversial reasoning.

The practice of *controversia* is especially interesting in Quintilian's chapters on the appropriate definition of rhetoric (2.15). Kennedy writes that this well-known discussion proceeds according to "Quintilian's basic method of reviewing and choosing or adopting what seems to be the best" (1969, 58). After clearing some etymological brush, he takes up the stasis question "*quid sit rhetorica*?" (what is rhetoric; 2.15.1) and immediately notes that "many definitions have been given" in the past (2.15.2). He begins with the common assumption that rhetoric possesses the power to persuade (*vim persuadendi*), but such a definition requires an interrogation of the genus (the general category) that is predicated by the *difiniendum* (the thing to be defined), the genus in this case being the category of "capacity" or "ability," what Aristotle called *dynamis*. Quintilian then launches into a detailed examination of the history of this definition, beginning with Isocrates and including references to Gorgias and Cicero (cf. *De Oratore* 1.138). Despite the imprimatur of authority, however, rhetoric as the power to persuade has drawbacks as a definition, since money, rank, and sex all have a similar power. So the *differentia* do not differentiate adequately. This first category of definition, then, doesn't meet the challenge of controversial scrutiny (2.15.8-9).

A second variation is drawn from the famous distinctions of the *Gorgias* (452e); in brief, the claim here is that rhetoric is the power of persuasion by speaking (2.15.10). This definition too has a history of adherents but is ultimately rejected because the orator is not always engaged in persuading and because persuasion is often engaged in by others "who are far removed from being orators" (2.15.11). Next, there are several possibilities drawn from Apollodorus, Aristotle, and Hermagoras that attempt to identify the goal or *telos* of persuasion (2.15.12-14), others that focus only on invention (2.15.16), still others that restrict rhetoric according to the topics it covers (2.15.17-20), and yet another group that sees the subject principally as a species of *ars* rather than as a power or science (2.15.21-22). Beyond these, there are "countless other definitions" (2.15.23), only some of which he can refer to directly (most notable is an extended critique of the *Phaedrus*; 2.15.25-31). According to Quintilian, the problem with this host of foregoing possibilities, especially those which approach rhetoric as a *techne* only, is that they tend to focus on victory, so that argument is reduced to eristic (2.15.32). Such definitions, however, do not address the stasis of quality or value (what is it good for?), a stasis central to Quintilian's own basic interest in molding an ideal orator "who is necessarily a good man" (cf. 1. Pr. 9 and 2.1.33). In response, Quintilian now turns to the more germane definitions that emphasize the connection between effective oratory and virtue.

The variant which best reflects this moral emphasis is, of course, the "science of speaking well" (*bene dicendi scientiam*; 2.15.34), with its famous play on the relation between effectiveness and ethics (speaking *well*), a pun which reflects rhetoric's traditional ambivalence between the demands of sound reasoning, on the one hand, and persuasion on the other. Having arrived at this formulation, Quintilian does not rest content, however; he proceeds, instead, to examine multiple variations of this privileged definition by "distinguished author[s] and professor[s] of rhetoric" until he can conclude that he has explored "practically all the most celebrated and most discussed definitions" available to him (2.15.36-37). As a result of this review, he is finally in the position to settle upon "what I believe to be the right view," a view based on the best criterion he has been able to discover—that practical wisdom combined with oratorical skill is the "ultimate goal of rhetoric" (2.15.38). He comes to this judgment, after a protracted period of "invention" which, in this instance, is taken up by a comprehensive review of all probabilities that seem worthy of attention (see 1.8.18). And, as noted, the method used here is never laid aside. Book III opens with "the necessity of examining a great diversity of opinions" (3.1.2), adds that all deliberation is based on comparison (3.8.34), and applies this method to such crucial questions as the role of the stasis categories in the invention process (3.6). In Book VII, Quintilian indicates that the critical method we have just examined is congruent with the important process of considering the point of view of one's opponent in forensic cases (7.1.4); and in the famous peroration on the ideal orator in Book XII, he once again places a legion of opposing positions in juxtaposition as the appropriate prelude to any judgment on the ends of oratory (see 12.1.14).

In sum, the critical method of *controversia* informs every major inquiry in this encyclopedic study of rhetoric and pedagogy because every thesis, every question involves an "infinite diversity of opinions" (3.11.2). In confronting this multiplicity, Quintilian reflects the breadth of interest advocated by Cicero's Crassus and samples "all the available" arguments as a prelude to judgment. And while the argumentative exercises that fill out the *Institutio* may not always rise to the level of theoretical insight imagined by Crassus, there is nonetheless an admirable congruence between Quintilian's own critical method and the practice of argumentation he will advocate for his students.[5] In the scope of his inquiries and in his lack of dogmatism, Quintilian endeavors to translate into his own argumentative practice the Academic, Ciceronian emphasis on controversial method as itself a serious theoretical stance. Central to this method are the related notions that knowledge of human affairs is inherently indeterminate, that deliberation requires an appraisal of conflicting opinions, and that no single position is either final or infallible. As Quintilian puts it, the domain of oratory is "extremely varied and develops some new aspect almost every day, so that the last word on the subject will never be said" (*nunquam dicta erunt omnia*, 3.13.17). In adhering to the practice of *controversia*, Quintilian not only advocates an argumentative methodology that contradicts any emphasis on canonical prescriptions, he also models in his own practice the methods he would promote for his students (see Kennedy 1969, 57). This self-reflexive effort is an act of professional integrity, a means by which teachers invest their lessons with authority as models in their own right. Centuries later, an unknown commentator writing of a different teacher (Geoffrey of Vinsuaf) remarked that Geoffrey "does what he teaches, which is the custom of the good teacher" (in Woods 1995, 76). Quintilian observes this custom, which is what we might expect since, by his own avowal, teaching "is the standard by which the greatness of my undertaking must be judged" (4.1.17).

3. QUINTILIAN'S PEDAGOGY OF RHETORIC AND ARGUMENT

Roman students, including both boys and girls of the aristocracy, began composition study at around twelve or thirteen with a *grammaticus*, a teacher responsible not only for grammar study but also for an extensive introduction to literature and literary style. The grammaticus would employ, on occasion, the elementary composition exercises of the *progymnasmata* (Marrou 1956, 274ff; Bonner 1977, 213-49; Clark 1957, 177-212; see also Quintilian 8. Pr. 1-5). These "preliminary" exercises in composition were, from the outset, fully involved with argument and, in particular, with argument in its dialogical form as *in utramque partem*. As the exercises progressed toward greater rhetorical sophistication, students moved (at about age sixteen) to the tutelage of the rhetor, who provided the equivalent of Rome's higher education and who, as his title suggests, was a specialist in oratory. At this point, Roman students (now all boys) came directly within the orbit of the Greek *paedeia*; i.e., within the pedagogical tradition of Protagoras, the Sophists, and most notably Isocrates, whose emphasis on oratory and practical training for public engagement had come to dominate Roman schools of advanced study.[6] It is oratory in general, and argument in particular, then, to which Roman students devoted themselves as they moved into the advanced levels of their curriculum.

Progymnasmata. These exercises are, literally, "pre-training" activities, preliminary in the sense that they lead to the full-dress arguments of declamation and, beyond that, to the kinds of deliberative, forensic, and epideictic speech that members of the Roman upper-classes could expect to perform in the course of their public careers. The systematic order of the exercises is of considerable interest to composition teachers not only because it emphasizes a reasonable sequence of writing challenges but also because each of the exercises in the sequence of a dozen or so effectively "reinforces" the lessons of the earlier stages (Murphy 1987, xxxii; see Nadeau 1952; Comprone 1985; Hagaman 1986; Clark 1957).

While still concentrating on matters of style and grammar with the grammaticus, students would begin with the composition of fables (mostly from Aesop) which teachers would first recite in verse and pupils would then paraphrase in their best prose (1.9). Hermogenes adds that students were often asked to "feign the words of given characters" in a fable and that topics were customarily contentious, as when the student was asked to paraphrase: "the monkeys in council deliberated on the founding of a city" (in D. L. Clark 1957, 182). So composition begins with imitation, impersonation, and occasionally the contention of mock-debate. Students would next pass to "fictitious tales" taken from literary sources, like Ovid, again employing impersonation and paraphrase, but this time revising familiar speeches (like Medea's justification for stealing the Golden Fleece) into more "correct and accurate" Latin (2.4.15; Cicero's *De Inventione* 1.19). These tales (the more realistic of which were called *argumentum*; 2.4.2) were followed by the *chreia*, exercises in the elaboration (*expolitio*) and interpretation of some well known maxim, such as "money is the root of all evil," or a saying by Isocrates. In this case, the student was composing rather than paraphrasing and, in effect, inventing the argumentative reasoning that supports the claim proffered in the maxim itself (see Hoch and O'Neill 1986; Bonner 1977, 258-60). It is notable that even in these early exercises, the young rhetor is routinely given a specific character along with some situational data, so that invention proceeds in relation to decorum, while the specific topics to be addressed are matters of probability and will customarily invoke opposing options.

Argument itself comes to the foreground in the next exercise, the narration, sometimes called "refutation and confirmation." Quintilian suggests that these exercises follow up on stories from literature and history that the student has already learned; but in addition to simply retelling the chosen incident, the student must "annex the task of refuting and confirming . . . from which no little advantage is to be derived" and for which "a quantity of arguments *in utramque partem* can be produced" (2.4.18-19). For example, Quintilian asks his students to consider the credibility of a raven settling "on the head of Valerius in the midst of combat" or whether or not Romulus was actually "suckled by the she-wolf" (2.4.18-19). At this point in the exercises, monological claims surrender to the dialogical interaction of competing possibilities, alternative options must be outlined and evaluated, decisions made as the result of comparative deliberation.

And because all students would recite their compositions aloud to the class, the group as a whole was exposed to "a quantity of arguments" on any and all sides of the case. At this relatively early point, then, exposure to alternative perspectives on controversial topics became a routine part of the student's educational culture, preparing the way for the prominence of *multiplex ratio* in school and beyond.

Exercise in *narratio* has the further practical advantage of helping the student prepare for the all-important section of an oration devoted to background (i.e., *narratio* is the second stage in a classical oration, coming just after the exordium, or introduction). And, in fact, the *narratio* is that moment in the arrangement of a topic (*dispositio*) that Quintilian considered "the most important department of rhetoric in actual practice" (2.1.10; see O'Banion 1992). The value of an ample *narratio* is not hard to calculate, and Quintilian's own critical practice serves as a model (see previous section). When students learn not only to tell a story but also to include as background to their own compositions a comprehensive array of arguments available on either/all sides of a question, invention is likely to be more thorough, judgment may well be suspended more willingly, and the orator's ethos becomes more expansive, less parochial. The *narratio* exercises generally involved questions of fact (did it or did it not take place), and from these considerations, the *progymnasmata* moved to "commonplaces" and then "encomia." Both of these were epideictic exercises in the praise of virtue and blame of vice, and they represented an increase in complexity over the comparatively simple calculations of fact or fiction that had come before. On such topics as adultery or gambling, "the mind is exercised," writes Quintilian, "by the variety and multiplicity" inherent in the topics because the rhetor must deal not simply with polarized conceptions of pro and con but with "degrees" of vice and virtue as well (2.4.21).

These increasingly subtle exercises in rhetorical reasoning progressed next to the composition of "comparisons" (between characters) and "impersonations." Comparison is, of course, the methodological backbone of controversial reasoning; and when asked to compare Ajax and Odysseus, simple recourse to praise or blame, either/or was clearly inadequate (Bonner 1977, 266-67). In addition, the juxtaposition of the two heroes constitutes another valuable opportunity for invention: we know Odysseus is clever; but, in comparison with Ajax, is he also loyal, and to what degree? In such a case, preparing both sides was again required, while judgment was a question of weighing contraries. The exercise in impersonation, or *prosopopoeia*, highlighted an historical figure faced with a singularly persuasive situation, such as Priam pleading with Achilles for the return of Hector's body (D. L. Clark 1957, 198). While such speeches were typically monologues, decorum was all: they required students not only to develop original representations of the feelings, reasons, and style of the chosen character (cf. "fictitious tales" above) but to cast these in light of the demands of a particular rhetorical moment. Augustine recalls the exercise this way:

> . . . the pupil won the greater praise according as he more faithfully represented the dignity of the character concerned, and more convincingly reflected her anger and resentment, clothing his ideas in a befitting style. (in Bonner 1977, 269)

Because young rhetors would move from impersonations of warriors to orators, from men to women, Greeks to Romans, familiarity with a range of psychological, social, and historical perspectives was also stimulated as a resource in overcoming the conceptual myopia referred to in the Introduction as the "my-side bias."

These various assignments proceeded in incremental steps to the final *progymnasmata* exercises of the "thesis" and the "laws." The thesis, or general question (*quaestio infinita*), involved the student in abstract, universal issues (whether one should marry or whether country or city life is best), questions such as Milton addresses in his "Prolusions" or Shakespeare in his first 17 sonnets (2.4.24; D. L. Clark 1957, 206). Whereas the outcome of Priam's interview with Achilles was basically a foredrawn conclusion, the thesis is specifically an inquiry into "matters still in doubt"; and although the topic may be, as Quintilian puts it, "remote from persons and times and places," it must still be argued *in utramque partem* (3.5.5; D. L. Clark 1957, 204-5; Bonner 1977, 270). The thesis (which had been developed in the Peripatetic and Academic Schools of the Hellenistic age) also stimulated philosophic considerations of justice, expediency, propriety, etc. and invited students to contemplate the relationship of a particular proposition (or hypothesis) to general, theoretical questions (*quaestiones philosophi*) which have a bearing on the case (3.5.6).

In sum, students were asked to employ dialectic as a resource for rhetorical deliberation. The "discussion of laws" also called upon students to confront a complex controversy and offer a persuasive judgment; but at this final stage of preliminary training, a finite point of law replaced the open philosophical question, though the law too was considered in the abstract. Quintilian is particularly interested in this exercise (2.4.33-41), perhaps because it so clearly addresses his commitment to preparing his students for the bar, perhaps because it is an effective preview of the legally-oriented declamation to come, perhaps because of the demand for rigor which these cases placed on the rhetor. In any case, while the "thesis" and "laws" take on added theoretical complexity in lieu of specific rhetorical contingencies, they nonetheless hinge on the operations of *controversia* and the interplay of equally good reasons on either or all sides (see Aristotle's *Topics* 101a).

It is a considerable leap from retelling fables to theoretical reasoning about complex points of law. But the pedagogical method remains the same: learning is best promoted through the practical application of the principles at issue. For Quintilian, the basic process is repeated, whether impersonating heroes in crisis, confirming or refuting some melodrama of history, comparing the virtues of robust characters, or arguing matters of ethics and law. No claim about these very human affairs exists uncontested—all such argument solicits *multiplex ratio disputandi*. But if there are always two sides or more to every question, there are, nonetheless, no rules for negotiating particular controversies. Decorum is, of course, always at play, but decorum is an improvisational faculty that must refashion itself according to circumstance. How, then, does one teach a process that only fulfills itself in particular actions? Imitation of effective models is one way, and the *progymnasmata* is replete with opportunities for *imitatio* (as we will see in sec. 4). But imitation is ultimately a form of practice, and the overriding principle of the preliminary exercises is that "[t]here are no subjects in which, as a rule, practice is not more valuable than precept" (2.5.15). Better still if this practice imitates the actual experience of oratory "in the broad daylight of the forum" (12.2.7-8). Of course, the actual practice of such argument "never did run smooth"; its questions are indeterminate, its personnel unpredictable, its procedures ever subject to surprise. So what students of the *progymnasmata* are actually practicing is, in a phrase, *preparedness for any eventuality*, or, more exactly, for arguments on any and all sides of the case.

At the outset of Book X, Quintilian digresses to emphasize the role of *facilitas*, the resourcefulness and spontaneity required for oratorical success (10.1.1). Later he refers to this facility as "promptitude and readiness for action" and adds that it "can only be maintained by practice" (10.7.24). The practice afforded by the *progymnasmata* prepares the student for argument in many ways, but no function is more important than cultivating *facilitas*, the capacity for easy adaptation to "varied and multiple" circumstances, a capacity that allows orators to move with an argument according to the flow of its development. Donald Leman Clark ends his own comprehensive review of the *progymnasmata* this way:

> The aim of the rhetorical exercises was not in general to make truth prevail, but to make one side of a debatable question seem as

plausible as possible and then turn around and make the other side of the question seem just as plausible. (1957, 212)

To make the "other side seem plausible" is to be prepared for both sides, to be prompt and ready for the twists and turns that characterize "the actual practice and experience of oratory" in the forum and elsewhere. By the time that Roman students arrive at the mock-debates of declamation, they have more than a passing familiarity with "the quantity of arguments available on either side of any question," the decorum required to match strategies and decisions to circumstances, and the *facilitas* of the capable orator who can adjust in the midst of evolving debate. These skills have been nurtured through "diligence of exercise" in an orderly but patient curriculum that is ultimately intent on incremental growth. The *progymnasmata* does not ask the novice to start by cooking an omelet; instead, it begins with how to break an egg, then how to prep the pan, then how to whip and spice, etc. Throughout it all, students are constantly adding new strategies to their own rhetorical menu while sampling some very different recipes for the same dish put together by other orators. All such experience, of course, is itself preparation for the next round which, if anything, places even greater emphasis on decorum, *controversia*, and the intersection between the two.

Declamation. Quintilian calls declamation "the most useful of rhetorical exercises [because] it includes practically all of the preliminary exercises [the *progymnasmata*] . . . and is in close touch with reality" (2.10.2). And, as the final stage in the Roman rhetorical curriculum, it operates as a capstone experience, summarizing and confirming all those skills required to participate in Roman rhetorical culture. The declamatory exercises themselves are mock judicial or deliberative debates on specific points of law or history in which the student orator assumes a particular persona and works within the confines of a specific situational narrative (either fictional or real). For Quintilian, the verisimilitude of the exercises was of particular importance, and he repeatedly emphasizes the need for declamatory themes to be "as true to life as possible" and for student speeches to "be modeled on the pleadings for which [declamation] was devised as training" (2.10.4). Indeed, if declamation is presented effectively, it should mimic as closely as possible the "real contests" of open, unpredictable debate that the student will encounter in public life (10.1.4). Declamation, then, is the classroom approximation of what Plato called "contending with words," a direct, if guided experience in the dynamics of practical argument "spoken well."

The lineage of these exercises is almost as old as the history of rhetoric theory. D. A. Russell notes the tradition of *melete* or mock speeches among the 4C Greek Sophists and points out their dual role as practical exercises and as literary expressions of oratorical skill (1983, 9-20; see 2.4.41). This latter role as literary models of discursive battle is on full display in Gorgias' *Palamedes* and Antiphon's *Tetralogies*, as well as in many of the paired speeches in Thucydides. Aristotle's *Topics* Book 8 assumes the existence in his own school of structured contests with rules and judges (see 159a 25-37); and Stanley Bonner, the preeminent authority on Roman declamation, notes that schoolroom exercises in debate are routine from Hermagoras onwards (c. 150 BCE; 1969, 1-26). Cicero refers to such exercises as *causae* and he too commends the practice of arguing these cases "in a fashion as near as possible to real life" (*De Oratore* 1.149).

By Quintilian's time, declamatory exercises had been divided into two distinct kinds: the *suasoria* or deliberative speech on a question of history or politics, and the *controversia* or forensic speech on a specific legal case. Sources listing both kinds are abundant, so we have a good idea of what they both were like (see Bonner 1977; D. L. Clark 1957; Nadeau 1964). As for the *suasoria*, Philostratus lists these themes from Greek history: the Spartans debate whether they should build a wall to fortify themselves against attack, and "Isocrates tries to wean the Athenians from their empire of the sea" (514 and 584, resp.). Seneca the Elder (c. 50 BCE - 40 CE), who chronicled a vast number of Roman declamations in the generation before Quintilian, records *suasoria* exercises in which Alexander's mother urges him not to cross the ocean and Agamemnon argues with the soothsayer Calchas about the sacrifice of his daughter Iphegenia (*Suasoria* 1 and 3). Quintilian cites Cato deliberating on whether on not he should marry (3.5.13) and Numa, a legendary figure from early Roman history, debating whether or not he should accept the Roman crown (7.1.24).

Cases like these built on a spectrum of skills students had already cultivated in the *progymnasmata*; but in declamation the scope of the rhetorical challenge was extended considerably. In most cases, students were asked not only to take into consideration specific historical circumstances they had studied in literature but also to take on the persona of a particular character and to address a particular audience. However, in addition to making decisions regarding ethos, audience, and situation, this time the student was arguing in the presence of a declamatory opponent, which gave considerations of *in utramque partem* greater currency and which brought the whole exercise into closer proximity with "the actual strife and pitched battles" of public argument (2.10.8). Before we turn to the specific form of the two declamatory exercises, it may help if we have some idea of the classroom principles and practices that characterized instruction in declamation.

Aristotle argues that "none of the arts offers theories about individual cases" (*Rhetoric* 1356b 28), but the study of declamation picked up at just that point in the rhetorical curriculum where students were asked to transfer their attention from matters fabulous, literary, and theoretical to considerations of controversy arising from concrete circumstances. In part, Quintilian seeks to effect this transfer through the critical analysis of working models, often Ciceronian orations, about which the teacher would question his students regarding various points of argument, organization, and style and whether or not the rhetorical choices made were suited to the circumstances of the case (2.5.1-26). Critical analysis was thus employed as an aid in identifying the kinds of decisions required by "the definite and specific case" (7.10.7). I will return to this relationship between criticism and composition in the next section. But the point at present is that declamatory criticism was focused on individual cases because the functional example allowed for consideration of concrete rhetorical practices. Or, as Quintilian describes the process (in a slightly different context), the specific example illustrates "the methods employed, whether wisely or the reverse, by individual [orators] under varying circumstances and conditions of time and place" (2.5.15; see 2.13.2-5).

This emphasis on individual cases, on what works in what circumstances, is, of course, consonant with the essential pragmatism of controversial pedagogy. As Cicero puts it, theoretical study would be "lame and defective were not practical results to follow" (*De Officiis* 1.153). No doubt for Quintilian the "practical results" of declamation would be measured in the character and careers of his students, with the ideal orator as the ultimate criterion for assessment (9.2.81). Given our own present focus on classroom practice, a less dramatic objective for this declamatory pedagogy can be theorized as follows: these simulated cases are intended to cultivate the related abilities to (a) prepare for all possibilities that may arise in the course of argument and (b) balance conflicting claims through appropriate rhetorical choices. Taken together, these abilities or capacities for action constitute the basic movement of the rhetorical process: invention and judgment, discovery and decision. Buckley calls these pairs "the double periods . . . [of a] single, ongoing process of thought" (1971, 147). In a nutshell, declamation is intended to instruct students in this two-part process by integrating invention and judgment in a single experience made concrete by appeal to the case narrative. Within the context of this simulated experience, students exercise their abilities by placing opposing positions in relation to one another, building a balanced understanding of probable claims, weighing these probabilities with rhetorical considerations of audience and context, and making decisions regarding what will best serve the comprehensive demands of the case. The ideal orator may be the intended outcome of Quintilian's *Institutio*, but in the classroom the ideal takes form in the exercise of particular capacities, an exercise declamation is designed to deliver.

To return, then, to specific classroom procedures for declamation: the teacher would present a declamatory problem, often in the form of a brief summary of the circumstances. In the case of *controversiae* or legal exercises, this narrative might also include a synopsis of the specific statutes in question (see Seneca the Elder for examples). The teacher would then provide some introductory analysis (*divisio*) of the case, often including opposing perspectives and how these might be arranged and presented (Clark 1957, 215; Edward 1928, xx-xxii; Bonner 1977, 320-21). The students were next assigned the same or a similar case as the subject of their own oratory and allowed to select a side (either side was available) (Edward 1928, xxxii). Students would then write out and read their initial draft to the teacher, who would question each pupil carefully; or, as Quintilian puts it, he would "test the critical powers (*iudicium discipulorum*) of his class" (2.5.13). It was further assumed that the young orator must have a significant grasp of both sides of the case and must deal with all its pertinent aspects, not just those which favor the chosen position (10.5.20-21). The student would then prepare a more polished composition for memorization and delivery before the class as a whole, and sometimes before the public at large. In these "public" orations, the audience would not hesitate to express its approval or disapproval of student performances, often going so far as to hiss or applaud vehemently (Sussman 1987, 11-12).

A distinctive feature of declamation, therefore, was that speeches were constructed with an antithetical response in mind. Typically, all students in a single group would declaim either for or against the same case, so that each declamation was subject to peer review and the inevitable diversity of opinion that would manifest itself amongst the students. Moreover, at this point in the curriculum, there is interaction *among* students on opposing sides of an issue, so that a well-crafted, but single-sided composition would not suffice. Students would now have to calculate both "confirmation" and "refutation" in earnest, as well as prepare for the serendipitous turns that were bound to arise as one argument follows another (2.23.1). If younger students working with the *progymnasmata* exercises were theoretically ready to understand the value of argument *in utramque partem*, the declamatory experience made preparation for any and all possibilities considerably more operational.

In addition, the performative context of declamation would quickly make students aware that their arguments were "addressed" to a critical audience, an audience that was not a passive proxy for authorial intentions. Instead, the audience was more likely to represent the "varied and multiple" positions that a bit of inquiry will undoubtedly reveal. The principle at work here is not the relatively simple one of recognizing the unique nature and situation of one's audience in order to calculate the best means for moving them into one's own camp. Such a motive assumes that the orator has the answer and that the role of the audience is, principally, to grant or withhold assent. According to such assumptions, the "other" is actually an extension of the self and argument remains essentially a monologue (cf. the discussion of oppositionality, Ch. 4. sec. 2). On the other hand, if orators are theoretically bound to invent alternative arguments as plausible as one's own, then the orator is similarly bound to extend the right of difference to at least some members of the audience and to grant these differences a purchase on probability equal to one's own. The preliminary exercise of "impersonation" is propaedeutic to this effort in the sense that it introduces the orator to alternative subjectivities and to reasonable grounds they may have for opposing positions. In addition, Quintilian shares the general Roman belief that between members of a common culture there is a reservoir of understanding, so that it is at least possible to project subject positions one doesn't know in person, so to speak.

From the rhetorical perspective, these efforts at "identification" not only serve the interests of conciliation discussed in Chapter 5, they also establish the basic conditions for dialogue. Once the audience or one's interlocutor is granted a voice independent of the orator, dialogue is at play and *controversia* can begin to tap its unique resource of invention and judgment. Ben Franklin remarks that "our critics are our friends because they show us our faults." Similarly, the audience and the interlocutor are one's partners in arguing that new ideas and new agreements are always available, that discovery lies at the heart of dissent, as long as we remain in dialogue. The theatrical constructs, *dramatis personae*, and role playing of declamation all call attention to the dialogical nature of controversial argument, but it is ultimately the rhetor who decides to invoke dialogue by granting the other side a voice of its own.

Under conditions of declamation, when the rhetorical environment is replete with oppositional voices, "the all-important gift of the orator," writes Quintilian, is a "wise adaptability" to "the most varied emergencies" (2.13.2). As we noted, Quintilian would foster this adaptability by employing models and promoting a critical reading that leads to effective *imitatio*. He also suggests that teachers provide their own sample declamations—both pro and contra—as a standard against which students could measure their own efforts (2.2.8, 2.4-5, 10.2.39). Roman students of

Quintilian's time could also observe professional rhetoricians who declaimed in public before large, enthusiastic crowds and who regularly displayed the strategies of *in utramque partem* (Bonner 1949, 51; Murphy 1990, 62, n. 35).[7] But ultimately, it is practice with the cases themselves that Quintilian relies on. We turn, then, to the *suasoria*, the declamatory impersonations of historical figures and to the means by which these cases, fanciful as they often were, nonetheless function as practical exercise in the rhetorical capacities that distinguish controversial argument.

As we've just seen, the Roman student had been prepared for *suasoria* by the *progymnasmata* training in *prosopopoeia*, the imitation of character. In this early exercise, the student might impersonate Achilles lamenting the death of his friend Patrocles or Dido pleading with, then excoriating, Aeneas. In the process, the young orator would practice the appeal to character (ethos) as a rhetorical proof, the appropriate deployment of style (*to prepon*), and the use of settings to heighten these hyperbolic moments ("description" or *ekphrasis*; cf. 3.8.59). But like the "encomium" or the "comparison of character," these impersonations were predominantly epideictic and focused mostly on the discussion of values, whereas *suasoria* and *controversia* were decidedly practical in their orientation toward particular problems and concrete action. More specifically, the *suasoria* involve deliberative decisions regarding future action, while the *controversiae* require forensic judgments of past events. Other characteristics add to this pragmatism: the *suasoriae* ask students for the first time to compose a complete oration, from exordium to peroration, not just a dramatic set piece. In addition, these assignments focus on a particular rhetorical question (how can Dido persuade Aeneas to stay?) rather than depict a character in general (Dido in lament). Such added focus encourages more precision in the analysis of character as well as a more comprehensive approach to invention.

In the *suasoria*, students may be faced with the following kinds of dramatic events: a Spartan general must urge his soldiers to fight or retreat at Thermopylae, or Alexander debates with his generals whether or not to ignore the oracles and enter Babylon (Seneca the Elder, *Suasoria* 2 and 4). In such cases, students were not simply acting *ex persona* (in the character of) and delivering a dramatic monologue, like Browning's Fra Lippo Lippi or Andrea del Sarto. Instead, they were attempting to persuade or dissuade an historically well-defined audience regarding a specific course of action, with considerable contextual detail available from the literature, details which the classroom audience would know. So flagrant departures from verisimilitude would be identifiable. Quintilian writes that in the process of these historical impersonations, the declaimer must

> bear in mind the fortune, rank, and achievements of each single individual . . . so that they seem to speak in their own persons. For a speech which is not in keeping with the man who delivers it is just as faulty as the speech which fails to suit the subject to which it should conform. (3.8.50-51)

He adds that all this is "the most difficult of tasks" as well as "the most useful exercise"; indeed, he describes the *suasoriae* as "absolutely necessary" for the developing orator, not simply because they punctuate the centrality of ethos to argument but also because they exercise the pupil's range of comprehension with regard to human motive and response (3.8.49; cf. Aristotle's *Rhetoric* 1356a, 1378a). Quintilian writes that his own students assume as many different roles in their declamations as comic actors on stage (3.8.51). Add to this role playing the fact that students routinely declaim on both sides of a *suasoria,* and the intent of declamation pedagogy becomes not only to extend the student's range of familiarity but also to prompt, in particular, an understanding of the role of opposites in argument. For the student, this exposure to alternative positions will tend to loosen dogmatically held allegiances, breed tolerance, and cultivate procedures of inquiry into areas outside their present grasp. The *suasoriae,* therefore, bring the theoretical requirement of "preparing for any eventuality" to dramatic life by asking orators not just to appreciate but to inhabit the position of others; to understand their motives and context; and, most importantly, to persuade third parties of a cause that is not one's own.

Similarly, the increased sense of rhetorical purpose in *suasoria* offers possibilities for cultivating the young rhetor's conception of audience. Quintilian writes that the *suasoria* may call upon the student to calculate the "considerable difference" between addressing such groups as

> the Senate or the people, the citizens of Rome or Fidenae, Greeks or barbarians, and in the case of single individuals, whether we are urging Cato or Gaius Marius to stand for office, whether it is the elder Scipio or Fabius who is deliberating on his plan of campaign. Further, sex, rank, and age must be taken into account, though it is character (*mores*) that will make the chief difference [in one's address]. (3.8.37-38)

Given this variety, one's rhetorical strategy in addressing audience will obviously be dictated by considerations that are local, temporal, and personal. What seems expedient to one group may be reprehensible to another; what is persuasive with Marius may be problematic for Cato. Once again, by taking on multiple parts in *suasoria*'s reconstruction of historical moments, the student is exposed to a panorama of conflict and required to adjust somewhat differently to each instance. In this pedagogy, knowledge follows from the lessons of practice, and students practice with a new audience for each problem. I have already addressed some issues of audience in the last section, but I would like to add here a note on the subject in connection with *suasoria* in particular. In cases like Dido's outburst or the Spartans at Themopylae, *suasoriae* ask orators to exercise more than a capacity for reason. By placing the student in the midst of someone else's crisis, these exercises seek, in Northrop Frye's elegant phrase, to educate the imagination. The fact that these logically rigorous exercises in forensic strategy are based on highly dramatic historical situations may indicate that the development of the imaginative faculties is preliminary to and necessary for the cultivation of critical reason.

We can say, then, that the *suasoriae* promote a comprehensive approach to the invention of deliberative claims by stimulating personal familiarity with a wide range of positions on all sides of the question. Of course, appeals to character and audience (which Quintilian tends to combine; 6.2.9-12) are only part of the process of invention. In an important passage, Quintilian argues that

> [t]here are three points which must be specially borne in mind in advice or dissuasion: first, the nature of the subject under discussion, secondly the nature of those who are engaged in the discussion, and thirdly the nature of the speaker who offers them advice. (3.8.15)

In *suasoria*, diverse appeals to ethos and audience can be practiced by role playing and by observing how various strategies fare with a proxy audience of one's peers. But the procedure for inventing claims related specifically to the subject (technically, appeals to logos or reason) is a much more complex matter in Roman pedagogy. Treated under the heading of status theory (*stasis* in Greek, also *constitutio* in Latin), this subject requires more time and attention than I can give it here (see Bonner 1977; Carter 1988; Clark 1957; Fahnestock and Secor 1981). But, in passing, I would note that in Quintilian's approach to the topic (to the extent that I can follow the discussion of 3.6) stasis theory is not simply an attempt to arrive at the crux of the issue, one's own particular stand; more exactly, it is a framework for insuring a comprehensive assessment of all arguments available on the topic (cf. Kennedy 1969, 61; cf. Ch. 4, sec. 2 above). Quintilian, as ever, introduces the topic by describing multiple variations of stasis theory and notes that "there is the greatest possible disagreement among writers about this as there is about everything else" (3.6.22). What we can say with some confidence, however, is that stasis theory assumes that invention is focused by asking specific questions in sequence and by suspending judgment while working one's way through the available options. Moreover, as Michael Carter has pointed out, the stasis questions naturally originate in the context of opposing positions; correspondingly, answers can be expected by investigating the issues *in utramque partem* (1988, 98-99).

Quintilian's own choice for the sequence of stasis questions is the same as Cicero's: questions of fact, definition, quality, and (in some cases) law. In specific connection with *suasoria*, Donald Leman Clark offers this approximate description:

> When anyone deliberates about anything, he asks himself certain questions. . . . In the effort to help the thinker and speaker find answers to these questions, ancient rhetoric suggested that he would find all the reasons, pro and con, if he would ask three specific questions about the issue being deliberated. Is it possible? Is it expedient? Is it honorable? (1957, 218-19)

So, as student orators prepare to exhort or dissuade the troops at Thermopylae, they would move diligently through opposing possibilities on each question, a daunting task given the number of potential responses (e.g., we will retreat/stand because it is expedient/honorable for the following reasons . . .). But complex as it may be, the process will clearly yield *copia rerum ac verborum* (abundance of matter and words) on multiple sides of the question and so prepare for informed rather than hasty, biased decisions (10.1.5).

According to Bonner, the "main problem" regarding the invention of arguments for *suasoria* was to find "the most relevant and cogent arguments," which students do by "the technique of arguing for or against a general proposition." This step is followed closely by "applying" these same methods "to build up arguments appropriate to the persuasion of particular people in particular circumstances" (1977, 280). D. A. Russell identifies the same combination of activities when he describes declamation as teaching people "how to *think up* relevant arguments and *employ* them in the most effective way, having regard to the prejudices of the audience as well as the facts of the case" (1983, 14; italics mine). Invention by *controversial* method supplies the rhetor with a host of ideas on either side; but, as Aristotle notes, we deliberate mostly about our actions and, because actions involve commitment, they require careful discrimination and judgment (*Rhetoric* 1357a). Invention by itself, then, is only part of the process. Our general at Themopylae will invent innumerable options on either side of each stasis, but decisions must also be made regarding "what is appropriate" and how ideas can be "employed in the most effective way." Such is the duty of decorum in rhetoric: to make decisions about what will prove persuasive "for particular people in particular circumstances." Under working conditions, however, a strict division between the production and discrimination of arguments becomes difficult to maintain, more a theoretical rather than a functional distinction. Are we inventing *copia* which are subsequently subject to discretion, or are all "relevant and cogent" options already scrutinized in the process of their making? How, then, do we distinguish invention from judgment so as to teach them effectively—or must we? These and related questions will require further attention. For now, we can say the *suasoriae* would instruct students in argument by engaging them in a compelling human drama that *does not segment the rhetorical process* but which does foreground the management of *multiplex ratio* as crucial to all stages of rhetorical action without exception.

When we turn from *suasoria* to *controversia*—the declamatory exercise devoted to forensic rather than deliberative cases—we turn also from cases steeped in drama and history to exercises in logical rigor, from the theatrical to the dialectical, if you will. These *controversiae* represent an increase in complexity over the historically-based *suasoriae* for a number of reasons. In the first place, they simulate more closely the give-and-take, the rough-and-tumble of the courtroom; secondly, the case narrative presupposes a more intricate "concatenation of circumstances" that emphasizes the messiness of argument in its quotidian rather than literary apparel; and finally, they require more detailed attention to the nature and cohesion of one's claims and proofs and so go beyond the demonstration of character and appeal to audience (Bonner 1966, 9). Seneca the Elder records his *controversiae* by first listing the title of the case, then by stating the law or laws that govern the suit, and finally by providing a brief narrative outline of the case. One popular conflict in the schools was as follows:

The Daughter of the Pirate Chief

Law: A parent can disown a child for disobedience.

> A young man captured by pirates writes his father for ransom. He is not ransomed. The daughter of the pirate chief urges him to swear that he will marry her if he escapes. He swears. Leaving her father, she follows the young man, who, upon his return to his home takes her to wife. A well-to-do orphan appears on the scene. The father orders his son to divorce the daughter of the pirate chief and marry the orphan. When the son refuses to obey, the father disowns him.

(from Seneca the Elder's *Controversia* 1.6.6; transl. Clark 1957, 231; cf. Bonner 1977, 314-15)

Although Roman law did require a son to obey his father during the parent's lifetime (*patris potestas*), the above law and the situation outlined are fictitious, even fantastic (D. L. Clark 1957, 231). Obviously, any defense of verisimilitude in the practice of this *controversia* would not be based on the events of the case itself. Rather, it was the intricate reasoning involved and the careful ordering of arguments required for effective presentation that defenders like Quintilian saw as essential to the realism of controversial practice; i.e., the argumentative rigor required from both prosecution and defense kept these cases from declining into mere theatricality and made them useful preparation for "the actual work of the courts" (2.10.8).

Students of declamation would begin their analysis of a case by first identifying its particular stasis as a question of fact, definition, quality, or law. Regardless of the point at issue, however, they would once again be asked to prepare arguments on both sides of the case (10.5.20). Quintilian notes that it is simply not adequate in forensic argument to take up only accusation or defense, that "sufficient acquaintance with the other side" is a prerequisite for effective argument (10.5.21). In the case of the "Pirate's Daughter," the controversy was likely to turn on a question of equity vs. law: is the law in this case universally binding, or is equity a higher virtue than the written statute? Strong arguments could be made on either side, and intricate reasoning would be required by either to carry the case. Bonner, in fact, indicates that the design of such fanciful circumstances is not necessarily a distraction because the primary concern is "to provide a situation in which one might plausibly argue either way" (1977, 312). Quintilian himself notes that a certain measure of fanciful complexity is suitable if it encourages students to think through problems with increased care (7.1.14-15).

As an example of the level of intricacy *controversia* could achieve, Quintilian cites a puzzler on inheritance law that is filled with provisions about bastards, the disinherited, and the proper sequence of legal heirs (3.6.97-103). In the case itself, a father has three sons, one of whom he gives up for adoption, one who is disinherited, and a third who is a bastard (the father here is a kind of Roman version of Karamazov *père*). At his death, the disinherited son becomes the heir but is barred by the law, which appears to disallow his claim. The disinherited son may have to argue on fact (is he actually disinherited?), on definition (what does disinheritance mean?), or on law (what is the law's actual intent?). The bastard faces similar difficulties: what is a "bastard" and under what conditions can his rank in the family supersede his brothers'? As for the son exiled through adoption, the other potential heirs will say that he had ceased to be a son at adoption, again a question of fact, definition, and perhaps of law. Because the statute has been made especially inconclusive, all arguments can claim no more than probability and will have to be set against other, competing probabilities. This comparison of probable claims is, of course, standard argumentative practice (2.17.35). What complicates the inheritance case is that all three claimants will also undoubtedly mount vigorous refutations of their brothers' positions. As a result, the case will turn not on the integrity of one's own proposition but on how well the student can position his claim vis-à-vis its competition. Which means that the ability to anticipate alternatives, prepare for any eventuality, and exhibit the requisite *facilitas* in debate (in other words, the skills of controversial reasoning) are what will determine the outcome.

A more down-to-earth case of *controversia* is referred to by Quintilian as "The Poor Man's Bees" ("Major Declamations" #13). In this case, there is a controversy between the rich owner of a flower garden and a poor neighbor whose bees invade that garden. The rich man spreads insecticide on his flowers, kills the bees, and the poor man brings suit. The *Maiores Declamationes* (attributed to Quintilian) fills out the poor man's speech in considerable detail, especially his refutation, which provides a comprehensive recapitulation of each point in the rich man's case before the poor man's detailed rebuttals (see D. L. Clark 247-50). What is interesting in this defense is that opposing positions (the rich man's claims, the poor man's refutations) are incorporated in a single speech (cf. Crassus in *De Oratore*, Bk. III); i.e., a single speaker has not only prepared both sides of the case but also given "double-voice" to the argument by including both sides within the boundaries of his own utterance. We can only guess at the effect of this strategy, but there is a certain ethos of *égalité* that goes along with such practice. When Quintilian treats debate (*altercatio*) proper (6.4), he reiterates the point that careful consideration should always be paid to "the arguments of the opponent" (6.4.14). Even when students find themselves in agreement, he says, it is best for them to practice their skills in *altercatio* by taking different sides (*diversas partes altercationis*; 6.4.21) and testing their ideas in the actual confrontation of positions. Such experience, writes Quintilian, may also convince student rhetors to avoid eristic and eschew the "coarse abuse of those who hold opposite opinions" (3.8.69); after all, student-declaimers will momentarily be arguing the opposing position themselves.

So, in the end, declamation is not a summary of the tricks of the trade, though it remains the capstone to the rhetorical curriculum. Rather, *suasoria* and *controversia* wrap practice in performance as a way of teaching what can't be known by precept. Quintilian likes to think of these performances as "mock trials," but actually the cases have been "fixed" to insure that there are options for sound arguments on either side. The performance of declamation thus spotlights oratorical and argumentative expertise: who can respond to the case by generating *copia* on contrasting sides, who can take stock of this abundant but contradictory evidence, combine it with situational concerns, and, ultimately, merge all with eloquence in a performance that coordinates all these capacities by "speaking well." A tall order, perhaps, but we reveal ourselves in our ideals, says Quintilian; and if we fall short, there's no shame (2.5.25; 12.11.26). A tall order, nonetheless, because the teacher will not tell you what to do, though he may demonstrate related practices and question you about what others have done. But when it comes to actually making a case, the students are in new territory and will have to respond to the indeterminacy of this specific situation with their own present command of controversial reasoning, practical intelligence, and *facilitas*, rhetorical capacities they have been rehearsing for some time.

Like the first Sophists of the Periclean age, Quintilian (a prominent member of the Second Sophistic) thought of himself as a teacher of virtue, though he conceives of virtue in especially practical terms. He asks "[s]hall we then, who are endeavoring to mold the ideal orator, equip eloquence not with weapons but with drums [*typana*]?" (5.12.21). His answer is that students need most what is most useful. Moments later, he laments that too often declamation is misused because "*utilitas* is not accounted one of the good qualities of eloquence" (5.12.23). Like Cicero, he would correct this situation by promoting in rhetoric an alignment of honor and practicality, *honestum* and *utile*. Also like Cicero, he sees this amalgam of virtue and effectiveness at work in the practice of *in utramque partem*. But perhaps more clearly than the great orator, the great teacher sees that knowledge follows from practice and that practice must be "as true to life as possible" (2.10.4).

There are, naturally, some problems with declamation, especially with declamation as the genre came to dominate Silver Age Roman letters. In particular, declamation was subject to excess. Professors of rhetoric began to invite the public more and more often to open *recitationes*, first to impress the parents of their students and so to attract more clients, later to display their own brilliance before ever-expanding crowds (Duff 1929, 10; D. L. Clark 1957, 215-16). And while it seems unlikely that this bit of history will be repeated soon with professors of rhetoric once again becoming public idols, we should note that the role playing of declamatory practice did give rise to self-conscious displays of rhetorical ingenuity by the professoriate in which performative affect considerably outweighed argumentative value. Quintilian is himself candid about the potential for extravagance that corrupted declamation. He notes that declamation became "so degenerated that the license and ignorance of declaimers may be numbered among the chief causes of the decline of eloquence in Rome" (2.10.3) and that this license threatens to corrupt classroom exercises (*domestica exercitatio*) as well (see 4.1.45, 4.2.29, 5.12.17, 8.3.23, 8.5.22-25, 9.2.42 and 48). Martin Clarke notes that the Roman schools in general and declamation in particular promoted an ornate style unsuited to real oratory and that the demands of imitation displaced the capacity for independent critical thought (1963, 162). Marrou complains that declamatory narratives became much too fantastic, "applying bizarre points of law which included nothing but tyrants, pirates, abductors, rapes, and disinherited sons." And while such cases were meant to sharpen the wit and accustom the mind to difficult cases, they sacrificed verisimilitude and "the actual conditions of the law courts" in the process (1956, 202-5). Marrou also points out, however, that declamation can be defended on historical grounds as an isolated opportunity for the practice of public eloquence during a period of steep decline in political freedom (288).

But Quintilian's defense of declamation seems to me the strongest and most relevant: for it is possible, he insists, "to make sound use of anything that is naturally sound" (2.10.3; cf. 2.16.10), and his method for ensuring the pragmatism of the rhetorical exercises was to make sure that they remained "modeled on the forensic and deliberative oratory," the "mimic battles" for which they were intended as training (2.10.8). Seen from Quintilian's perspective, then, as "a foil wherein to practice for the duels of the forum," declamation becomes not simply the "crowning exercise" that summarizes the entirety of the Roman rhetorical curriculum (5.12.17); more importantly, the *progymnasmata* and declamation taken together represent a rite of passage designed to ease the transition from theory and exercise to responsible practice, from cloistered experiment in the basic features of argument and the process of reasoning *in utramque partem* to the unavoidable demands on advocacy in an environment conditioned by difference, disagreement, and change.

Certainly there is not much doubt about the pedagogical attractiveness of the *progymnasmata* and declamation throughout much of Western educational history. As noted, Marrou refers to the rhetorical curriculum inaugurated by Protagoras and the Sophists and advanced by neo-Sophists like Quintilian as "astonishing" in its effectiveness and its influence (1956, 51). Declamation is the direct ancestor of medieval *disputatio* (Murphy 1990, 67); it was embraced throughout the Renaissance (as was Quintilian in general) and referred to by Erasmus as "the most fruitful" of all rhetorical exercises (D. L. Clark 1957, 260); it remained alive and well in American colleges through most of the 19C (Graff 1987, 41ff). The issue before us now, however, is whether or not these classical ideas are transformable into something of use for us; or, alternatively, whether or not there is space in contemporary theory and pedagogy for such a holdover. Such questions are the burden of the last chapter of *Many Sides*. At this point, however, it may help to look forward to that task by briefly glancing at some central features of controversial pedagogy in the context of Quintilian's own teaching.

4. QUINTILIAN AND PEDAGOGICAL TRANSFORMATION

According to the argument of this chapter, the rhetorical curriculum of the *Institutio Oratoria* is informed at every turn by controversial reasoning, a form of argument distinguished most obviously by the effort to prepare claims *in utramque partem*, on both sides of a question. The immediate goal of this practice is to develop a comprehensive view of the subject that readies the orator for all possibilities in debate. The theoretical motives for *controversia*, however, extend well beyond this initial purpose. For his part, Quintilian does not much comment on such motives, preferring instead to cast *controversia* in instrumental terms as an enabling practice that informs and structures particular rhetorical exercises. Nonetheless, the *Institutio* belongs to a sustained intellectual tradition that provides background and ballast for its detailed efforts to render controversial reasoning in concrete pedagogical form.

As outlined in *Many Sides*, this tradition is initially Protagorean, fundamentally Sophistic, and ultimately Ciceronian in terms of its rhetoric. Its key concept lies in the convergence of knowledge theory and discursive practice, which in Cicero and Quintilian becomes the "equation" of right thinking with speaking well (Struever 1970, 29). At the philosophical level, the Protagorean tradition avoids ideal conceptions of knowledge in favor of the indeterminate field of human perception. In turn, its language theory assumes the contingency of all claims to human knowledge and avoids the allure of universal propositions. It focuses instead on inevitable contradictions in the realm of social knowledge and on discourse as a practical means for transforming disagreement into action. As the operational method for such theories, antilogic/*controversia* gives voice to opposing parties and in so doing lends discursive form to social differences. But, as noted, *controversia*'s ambitions are not confined to listing *dissoi logoi*. For in the process of articulating difference, *controversia* also seeks to discover meaningful content through the dialogue of opposing positions.

Taken yet a step further, *controversia* employs opposing arguments not only as a subject for reflection but also as matter from which to engineer agreement and, hence, build community. Seen thus (as a combination of epistemic and practical motives), *controversia* becomes a uniquely rhetorical system of thought that employs the dialogue of contraries as a creative means for finding ideas that work in the world of human affairs.

Cicero is fully alert to controversial theory and to the tradition from which it develops; and through his own philosophical works, he makes this theory and tradition available to later Latin thinkers. As Nancy Struever puts it, Cicero's works provide "a treasury of the vocabulary and attitudes which make up the serious content of the rhetorical tradition from the Sophistic age on" (1970, 28). With the possible exception of Tacitus, no other Latin figure is more fully attuned to the Ciceronian rhetorical legacy than Quintilian (see Kennedy 1969, 111). So we can assume with some confidence that the "serious content" of the rhetorical tradition is both background and context for the consistent emphasis on "controversiality" throughout the *Institutio*. In Quintilian's text itself, however, these theoretical ideas appear only in mufti; i.e., in civilian attire as the working method which animates "the practicalities of instruction." If we are going to comprehend the full import of this pedagogy, we will have to negotiate for ourselves the relationship between the concrete practices of Quintilian's pedagogy on the one hand, and the theoretical concepts of Sophistic tradition on the other. Moreover, Quintilian's example indicates that the recipe for this particular pedagogy will not include a thick roux of rhetorical theory in a thin broth of classroom practice. Protagoras and Cicero provide necessary theoretical ingredients, but a fully realized pedagogy for *controversia* awaits Quintilian and the teacher's attention to the concrete practices that guide learning. If anything, the model presented in the *Institutio* indicates that—for the student—practice is primary and rhetorical theory is "dumb science" if not embodied in practical experience that is (at one and the same time) realistic and imaginative (5.10.119-20).

In the notes that close this chapter and in the discussion of the next, I will try to keep the congruence of theory and practice in mind. As for Quintilian, I will summarize three elements of his pedagogy that seem to me intriguing subjects for potential adaptation to present use. We can begin with the controversial approach to invention, which we encountered first in connection with Crassus, Antonius, and the purposeful collision of opposing ideas. Like Cicero, Quintilian identifies invention with dialogue and, in particular, with the pro and contra reasoning of argumentative exchange. In the dialectical tradition of argument, of course, invention proceeds differently; most often, students seek to "find" (*heurisis*) ideas by scouring the catalog of Aristotelian topics and/or by giving formal precision to their own best thinking as a proposition to be defended against rivals. On the other hand, invention by *controversia* asks students to complicate rather than confirm initial presumptions by investigating and defending positions other than their own. Quintilian does list the *topoi*, but he objects to "knocking at the door of each with a view to discovering whether they may chance to serve our point"; rather, he encourages "constant practice" as the appropriate avenue to inventive facility (10.5.122-25). And, in his curriculum, one "practices" through dialogue of all sorts: intersubjective dialogue with others who offer competing positions, "internal" dialogue by a single rhetor alternatively "speaking" for and against a particular claim, dramatic dialogue in which each student takes diverse roles in coming to grips with the question under discussion. In particular, Quintilian seems attracted to role playing as a way to give form to the dialogical view of invention; and in the *progymnasmata*, students begin at once to interact with alternative views, not as foils for eventual refutation but as new possibilities for one's own use. Once established in the preliminary exercises, dialogue becomes the standard procedure in declamation, where it is built into the scenarios (Alexander debates with his generals that . . . .) and constitutes the core of the final performance. This emphasis on dialogue, as we will see, follows directly from Protagorean theory and the notion that knowing a subject means knowing what others think about it (see Ch. 2, sec. 2).

There is, of course, a great deal more to say about the theory and procedures of controversial invention, much of which I will reserve for the discussion of dialogue in the next chapter. For now, I would note two implications for our own pedagogy suggested by controversial dialogue. First, Quintilian's example indicates that dialogical contact between students in response to controversy should begin early and be repeated often. One can, of course, survey diverse pubic opinion via research or on-line investigations; but the unpredictable, dynamic nature of direct argumentative exchange is an invaluable stimulus to rhetorical invention (cf. 1.11.9). For teachers of composition, "contact" with interlocutors may well involve an increased oral component in argument training. The second implication of an emphasis on dialogue is that we must be more inventive in simulating the continuous give-and-take that constitutes real-world argument. Argument *in utramque partem* implies repeated reversals: first one side speaks, then another, then the first again and so on, with each iteration modifying the way in which the case is framed. Unlike the statement/rebuttal procedures of forensic debate, the *refutatio* of classical arrangement, or the "reservations" of the Toulmin scheme, preparing for any and all eventualities requires direct exposure to the ongoing flux of dialogical transaction. There is no evidence in the *Institutio* that Quintilian was fully able to simulate the dynamic nature of argumentative dialogue; but ongoing dialogue remains a goal to be pursued, not only because routine argument is naturally conversational and unbounded by formal rules, but also because extending the scope of the dialogue may extend the range of our *inventio*.

A second pedagogical element drawn from Quintilian is *imitatio*, perhaps the principle candidate for contemporary transformation. Introduced in theory in the *De Oratore*, imitation pedagogy achieves a more practical form in Quintilian's *progymnasmata*. For example, in the first exercise involving the translation of fables, instead of simply reproducing the original in a different idiom, students were expected to amplify the tale through interpolated speeches and descriptions in the spirit of the original. In his commentary on imitation theory, Quintilian indicates that students were encouraged to modify the text according to their own judgment:

> I would not have paraphrase restrict itself to the bare interpretation of the original . . . .For if there were only one way in which anything could be satisfactorily expressed, we should be justified in thinking that the path to success had been sealed by our predecessors. But, as a matter of fact, the methods of expression still left us are innumerable, and many roads lead to the same goal. (10.5.4-7)

Accordingly, the student's copy is not so much a transcription (what Rita Copeland calls "the copious likeness of the original") as it is a translation, a rendering of the student's understanding of the source, both an interpretation of the text itself and a re-presentation of the text that draws on the young orator's own resources (1991, 37). The same standards guide all the ensuing exercises of the *progymnasmata*, from amplifying a *chreia* or maxim, to improvising on a moment of history in the *narratio*, or impersonating a character in *prosopopoeia*. All are founded on the pedagogy of *imitatio*; and, as Quintilian makes clear, there is a good deal more to this pedagogy than the simple reproduction of the source.

In the first place, instruction based on *imitatio* subordinates the role of systematic method in rhetorical training. "In everything we teach," writes Quintilian, "examples are more effective (*potentiòra*) than the rules which are taught in the schools" (10.1.15). What Quintilian teaches is oratorical practice; and as such, his subject differs, he argues, from the study of grammar insofar as grammar requires only understanding of and obedience to the rules (2.1.45). Rhetoric, on the other hand, is a practical art, concerned primarily with action arising in the context of varied and changing circumstances. The detailed study of diverse models is intended to reveal the inherent relation of rhetorical practice to the prevailing set of circumstances. When faced with compositions of their own, Quintilian's students are consequently encouraged to consider the variety of models to which they have been exposed before choosing appropriate resources for the present circumstance (10.5.3). Moreover, as students develop, imitation alone becomes inadequate if it does not appropriate the inventive faculty of the models, the *facilitas* required for the effective adaptation of available resources to particular settings (10.1.1-4). But because *facilitas* is everywhere original, it cannot be abstracted in rules. It can, however, be instantiated in the actual functioning of well-chosen exempla, and it can be imitated if invention intervenes to reframe the original according to the demands of a new time and place. The emphasis on examples, therefore, serves to introduce the situated nature of all rhetorical practice, but it does so in the context of the student's own practice, not as extrapolated in abstract principles.

In the second place, *imitatio* also highlights the role of interpretation in the basic economy of rhetorical production. Indeed, the pedagogical reliance on examples assumes not only that they provide what Michael Leff calls "the equipment for future production" but also that the student possesses the hermeneutical resources to appreciate that equipment and to recognize what is potentially portable to another rhetorical space (1997, 201). Quintilian argues that a crucial role in this regard is played by the teacher, who can choose to follow the recitation of model passages with an initial interpretation which assesses its properties in terms of the rhetorical project at hand (2.2.8, 2.6.1-7). In this way, the teacher illustrates *the ability* to translate the model into a usable idiom through the exercise of interpretive skill as well as *the authority* to transform its resources by displacing them in the direction of present use. Quintilian addresses this process of transformation in a remarkable analogy between reading and digestion:

> [J]ust as we do not swallow our food till we have chewed it and reduced it almost to a state of liquefaction to assist the process of digestion, so what we read must not be committed to the memory for subsequent imitation while it is still in a crude state, but must be softened and, if I may use the phrase, reduced to a pulp by frequent re-perusal. (10.1.19)

The primary role of reading in oratorical training is, of course, to supply the orator with *copia rerum ac verborum* (a full supply of words and matter; 10.1.5). But as the doctrine of imitation would have it, the value of the model depends upon the rhetor's ability to transform it into something useable in other circumstances, a transformation made possible through the command of hermeneutic procedures. In sum, imitation pedagogy recognizes a firm connection between the consumption of texts and the production of new arguments, a connection that expands the scope of the imitation exercises, deepens their implications, and enhances their potential for contemporary adaptation (see V. Kahn 1997; Leff 1997).

A third element of pedagogy I draw from Quintilian might well be considered a continuation of imitation pedagogy. In discussing the rhetorical exercises, Quintilian indicates that assignments like the *narratio*, impersonation, and declamation require the student to inhabit unfamiliar roles and to speak in the voice of strangers (11.1.41, 3.18.51). In the language of our own times, such exercises ask us to decenter our own presumptions and examine the territory of the other. Daniel Bender adds that such exercises extend in social, ethical, and political directions as well as psychological ones by acquainting students with unique paradigms of community and commitment (1993, 115). This potential is especially prominent in the declamatory exercises, which amount to simulated cases in circumstantial argument. Case studies in argument have, of course, been popular for some time in professional writing and legal studies, but they run counter to the emphasis of most argumentation texts on what Richard Fulkerson calls "comp-logic" and the role of ethos and audience as argumentative "proofs" (1988). What declamatory exercises provide is a dramatic evocation of the multiplicity, ambiguity, and contingency that characterize argument with its social dimensions intact. Michael Billig points out that the nuance of human affairs can never be reduced to method, so "finite laws [or rhetorical precepts] are likely to be embarrassed . . . by novel particulars" (1987, 62 and 68). As Quintilian recognized, the well-conceived declamatory case is the capstone of rhetorical training because it immerses the rhetor in the complexity of novel particulars and requires the full array of rhetorical skills. Quintilian's advice on how best to employ these "most useful" of exercises is abundant and has much to say to us still.

In Aristophanes' *The Clouds*, students go to the school of Sophistic thinking to learn to bicker with their parents and import corruption into the body politic. Quintilian reverses the moral orientation of advanced rhetorical education, but he continues to place argument at the heart of the curriculum. Only through the prudent management of controversy can the student become what Quintilian terms a truly Roman "wise man;" i.e., one who reveals his virtue "in the actual practice and experience of life" (12.2.7). Throughout the *Institutio*, it is exercise in *controversia*/antilogic/*in utramque partem* that Quintilian calls on most often as the pedagogical means for realizing his practical, if ambitious goals. The student learns to generate two or more positions in conflict and to simulate a productive dialogue among these sides because the problems one is likely to encounter in the "actual practice and experience of life" invariably display *multipex ratio disputandi*. Preparation for rhetorical culture, therefore, is not complete when one's rhetoric is confined to monological claims, nor is judgment adequate when the standard is merely formal and abstract. According to this pedagogy, direct experience with complex, if fanciful cases of *multiplex ratio* is the best means for developing the prudence necessary to respond well to specific situations and the eloquence required to articulate one's response with effect. Classroom exercise *in utramque partem* was, for Quintilian, the principle method of preparing students not only to speak well but also to think and act wisely.

Looking back, the *Institutio* and the Ciceronian rhetoric on which it is based give comprehensive, coherent form to the "serious content" of the Sophistic rhetorical tradition. Looking forward, they are abundant with clues to further developments in that tradition. In the next chapter, I will also do my best to look both ways: to keep the precedent of Protagoras, Cicero, and Quintilian in mind while reframing that precedent for a very different time and place. A very different time and place. Whatever the outcome, I would like my own effort to be measured by Quintilian's standard: does it aid teachers; does it contribute to good classrooms; is it a benefit to students?

Epilogue

An Appropriate Pedagogy for Antilogical Argument

[In learning to speak and judge well, it is] "wonderfully useful . . . to rub and polish our brains by contact with those of others."

Montaigne, "The Education of Children"

In the closing pages of this study, I contemplate the basic outline of a contemporary pedagogy for antilogical argument. Many of the principles of this pedagogy have already been introduced, most explicitly in my discussions of Cicero and Quintilian. But while the educational innovations of Greek and Roman Sophism are obviously the inspiration for what follows, their precedent does not provide a blueprint for contemporary instruction. Put another way, *imitatio* at this distance would require not only the vigilant mediation of historical consciousness but also monotonous disclaimers to "please note the vast differences between . . . ." Instead, it seems best to declare at the outset that the classical conception of antilogical argument has been refashioned in attire suitable to the practical conditions of contemporary pedagogy. The *Institutio Oratoria* is an invaluable model for the teaching of argument; but two millennia of demographic, social, and material transformations, along with a sea-change in conceptions of social justice, simply don't admit blithe adaptation. As a scholar investigating the history of my discipline, I have tried in preceding chapters to shift my attention between past and present, theory and practice; as a teacher, however, my commitments are weighted in favor of the present. Indeed, the arrival of new students every term requires teachers of rhetoric not only to adapt to ever-changing circumstances but also to reconsider the fundamental pragmatism of our discipline: i.e., if knowledge happens in the process of making ideas functional, then our attention as teachers is appropriately focused on individual human achievements, on making things happen for this class of students, for this particular student. It may help to glance in the rear-view mirror; but our steady gaze ought to be trained on the people and prospect before us.

Another caveat is in order. The discussion that follows draws heavily on my own experience in the classroom. Perhaps this is unavoidable; for regardless of my reading about this pedagogy or that learning style, or memories of other teachers, or discussions with colleagues and students about their classes, or visits I've made as an observer, my own experience in my own classrooms will always be more vivid, comprehensive, and accessible. Moreover, as a rhetorician I would contradict my own basic principles if I were not alert to the personal and contextual nuance of particular situations. So, despite any claims I might make about the adaptability of my own pedagogy, I simply cannot presume that my own two decades teaching rhetoric, composition, and argument in a large, Midwestern university are comparable to the teaching experience of others. Differences in teaching circumstances may not be as dramatic in synchronic comparisons as they are in diachronic ones, but they are nonetheless significant. In consequence, making pedagogical recommendations will always be "tricky" because the transformations required to move from one classroom to another will invariably resemble a game of telephone: we may attempt to repeat a story we've heard, but differences in delivery, setting, and audience will likely produce a quite different narrative (cf. Lynch et al., 411).

These historical and situational considerations lead to certain decisions regarding the scope and purpose of this Epilogue. My goal here is to complement the preceding pages by developing the most promising features of the antilogical tradition in ways useful for contemporary teachers. Instead of a prescribed pedagogical plan, however, I offer an outline of the major theoretical and practical considerations that teachers may want to keep in mind when thinking through antilogical argument for themselves. In brief, this Epilogue presents, first, a summary of Protagorean theory for teachers; then, an outline of antilogical pedagogy organized around its primary features; and, finally, some notes on an antilogical approach to course design and classroom activity. More specifically, I begin with a review of the basic principles of neo-Protagorean philosophy because these ideas not only motivate a specific approach to argumentation, they also serve to orient classroom practice in relation to theories of knowledge, discourse, and learning. It is easy enough, of course, to jerryrig one's particular pedagogy without giving much thought to philosophical matters, especially for teachers of composition pressed for time by class prep, paper grading, student conferences, committee meetings, and the arrival of another new journal issue with articles one really ought to read. In my own experience, however, time spent on what one might call the teacher's "first philosophy" (i.e., questions of epistemology and ethics) not only contributes to a coherent orientation for decisions regarding "the practicalities of instruction," it also rekindles one's engagement in teaching by recalling us to our own excitement in the process of learning. With regard to the basic argument of *Many Sides*, I have already asked readers to contemplate Protagorean theories of knowledge and ethics as a prelude to analyzing antilogical practice; I would now ask teachers to keep in mind the rudiments of this same philosophy, not only because its general principles inform particular approaches to instruction but also, and more importantly, because Protagorean epistemology motivates fundamental adjustments in our perspective on what counts as student learning, a perspective I will try to clarify in the course of this Epilogue.

The philosophical prelude is followed by what amounts to a reconception of antilogical practice under contemporary conditions and for pedagogical purposes. The distinguishing feature of this account is its dual emphasis on the dialogical dynamics and the pragmatic intentions of contemporary antilogic. The dialogical nature and function of antilogic has been both a leitmotif of *Many Sides* and the primary constituent of an evolving tradition that is articulated in each new age with a somewhat different accent. I would argue that in our own time this accent is on fully bilateral relations, on the transformation of partiality into partnership, and on the opportunity afforded by egalitarian relations for arriving at good reasons. In this second section, I review some modern approaches to dialogue and their connection to argumentation; then, in keeping with the present emphasis on pedagogy, I reconceive these ideas in direct relation to rhetorical invention. In so doing, I hope to show that the conjunction of dialogue with invention opens up new options for instruction in argument. Similarly, antilogical pragmatism has already been the subject of considerable discussion in *Many Sides* (see Ch. 3, sec. 1); in these final pages, however, I revisit this topic by examining its relation to rhetorical judgment. It was customary from Rome to the Renaissance to divide the study of rhetoric into invention and judgment, the discovery and development of arguments followed by assessment and the decision-making process (Cicero *Topica* 2.6; McKeon 1966). I have borrowed this framework and am using "judgment," the less familiar term, to signify the procedures by which we test claims, weigh options, and commit to the most useful course of action. In sum, I will argue that dialogue promotes invention, that pragmatism directs judgment, and that an antilogical pedagogy would revolve around these two not-entirely-separate phases of the argumentative process.

To a significant degree, this constellation of major concepts—dialogue/invention, pragmatism/judgment—is a digest of the basic themes developed throughout *Many Sides*; at the same time, this same set of relations operates as a springboard for some closing speculations on how teachers can adapt all this theory to the classroom. These "practical speculations" are cursory by design, a gesture in the direction of a workable pedagogy that teachers can consider, adjust, or transform according to their unique needs and circumstances. I have organized these classroom practices under the headings of critical interpretation, rhetorical reinvention, and case simulation, a trio which can be taken as a sequence or rearranged, again depending on the teacher's goals.

However these gestures, hints, and speculations are used, it does seem fitting to me to end *Many Sides* with considerations of practical pedagogy. You will recall that Protagoras saw himself principally as a teacher (*Protagoras* 317b). Antilogical argument was, of course, the distinguishing feature of his teaching method, and Marrou asserts that this method was "astonishing in its practical effectiveness" (1956, 51). In these closing pages, I intend to take stock of antilogic by considering how best it might be adapted to the contemporary needs of teachers and students. In the end, the continued relevance of this ancient and obscure form of argumentation will be confirmed (or not) in the classroom, where outcomes are seldom astonishing, but where effective methods, new or old, should always be welcome.

1. A PHILOSOPHICAL PRELUDE

What, then, does the teacher need to know about neo-Protagorean epistemology? In a nutshell, there are four related principles or tenets to keep in mind:

1. human knowledge is relative to knowing subjects and their place in culture and history;
2. because individual knowing subjects are all "placed" somewhat differently, questions of knowledge arise within a field of competing options;
3. moreover, because all knowledge claims issue from distinct subject positions, no single claim can assume authority as universal and absolute;

4. nonetheless, competing claims can be negotiated and collective agreements achieved by appeal to mutual patterns of understanding and practice and practical standards arrived at through argument.

The first of these "principles" is a modern variant of the Protagorean dictum that "humanity is the measure of all things." At once famous and neglected, this gnomic maxim can be cast as both a theoretical hypothesis regarding the production of human knowledge and an empirical observation on the way knowledge operates in actual practice. The nucleus of the doctrine posits human cognition as the "precondition" of any claim we make about the world; or, to reverse the terms, what we claim as knowledge follows from and is structured by the cognitive practice of knowing subjects. Accordingly, our knowledge of the world is not an immediate reflection of things as they are in themselves; rather, our diverse "ways of knowing" invariably leave their impression on what we take to be real and true. Consequently, when we claim *knowledge of* this experience or that phenomenon, such a claim is also and unavoidably a matter of *knowledge from* a particular human perspective. Seen in these simple, unqualified terms, the human-measure doctrine may seem almost too obvious to many modern readers or, alternatively, too relativistic for those with a stake in objective certainty. Nonetheless, the human-measure persists, not simply because it is the *fons et origio*, the base and origin of Western humanism but, more importantly, because it articulates so succinctly a basic theory of what happens when human beings "understand" their world.

As a theory of knowledge, the initial Protagorean proposition is relatively straight forward: the human-measure doctrine argues that what we know is contingent on the cognitive operations of the knowing subject. However, the basic notion of the human-measure doctrine quickly becomes more subtle when modified by the historical/social caveat that fills out the text of tenet #1. For if the world appears to us as structured by human agency, and if individual agents are all placed somewhat differently along the axis of culture and within the flux of history, then each agent is likely to know the world in somewhat different ways. As a result, what counts as knowledge will in most cases find expression as a plurality of positions, a range of approximations that reflect the deep contingency of all human efforts to "measure" or interpret the world.

To expand slightly, each individual is immersed in a network of specific relations (familial, social, political, economic, linguistic, etc.) which—despite certain commonalities across time and within communities—are unique to that person. Or, more exactly, we come into self-consciousness amidst a mass of conventions, distinctions, and interpretive strategies available at a particular time and place, and we draw on these "availabilities" in our own attempts to understand the world. The knowing subject is really, then, a social subject whose cognitive practice emerges from the collective resources of time and place. Even among contemporaries and colleagues, however, we all exist in somewhat different relation to the constellation of possibilities that characterize a particular historical moment. All of which means that we are likely to develop patterns of hermeneutical practice, or ways of knowing that are distinct in varying degrees from those of others around us.

In accordance, therefore, with tenet #2 above, what we refer to as human knowledge will manifest itself as an expansive spectrum of historical, cultural, and individual differences. Put another way, what counts as knowledge in this time or that place, for this or that individual, will be calculated according to varied canons of judgment. The heavens are viewed differently by professors at MIT and the priests of Chichen Itza, questions of social justice take different forms for Georgetown attorneys and scholars of the Koran; indeed, as we all know, scientists, scholars, and attorneys with similar backgrounds can have very dissimilar perspectives on the same subject of study. In sum, the human-measure doctrine assumes that what we know is in large part a product of human interpretive practice and that, at a deeper level, such practice is an ontological feature of human experience, something we do *by nature*. And yet, according to the caveat (i.e., according to Protagorean doctrine in contemporary form), that nature is always and already historicized, always located within a network of particular relations. So, while human-measure in its original form posits only that what we know is relative to how we measure, a growing sense of historical and cultural difference asks us now to consider how we measure as relative to when and where we are.

Of course, such relativism stands in direct opposition to the master tradition of Western philosophy. To cite a single example: the Cartesian *cogito* posits the cognitive self as potentially free from attitude and belief, outside of history and culture, and devoted solely to reason. Thus purged of subjective bias, the rational self operates by invoking formally consistent methods that stand in fixed relation to "objective" reality and reveal that reality as a timeless, universal structure. In turn, this transparent, invariant structure serves as an unimpeachable foundation for knowledge claims and, by inference, a warrant for justified action. By contrast, the human-measure doctrine emphasizes the role of human agency in the epistemological process rather than the transcendent authority of a knowable reality. More specifically, neo-Protagorean theory assumes that individual interpretive strategies condition the form and content of what we know, that such strategies vary according to the "location" of the knowing subject, and that different constructions of the known world will arise accordingly. Under such conditions, the Cartesian claim to certain knowledge is simply unavailable because there is no foundational structure to operate as an absolute guarantee. Instead, what counts as human knowledge for the neo-Protagorean relativist is a matter of human conjecture—which means, of course, that it is also subject to argument.

The concept of a foundational structure or overarching framework of universal truths is at issue in tenet #3. Protagoras himself addresses this concept in another surviving fragment, the *peri theon*, in which he declares that "concerning the gods, I am unable to know whether they exist or whether they do not exist, or what they are like in form" (see Ch. 1, sec. 1). The gods may be non-existent, or they may simply be unavailable; the fragment does not directly challenge the presence of a fixed and transcendent reality. Rather, it argues that ultimate authority is unknowable as such by humanity. And, without recourse to an unconditional foundation on which to ground our claims about the world, what we know must remain sublunary, below the level of the absolute, conditioned by contingency and multiplicity rather than permanence and universality. From the perspective of its critics, the absence of foundational principles forces the relativist to approach knowledge in ways that are at best subjective, at worst nihilistic, incoherent, and immoral. Indeed, how are we to winnow truth from falsehood if ultimate principles are unavailable and if we accept as valid different epistemic standards from different times, places, and people? Yet, from the Protagorean perspective, we simply have no purchase on indubitable standards and must learn to grapple with the "irreducible pluralism" of human interpretive positions. Or to paraphrase Jane Austen, "we have no talent for certainty." Nonetheless, the absence of absolutes to which our knowledge claims can be anchored does not leave us entirely without epistemic resources. Indeed, tenet #4 expresses confidence in the availability of useful knowledge, a confidence based, in particular, on our capacity for argument.

The reasoning here is detailed (see Chs. 1–3); but in sum, we know from the first two tenets that human knowledge is relative to the place and person of the knowing subject and that in the absence of a consolidating authority knowledge claims are likely to proliferate in an array of interpretive options. However, even if knowledge at root is "knowledge for A," there is no corresponding reason to assume that all knowledge claims are necessarily equal; nor, in the absence of a neutral standard, is there cause for despair over the subjectivity of judgment. Protagoras addresses these important issues directly: though each of us is the measure of what is or is not the case, and though one argument or logos may not be "truer" than another by any absolute standard, one logos may still be "stronger" in the sense of more advantageous or useful (*Theaetetus* 166c-68c). Accordingly, the Sophist or wise person is someone who is able to distinguish among opposing arguments and discern that which will prove most beneficial under prevailing circumstances. So, while it originates in an irreducible pluralism, Protagorean reasoning is oriented toward practical benefits and requires the ability to juxtapose alternatives in preparation for making choices. In practice, the search for useful knowledge relies upon the operations of argument. Put briefly, antilogical exchange immerses us in the dialogue of competing claims (anti-logoi) and, in the process, works to extend our conceptual horizons. In turn, the dialogue of antilogoi creates opportunities for new arguments at the same time it provides a comparative context for reasoned judgment. The knowledge that issues from this intersubjective exchange will be neither certain nor final because other arguments will inevitably arise to challenge the status of an ascendant logos. Nonetheless, understandings achieved through the consideration of all options and the authority generated through the process of dialogical exchange provide a reasonable standard for human judgment, a practical framework for collective agreement, and ample authority for justified action, even amidst the flux of opinion.

To distill this reasoning yet further: knowledge is something we accomplish by submitting the many sides in controversy (*multiplex ratio disputandi*) to the challenge of well-conducted argument or—more exactly—to the collective search for the strongest possible logos within and through the dialogue of opposing parties. Seen thus, argument is an affirmative act that transforms the clash of opinion into justifiable commitments by forging relationships amidst difference. As such, Protagorean theory achieves its pragmatic fulfillment in antilogical practice, a practice that imagines hermeneutics and rhetoric, measuring and arguing as inseparable features of a single process.

Naturally, there are qualifications that attend the Protagorean epistemic. In the first place, I should clarify the extent to which what we know is contingent upon human cognitive practice. To its critics, relativism traps us in an inescapable solipsism, divorced from the world, confined to a cell of private opinions. And yet, there is nothing in neo-Protagorean theory to suggest that the materiality of the world is unavailable for analysis and understanding. Indeed, as modern relativists have pointed out, the world itself not only provides raw data for observation, it also limits the range of acceptable interpretations. "I cannot," writes Lorraine Code, "*choose* to experience sunshine when it is cold and snowing" (1982,168). Moreover, given the pragmatic nature of Protagorean knowledge, any view which does not conform to the raw data of the world adequately enough to be useful will eventually be rejected as unpersuasive. Consequently, while the known world remains contingent on the conceptual practices of the knowing subject, the extent of the subject's influence is bounded by the presence of knowable data and the collective agreements we make about this data.

A second qualification follows from the first: i.e., relativism is not equally at play in all sectors of knowledge. In the arena of "technical" knowledge, strict inquiry procedures seek to circumscribe the constitutive power of the individual subject and comprehend the state of affairs which obtains outside subjective influence. Of course, once translated into any symbol system, technical data are reconceived in a framework of human construction. Nonetheless, technical inquiry works to minimize the extent to which subjective intervention can alter the object of study. The case is obviously different with "social" knowledge, where data are ambiguous and alterable according to the practice and perspective of the subject. It is only fair to acknowledge, then, that relativism will be more fully in force in the latter category than in the former though, as noted, human interpretive influence is never reduced to zero.

Even with qualifiers, however, the basic proposition of Protagorean thought remains intact: humanity is the measure of all things because knowledge about the world proceeds from the cognitive operations of active human agents. In the ancient Greek world, the human-measure doctrine marked a shift from an ontological concern with the cosmos as the origin of abstract, foundational principles to a hermeneutical interest in human practice as the source of both diverse interpretations and defensible belief. This humanism has found diverse expression over the ages; but, as Joseph Margolis puts it, "there has never been such a Protagorean age as our own" (1991, 119). For example, human-measure now finds expression in the growing recognition that historical placement conditions perception and that cultural positions confer tacit prejudice, in the rejection of foundational principles that claim invariant status and the bivalent truth values that attend them, in the corresponding endorsement of multiple ways of knowing and the acknowledgement that reason itself can take many forms, and in the admission that—in the context of many, local, conflicting beliefs—the truth is a matter of discourse and a topic for argument. From the perspective of the dominant Western philosophical tradition, these Protagorean developments mark an epistemic tragedy, a falling away from the certainty conferred by absolute, universal, and invariant ideas. But for the neo-Protagorean, an historicized relativism enables the proliferation of *petits recits*, of many sides and *multiplex ratio*, all of which have something to contribute to comprehensive understanding. Of course, we must still seek convergence and make judgments, but we do so by invoking the flexibility of contingent standards rather than by submitting to the prescribed methods of formal reasoning. Consequently, in a Protagorean world, the way we measure is no small matter.

2. THE DIALOGUE OF ANTILOGIC AND THE PEDAGOGY OF INVENTION

The philosophical framework outlined above follows from the human-measure doctrine and its proposition that knowledge claims are relative to knowing subjects. The additional tenets fill out this basic theory by arguing, among other things, that distinctions in subject placement result in incongruent subject horizons, so that what counts as knowledge on issues of substance is routinely a matter of controversy. Put another way, the search for knowledge is inseparable from the practice of argument since the adjudication of differences depends upon procedures for inter-subjective exchange. The Protagorean theory of knowledge therefore assumes a rhetorical component—antilogical argument—as the pragmatic extension of its philosophical tenets. For its part, antilogical practice is based on the Protagorean assumptions that knowledge is multivalent, that any single position in controversy can be complemented by a well-reasoned alternative (an antilogos), and that the interaction among opposing positions in argument will yield new questions rather than determinate results (see Ch. 4, sec. 2). Given the limitations of any single logos, the strategy of antilogic is to seek understanding through productive exchanges among opposing positions, exchanges in which all parties stand on equal footing. In brief, the defining feature of contemporary antilogic is its emphasis on dialogue, on bilateral relations between partners in controversy, each of whom is asked not just to take turns speaking and listening but to participate in the comparative examination of opposing claims and the joint pursuit of the best available logos.

This dialogical emphasis will naturally influence antilogical pedagogy and, in particular, the antilogical approach to the invention of useful ideas. But before I turn to this pedagogy, some clarification is in order because in modern parlance "dialogue" has become a portmanteau term, an open lexical space waiting to be filled according to the inclinations of its users.[1] For our purposes, I will argue that the main lines of modern dialogical theory can be traced to the work of Martin Buber, Mikhail Bakhtin, and Hans-Georg Gadamer. The legacy of these important thinkers has been especially influential in modern rhetorical studies, with a host of contemporary scholars reviewing, revising, and extending original insights to the point that dialogical rhetoric can be justifiably considered a coherent, if not exactly uniform body of study. My own list of major contributors to a theory of dialogical communication includes Douglas Ehninger, Maurice Natanson, Richard Johannesen, Wayne Brockreide, Joseph Wenzel, Wayne Booth, John Stewart, Thomas Sloane, Anthony Blair and Ralph Johnson, Kenneth Bruffee, Michael Billig, Frans van Eemeren and colleagues, Don Bialostosky, Gregory Clark, Sally Jackson, Thomas Kent, William Covino, Tim Crusius, Victor Vitanza, Mariolina Salvatore, and James Crosswhite. At this point in *Many Sides*, I can only outline what I take to be the basic features of this emergent rhetorical tradition; but it is nonetheless possible to glimpse something of its expansive nature and revolutionary promise by noting lines of inquiry with particular relevance for argumentation theory, beginning with the Gadamerian conception of knowledge itself as a dialogical process.[2]

According to the tenets outlined above, the unique cognitive patterns of each knowing subject are shaped by a network of relations that make up our social, historical, and linguistic heritage. Gadamer refers to this heritage as "tradition" and writes that "to be situated in a tradition does not limit freedom of knowledge but makes it possible" (1994, 361). And, because we are all situated somewhat differently in relation to our tradition(s), we interpret the world as independent agents conditioned by "historicity" but distinguished nonetheless by a unique set of cognitive patterns and "prejudices" (360). Given the prejudicial nature of human consciousness, it follows that no single position will be either "objective" or "definite" and that the "infinite web of motivations" that characterizes knowledge in its multivalent complexity will customarily transcend the "finitude" of the individual subject (372-73). How, then, does dialogue allow us to address our partiality and extend our horizons? Gadamer answers this question by appeal to the *docta ignorantia,* the ancient Socratic principle that knowledge itself begins with the admission that there are things about any topic that I do not know (362-63). Once I acknowledge the bounded nature of my conceptual field, I can begin to address my limitations through the consideration of opposing positions. Indeed, Gadamer is emphatic in declaring that "Knowledge means, precisely, considering opposites" (365). I would gloss this deeply Protagorean concept by noting that the juxtaposition of opposites transcends the boundaries of any single claim and, as such, locates knowledge in the shared space "between" subjects with opposing views. Put somewhat differently, knowledge is an experience we participate in with others rather than a commodity we acquire on our own (see R. Palmer 1969, 165).

In actual practice, this dialogical experience can operate in a number of ways that foster productive argumentation: e.g., the interrogation of any single logos by contrary positions may help to identify inadequacies in the nature of the claim in question; alternatively, dialogical encounters may reveal unfamiliar perspectives hidden by our "prejudices" but available to others whose "placement" gives them a different perspective. Even more interesting is the potential of dialogue to develop in ways that transcend the existing horizons of those involved and extend the scope of knowledge through the gestalt of opposing positions. In this latter case, dialogical experience is a crucible for the transformation of discrete logoi into something unprecedented, a medium out of which new meanings emerge. In all cases, the focus is on the relationship between/among participants, with knowledge itself imagined as an interpersonal phenomenon, something that happens when knowing subjects with alternative views are in dialogue with one another (cf. Stewart 1978, 196).

For dialogue to function as an epistemic process, however, its role in argument must involve something other than pressing one's own position or correcting the mistakes of others. Indeed, antilogical dialogue requires a decisive shift away from common conceptions of argumentative exchange; i.e., away from the discursive product and the formal nature of the argument itself, away from the methodology of thrust-and-parry that has dominated argument instruction, away from such objectives as conceptual certainty and audience assent. Instead, dialogical communication will focus first and foremost on *the relationship between and among the subjects involved* and the protocols that support a multilateral distribution of emphasis, responsibility, and respect. It can be difficult to generalize about such relationships since the local, situational, dynamic features that predominate in actual dialogue resist systematization. But if we acknowledge that dialogue in practice is always beyond method, we can (with caution) theorize certain features of the dialogical relationship that provide guidelines for instruction. Most obviously, such relations are egalitarian in their emphasis on the two-sidedness of dialogue, on shuttling back and forth so that all sides receive a hearing and no claim escapes challenge. This procedural parity, however, is based on more fundamental assumptions about the relationship between interlocutors in dialogue, assumptions that emphasize what Protagoras refers to as *aidos* or respect for one's partners (Ch. 3, sec. 1).

In contemporary parlance, Protagorean *aidos* will manifest itself in the recognition of one's interlocutors as free and independent subjects whose world is distinct from but no less real than one's own. That is, the dialogical rhetor is committed to comprehending not only the topic in general but also—and in the first place—the claims of one's opponent and the context or worldview out of which these claims arise. For Buber, this commitment to others is crucial to human development: we enter into inter-subjective contact because the independent "I" we all seek to cultivate arises only in the conduct of our relations with others. Or, as he puts it: "the inmost growth of the Self is not accomplished, as people like to suppose today, in man's relation to himself, but in the relation between the one and the other . . . " (1947, 21). And, at another point, he adds that the "individual is a fact of existence insofar as he steps into a living relation with other individuals" (1947, 202).

To accomplish these "living relations," participants in dialogue must extend themselves beyond a cursory acknowledgement of one another's presence and seek instead to comprehend their interlocutors as fundamentally different from themselves. Buber writes that the other "does not have merely a different mind, or way of thinking or feeling, or a different conviction or attitude, but has also a different perception of the world and a different recognition and order of meaning . . ." (1947, 61-62). Dialogical argument responds to these differences by encouraging the unfettered expression of the "other's" views. Moreover (and perhaps more importantly), in the process of attending directly to my interlocutor, to the distinguishing features of her ideas and attitudes, I confirm this same person as a free subject, a "Thou" rather than an "it," a person with her own convictions and claims. In turn, by confirming the independence and authenticity of my dialogical partner, I make it possible for this interlocutor to reciprocate in kind; i.e., as a free subject herself (rather than the object of my rhetorical intentions), my dialogical partner is able to recognize and confirm in return those "differences" that distinguish my own unique personhood. From the dialogical perspective, then, the very notion of personal identity takes on interpersonal dimensions: I emerge as myself in and through my relations with others and, in particular, through the process of attending to someone else.

For Buber, as for Gadamer, the interpersonal dynamics of dialogue have epistemic implications, for it is only in the open exchange of views between free subjects that human beings experience their unique capacity as creators of meaning. That is, "[w]e do not find meaning lying in things nor do we put it into things"; similarly, we do not find it strictly on our own or solely in the possession of someone else; rather, meanings "unfold" in the inter-human space we can only inhabit together, a space greater than the "mere sum of two individual entities," a space in which we achieve an unprecedented wholeness of experience and capacity for insight (1947, 203-205; cf. Friedman on Buber, 78-79). Seen thus, dialogue is not simply a means to something else; it is, for Buber and Gadamer, the constituent feature of human consciousness—the way in which we know both ourselves and our world.

Lest the foregoing perspective seem too abstract to support specific protocols for dialogical practice, let me add some clarifications. In the first place, while the emphasis of dialogical argument falls predominantely on the participants involved (rather than on the discursive product) and while the attention of the dialogical rhetor is oriented initially toward the ideas of one's partners (rather than the demonstration of one's own claim), it does not necessarily follow that to express respect for others requires us to agree with their arguments (cf. J. Poulakos 1974, 206). Our first responsibility in dialogical argument is to do what we can to comprehend the views of others; and because our partners may understand the topic in considerably different terms than we do, we are obliged to pursue those differences as they exist in their own right. If, on inquiry into the views of my partner, I find I disagree with this or that position, I am free to resist the claims of the argument without rejecting the relationship with the person.

Second, rhetors in dialogue will typically fulfill their obligation to understand their interlocutors by asking questions. That is, if each interlocutor "measures" the world somewhat differently, then dialogue is inherently a learning experience in which rhetors seek to discover from others what they themselves do not yet know. To comprehend the alternatives offered by others, to inquire into their defining features and test their viability, the rhetor must seek to know not only what the interlocutors claim but also what questions in particular an interlocutor is responding to. As Bakhtin makes clear, all claims take the form of rejoinders to previous questions; so that to understand the nature of a claim correctly, we must understand the question to which it responds (1986, 75-77).[3] Consequently, when I question an interlocutor in dialogical discussion, my questions are neither "rhetorical" (posed with the answer in mind) nor "elenchic" (designed to discredit the respondent); rather, I ask questions of my partner with the sincere purpose of coming to terms with a "different order of meaning."

Third, when I ask questions in authentic dialogue, I declare my openness to new ideas; i.e., I announce that, in my opinion, the topic at issue is undecided, that our conversation is dynamic and may lead in unexpected directions, and that, at present, inquiry takes precedence over closure, discovery over judgment (cf. Gadamer 1994, 367). Indeed, in a dialogue based on mutual respect, all participants are at liberty to pursue any option—to accept or reject, to affirm or deny, to rephrase or revise—because the goal is understanding rather than victory, which opens up the range of response.

Fourth and finally, it follows that if I open myself to any and all developments in dialogue, then my own position and the a priori assumptions that support it are themselves open to challenge. For Maurice Natanson, the rhetor's commitment to openness in dialogue amounts to the assumption of personal risk (1966, 15-16). On the one hand, when I offer claims for dialogical consideration, I invite my interlocutor to interrogate these ideas from her own distinct perspective. In brief, to make a claim is to ask a question, the answer to which we can not control (cf. Crosswhite, 51-83). On the other hand, when I commit myself to understanding the views of others, I acknowledge their potential to displace my own present thinking. To enter into a dialogue of mutual respect is, therefore, to open one's self to the possibility not only that I could be wrong but also that I may have to change in response to a stronger case. Given the dynamic nature of dialogical interactions, however, the personal risk I undergo in argument is offset by this opportunity for discovery: we engage in conversation, we inquire into alternatives, we commit ourselves to others with the understanding that along with the possibility of change comes the potential to pool resources and jointly extend our horizons. So, while dialogue demands that we risk change, it compensates with the chance that the risk could be worth it.[4]

In brief, dialogue operates as a "covering principle" for antilogic, an organizing concept comprehensively distributed to all facets of neo-Protagorean practice. By definition, this covering principle is everywhere in force, but nowhere is it more decidedly manifest than in the antilogical approach to invention, the art of discovering new ideas through dialogical interactions with others. Of course, any approach to invention organized around the local, personal and, hence, unpredictable rhythms of human dialogue is likely to be a little messy. Certainly, in comparison with the orderly framework of the Aristotelian topics or the systematic progression of the stasis questions, dialogical inquiry will seem less stable, more indeterminate, but also less prescriptive, all of which follows from dialogue's allegiance to the shifting, uncertain nature of human relations rather than the formal purity of well-tempered propositions. On the other hand, dialogical invention does not preclude the employment of either the *topoi* (most notably, antithesis, analogy, contrast, and parallelism) or the stasis questions (in either ancient or modern form) as useful tools for defining the issues and identifying data relevant to them. In sum, for all its immersion in topical exigency, antilogical invention remains a calculable *praxis*, distinguished by an identifiable set of standards which inform its practice and so can be taught.

For these standards, I return to my earlier outline of antilogical form and the roles of multivocality, oppositionality, and dynamism as the constituent features or "virtues" of Protagorean rhetoric (see Ch. 4, sec. 2). This time, however, I review each of the three in the context of practical pedagogical concerns: specifically, how can rhetors appropriate these "virtues" as aids to their own invention and how might students understand their contribution to effective practice? Of course, the subject of dialogical practice, even when restricted to invention, could fill volumes, probably with paradigmatic examples, as in a case book, because the whole process is so closely tied to particular situations. We can, however, limit this expansiveness to teachable proportions by concentrating on the disposition or attitude of the rhetor, what Cicero calls "habits of mind," rather than on generic techniques. Such a focus reflects the emphasis on persons dominant in both the human-measure doctrine and in dialogical theory; it also reverses the pervasive emphasis of argumentation textbooks on productive instruments (e.g., the enthymeme or the GASCAP *topoi*) to the neglect of matters of ethos. In particular, then, the dialogical rhetor in the act of finding ideas for argument will (a) seek out multiplicity or *copia* by exercising a fundamental interest in and respect for the views of others, (b) promote productive opposition by invoking a critical scepticism in response to all initial claims, and (c) foster the dynamic nature of dialogical inquiry by reserving judgment until the inventory of ideas in response to controversy is full.

To begin with multivocality, antilogical invention is grounded in the belief that the most dependable meanings are negotiated within the widest possible field of options. Put another way, antilogic aims to fulfill the Aristotelian desideratum by discovering all available hypotheses, all those ideas that might contribute to the case at hand (*Rhetoric* 1355b 27-28). This "Protean variety" comes initially from a careful inventory of existing opinion, with additional logoi that follow from protocols for juxtaposition and critique. The first order of business for the antilogical rhetor, however, is to uncover the plenitude of competing logoi, the *multiplex ratio disputandi* that inevitably surround and proliferate in controversy.[5] The rationale for this inclusiveness comes directly from the Protagorean notion that each individual stands in a new horizon and so can contribute something original to the heuristic process. Any effort to comprehend controversy as it exists within an actual community of interested parties will, therefore, cast its net widely, not only soliciting prominent logoi but also promoting seemingly "weaker" positions whose claims and critiques may hold useful insights not yet generally known. So, the initial goal of dialogical invention is to survey the landscape of opinion in a particular controversy.

The appropriate rhetorical stance for promoting this multivocality is the *attitude of respect* for one's dialogical partners outlined in the last section. Since its beginnings, antilogical practice has sought to draw out and give shape to controversy by making sure that both/all sides are fully articulated (Cicero, *Academica* 2.7-8). As a result, dialogical rhetors are especially responsible for understanding positions other than their own. In practice, this means that transactions among dialogical partners are the order of the day and, in effect, invention is a collaborative process; i.e., the generation of acceptable logoi proceeds in the context of relationship and inter-subjective exchange. Admittedly, coming to grips with the claims of others is an interpretive act in which rhetors reconceive an original idea according to their own criteria. Nonetheless, authentic dialogue, with its respect for difference, provides a supportive medium for the collision of contraries, for productive contrasts that hold the promise of contributing in more or less meaningful ways to the expanding field of comprehension. Under these conditions, the rhetor's ability to access the bounty of other positions originates in a commitment to the "personhood" of the interlocutor, to conferring authority on others by acknowledging their capacity to see the subject in meaningful ways. This essential respect can be communicated through deliberate strategies for listening to and asking questions of one's partners; and in a comprehensive pedagogy, the arts of listening and questioning will demand considerable attention. But the crux of the matter is this: if the rhetor's inquiry process is not motivated by respect for the authenticity of other positions, it is unlikely that ideas outside the orbit of one's own views will take on the status of viable contributions to the accumulating *copia* of argumentative data. On the one hand, the attitude of *aidos* prompts a willingness in one's dialogical partners to articulate what they know about the topic without undue inhibition; on the other hand, the same attitude can liberalize the thinking of student rhetors by revealing new possibilities unavailable from their own cognitive context. The desired result is that "widest possible field of options" which Aristotle, Cicero, and Quintilian all see as the reasonable, if unreachable point towards which any systematic approach to invention will aspire.

The attitude of *aidos* is, consequently, inclusive: it seeks out dialogical interactions with any and all potential contributors to the multiplicity that is the initial objective of antilogical invention. The resulting proliferation of potentially interesting and useful ideas is further advanced by the alliance of respect with oppositionality, the readiness to contemplate each idea or logos in direct relation to the strongest available opposing position. This defining feature of antilogic finds its basic motivation in the Protagorean recognition that, while all claims are partial, the comparison of each with its appropriate counterpoint is an epistemologically productive way to think through the relative merits of competing probabilities. Put another way, the juxtaposition of a claim with its opposite places the subject of study in relief and, as a result of the contrast, provides a more vivid outline of its proportions. Moreover, from the dialogical perspective, the counterpoint of opposing positions in argument is crucial because the lack of someone else's response makes the dialogical relationship incomplete and thus inert as a resource for invention.

To operate as an effective aid to invention, however, opposition also requires its appropriate rhetorical attitude. For example, new ideas will hardly flourish if rhetors simply attack the claims of their interlocutors. Such an attitude is the rhetorical ethos of eristic, which seeks victory in argument rather than discovery. Instead, what is required to put opposition at the service of invention is a *productive scepticism* which, when combined with *aidos*, works to rescue contention from contentiousness and allow the collision of opposites to assume its inherently creative potential. More specifically, the attitude or ethos considered here does not direct its opposition against persons who, by virtue of their willingness to risk relationship, necessarily merit respect. As a contribution to this relationship, however, the dialogical rhetor will approach all claims with a robust scepticism that assumes not only that the claim itself is partial and in need of its antilogical complement but also that—out of respect for both the person and the process of argument—it is an act of goodwill to raise objections. You will recall Burke's motto for this unusual combination of respect and contention: *ad bellum purificandum*, towards the purification of war. The practicability of this paradoxical effort is illustrated by the "yes, but" strategies of Cicero's orators in the *De Oratore* who combine sincere expressions of goodwill, admiration, and even partial agreement with a strict commitment to making the opposing case as forcefully as possible (Ch. 5, sec. 2). In Cicero and for antilogic generally, to oppose with respect is to prompt the discussion a step closer to the strongest possible logos. Consequently, under the aegis of *aidos*, contention and rejection are not to be confused.

If dialogical scepticism is not purely negative, then the dialogical rhetor is in the position of the loyal opposition, not only raising objections which probe the strengths and weaknesses of the logos at issue but also articulating the positive features of the opposing side.[6] In other words, dialogical scepticism will approach an argument *in utramque partem*: i.e., by considering both/all sides as the most productive approach for uncovering the nuance of the case at hand. For purposes of invention, generating meaningful opposition not only adds an opposing logos to the inventory of available options but, more importantly, the juxtaposition of opposing arguments (anti-logoi) also prompts revision/refinement in existing claims and provides the catalyst for further heuristic developments, including new discoveries by gestalt. All of which serves to advance the accumulating stock of useable resources. For the student, the adoption of a sceptical ethos promotes both theoretical and practical insights; i.e., to challenge ideas in argument is to approach claims as human constructs conditioned by specific circumstances, subject to the partiality of the rhetor, and so in need of investigation; to challenge *all* claims—including those to which they are themselves inclined—emancipates students from the confines of their own partiality, often a difficult achievement for younger students. Admittedly, the purification of conflict through the invocation of a pervasive scepticism may seem a significant departure from what most students think of as standard practice for argument; but the value of dialogical opposition and its supporting attitude will be judged by its capacity to produce the desired results. Students can assess for themselves how effective this combination of respect and scepticism can be as a protocol for the production of *copia*.

One additional note on this rhetorical attitude: the sceptical opposition to all claims need not be limited to oral exchanges but should naturally extend to text sources as well. I raise the issue of dialogical reading under the heading of scepticism because many students only reluctantly accept the notion that authorities in particular and texts in general are subject to dialogical interrogation. Nonetheless, as a major constituent of antilogical invention, scepticism assumes critique; i.e., students practicing oppositional protocols will learn to "speak back" to the text, to demand that it defend itself, to seek out weaknesses as well as acknowledge strengths, to restate and revise its ideas in their own voice and context. For the student, the difference between conventional approaches to texts as resource sites and dialogical transactions with texts as "living relationships" may take some time to comprehend. But to transform the act of reading into a productive means of invention, students need to come "off the sidelines" and engage the text as an actual presence, a human voice to which they can relate and respond rather than an abstract authority to which they must attend in silence.

Finally, dialogical invention will cultivate a dynamic approach to invention that does not stop simply because all voices have been sampled and a critique of each has been mounted. Rather, the dynamism of antilogic assumes that new ideas are always potentially emergent because all existing claims are presumptive and so subject to revision. Similarly, dialogue itself is always potentially dynamic, with each new addition, comparison, critique marking a more or less new turn in the progress of argument. For purposes of invention, in particular, this dynamism implies that each rejoinder in argument is a gesture towards discovery; and, in that sense, dynamism is the natural extension of antilogic's oppositionality, which seeks to move argument forward by introducing respectful contention. Dynamism advances this commitment by extending the scope of argument well beyond the "single-message encounter," beyond the simple claim and response that characterizes the accepted unit of practice for much argument theory. Instead, from the antilogical perspective, the potential of dialogue to produce new ideas is perpetual, resistant to closure, committed to the future, to speculation, and to the unpredictable dynamics of dialogical relationship. And while *aidos* insists that invention continue until "the last person speaks," there is—for invention and for argument—no last word because there is always an antilogical rejoinder that may well mark a new start (cf. Billig 1987, 105-110).

If the multiplicity that antilogic seeks is fostered by the adoption of *aidos*, and if oppositionality is productively enacted through scepticism, then the dynamism of dialogical invention is accomplished through what the ancients called *epoche*, the *suspension of judgment*. In one sense, *epoche* is a refinement of *aidos*, a more concrete practice that extends the respectful presumption of feasibility that one assumes for one's own ideas to other initial claims (cf. Booth's "golden rule," 1974). In another, *epoche* complements the efforts of scepticism by opening up for critique any logos capable of stimulating a response. More to the present point, however, *epoche* allows free reign to antilogic's dynamism by holding *aidos* and scepticism in creative tension, thereby sustaining the potential of dialogue to evolve in unforeseen and useful directions. On the other hand, separating the development of a new idea from the determination of its merits can prove a difficult cognitive challenge, especially in matters that prompt strong emotions, given our inveterate tendency to conflate "thinking about" with "deciding upon." Moreover, there are many argumentative circumstances that demand a decision on schedule, circumstances that don't allow the luxury of an open-ended approach to invention. For these reasons, *epoche* presents special problems for instruction, problems that will require more study and space for development. Nonetheless, the heuristic goal of *epoche* as a rhetorical "habit of mind" is to keep the conversation alive, the value of which for invention is clear enough: sometimes, it is only after the pot has simmered that the best flavors begin to develop.

I would note in closing this discussion that, for the purposes of invention, the perpetual openness that follows from the successful suspension of judgment gives rise to a certain redundancy in the field of options, a network of claims and reasons in which similar arguments proliferate and overlap as contributors to the ongoing process of inquiry. In this context, all claims are complementary because all add to the plenitude of resources from which the best possible logos will be determined. This plenitude and the freedom of discussion from which it develops are central to the heuristic enterprise, and fostering them in the classroom will certainly require a more detailed pedagogy than I have outlined here. In this Epilogue, my approach to the pedagogy of invention has been predicated on the notion that "techniques" are the extension of certain rhetorically productive attitudes. Without the appropriate "ethos of invention," technique alone is a rough instrument, generic rather than focused in its practicality, bereft of its own rationale, dependent on its user for adjustments to circumstance. So, when it comes to the practice and pedagogy of invention, I am arguing that ethos comes first, not because rhetors are the ones who employ and adapt technique, but because invention thrives on productive relationships and relationships, in turn, prosper when they are supported by appropriate attitudes among the people involved.

3. PRACTICAL JUDGMENT AND ITS CHALLENGE TO INSTRUCTION

Protagorean pedagogy is based on the notion that by entering into dialogue with opposing positions students will liberalize their thinking, cultivate their skills at argument, and thereby prepare themselves to contribute with confidence and authority to the controversies that inform their public and private lives (*Protagoras* 318e-319a). The process begins by reconnoitering among available opinions, it develops by juxtaposing claims and counter-claims, and it is sustained by suspending judgment. But if knowledge of these practices is to have an effect on the affairs of people and groups, then decisions must be made. As a theory of discourse that seeks workable knowledge and useful courses of action, antilogic must therefore respond to the content it generates in invention with measures for its rational assessment and just disposition. That is, as a systematic theory of discourse (*ratio diligens disserendi*), antilogic would be incomplete without guidelines for effective deliberation. In the ancient world, this crucial part of the rhetorical process was a highly focused activity; i.e., to deliberate on indeterminate questions (rather than universal ones) involved the capacity to identify suitable courses of practical action given the resources available in particular situations (see Aristotle's *Nic. Ethics,* Bk. 4, 1141b). The case remains much the same today: rhetorical arguments call for practical judgment, a process that invokes some of the more subtle features of the rhetorical art. In the rest of this section, I hope to prepare the way for a pedagogy of judgment by (a) reviewing a prominent classical model for rhetorical deliberation, (b) considering problems for instruction inherent in this model, and, (c) defining rhetorical judgment in pragmatic terms intended for the classroom. As a framework for classroom practice, this approach is highly speculative; but it does at least call attention to *krisis*, or judgment, as a neglected subject for pedagogical attention, a subject which, regardless of its resistance to "system," will eventually require a role in any conscientious approach to instruction in argument.

The general pragmatism of Protagorean argument was the specific subject of earlier discussion (see Ch. 3, sec. 1). So, at this point I will only sketch some ideas that relate pragmatism to judgment and its pedagogy. To begin, the antilogical/dialogical approach to judgment has a strong family resemblance to Aristotelian *phronesis*. Both deal with questions of practical conduct arising from the contingent details of dynamic situations (*Nic. Ethics* 1095a 5). Because of their engagement with particulars, both also operate with minimal recourse to either idealized precepts or formal logic. Moreover, both blend functional and moral motives in their effort to identify action that is, at the same time, "advantageous" and "good," capable of producing tangible benefit but doing so justly *(Nic. Ethics* 1140a 25-30). This said, there are also important distinctions: *phronesis* is the work of a single person, the *phronemos* who reflects on the means for concrete action (*Nic. Ethics* 1144a 7). By contrast, dialogical reasoning locates deliberation in the midst of relationship, either in the direct exchange of dialogical partners or mediated through protocols that represent this exchange. In addition, *phronesis* assumes a categorical separation of thinking from acting not shared by antilogic: i.e., the *phronemos* contemplates the means for actions to be taken at another time; whereas dialogical deliberations undertake reflection from within the field of controversy and (in the process of semiotic exchange) generate argumentative content in dynamic form. Seen thus, antilogic becomes what I imagine it was for Protagoras: a form of practical action in its own right rather than a speculative preliminary to action. In turn, antilogic's standards for judgment will reflect this pragmatic orientation and seek, above all, to identify conduct appropriate to these particular people in this specific circumstance.

And yet, if practical argument is, by definition, anchored in the concrete details of specific cases, how do we generalize about effective practice or, more to the present point, how do our students learn to make sound judgments under conditions that are infinitely variable? Aristotle offers limited aid on this point (see *Nic. Ethics*, Bk. 3). But for Cicero, the linked topics of practical judgment and right conduct are of consummate importance, not least because they are also replete with challenges to effective practice. Not only are the contingent details of each case unique, but judgment in an antilogical/*controversial* context must also grapple with the *multiplex ratio* which successful invention has produced.[7] To further complicate matters, moral guidelines to sound action—while standard criteria for judgment—nonetheless present their own difficulties: there are likely to be disagreements over the nature of any particular virtue or "duty," virtues like "goodness" and "utility" are routinely in conflict, and, like all else in practical argument, values are subject to contingency and so will change with changing circumstances (cf. *De Officiis* 3.18).[8] Given these complexities, the operating procedure for practical judgment in Cicero, and for the rhetorical tradition generally, has been to consider "what is most needful in each individual case . . . and by calculation perceive where the weight of duty lies" (ibid. 1.59).[9] In other words, the exercise of practical judgment calls for *propriety*: for what is most needful, useful, advantageous, and/or fitting under conditions of competing probabilities and for the purposes of effective social action. For modern readers, propriety may connote a rather stuffy, conservative standard, the norms of narrow convention, concerned primarily with restraining behavior within the boundaries of the "proper." In its Ciceronian context, however, propriety is a powerful, even dominant moral and rhetorical force. Yet, despite its centrality to the deliberative process, Cicero writes that "nothing is harder to determine than what is appropriate" (see *Orator* 70). Of course, this difficulty doesn't make propriety any less pertinent, or get us off the hook.

Though there is some variation in its deployment, propriety is dependably at play in actions well suited to their situation and to the interests of those involved. In specifically rhetorical terms, discourse is "appropriate" when the subject is treated in ways that serve the needs of the audience and the purpose of the rhetor. And yet, while propriety is undeniably functional, concerned with what works, it requires that utility be complemented by a concern for what is "right" in the moral sense. Indeed, according to Cicero, propriety cannot exist outside the boundaries of moral goodness (*honestas*); so as an agent of right conduct, propriety will always seek to integrate the functional and the ethical, to blend "Does it work?" with "Is it honorable?" (see V. Kahn 1985, 32-34). In turn, this balance gives propriety unrivaled authority as a criterion for practical judgment: "The universal rule, in oratory as in life, is to consider propriety" (*Orator* 71). In fact, when it comes to oratory and argument, the rule has special force, though propriety in rhetoric typically trades under the name of *decorum* and, with this title, carries an impressive portfolio.

Like propitious moral or political judgment, decorum is fundamentally concerned with utility, with the adaptation of discursive form for persuasive effect. But Cicero argues vigorously that rhetoric's propriety is neither opportunistic in motive nor singular in focus. That is, decorum is that element of discursive style which not only adjusts the formal elements of language in conformity with rhetoric's practical goals, it also simultaneously considers what language is most suitable to the intrinsic nature of the subject itself. Or better yet, decorum works to give aesthetic form to the subject in a way that reveals its nature to those whom oratory would influence. Seen thus, decorum coordinates the internal or formal dimensions of language with the contingent, persuasive interests of rhetoric, complementing a dominant pragmatism with a literary appreciation for the appropriate expression of a particular idea (see Ch. 5, sec. 3; cf. Leff 1990, 117-21). In sum, decorum fulfills the antilogical imperative by holding contrary impulses in productive tension, this time as a way of arriving at a "reasoned judgment" regarding the articulation of meaning in ways that are both accurate and purposeful.

There is an obvious analogy, a homology actually, between the integration of moral and functional criteria in prudence and the balance of literary and persuasive interests in decorum. The similarity is more than formal, the kinship more than casual; i.e., both operate within the scope of prudence. To Cicero, prudence is "the foremost of all virtues," the incarnation of wisdom (*sophia, sapientia*) in practical form, the capacity for sound judgment in those questions regarding human conduct in which the issues are immersed in context, deliberation must contend with opposing probabilities, and decisions lead to concrete action (*De Officiis* 1.153). It is prudence that is the superordinate category, comprehending both propriety as its principal criterion in deliberation and decorum as its operating procedure in rhetorical matters. It is prudence, then, with its balance of insight and pragmatism, that is the fundamental component of reasoned judgment regarding human conduct in any sphere, including rhetoric and argument (see Garver 232-48). But these distinctions, while important, are among cognates that share a uniform nature; and while the terms themselves may seem a bit obscure or antique to many readers, they do explain the process of judgment in a way adaptable to the contemporary classroom.

More importantly, these cognate terms hold what I take to be a crucial implication for rhetorical pedagogy. That is, while prudential wisdom may seem a somewhat rarified capacity, one can, in fact, exercise one's potential for decorum by making particular decisions about this or that feature of composition. Admittedly, decorum, like its conceptual relatives, is deeply embroiled in contingency, always a matter of discretion, and ultimately beyond rules; so that the propriety of this rhetorical choice may not hold in that rhetorical situation. But, the situated nature of decorum is essential to its utility. As a synecdoche of prudence—the rhetorical expression of more general action—decorum calls for the same ability to make choices in the face of contingency that characterizes other forms of practical judgment regarding human conduct. Consequently, to practice decorum in composition is to cultivate prudence in general, a causal relation of considerable importance (though, of course, open to dispute). Filling out this provocative relation, however, would require more time and space than are available. Instead, I will offer a Cook's tour of topics related to instruction in rhetorical judgment, followed by an antilogical construct for such instruction. It all goes by quickly, but I will do my best to stake out the central issues for further inquiry.

What else, then, might we consider that would contribute to a workable pedagogy for rhetorical judgment? More specifically, are there topics of general theory that would aid students' understanding of the subject and capacity for successful practice? Here is a miscellany of considerations. Traditionally, students of rhetoric have been supplied with systems of classification that tame the infinite variety of situations, schematize categories of analysis and response, and limit the options for action to feasible proportions. The study of judgment is naturally subject to such efforts and has a number of likely suspects for application. There are the traditional "genres" of classical rhetoric, most of which are already supplied with criteria for judgment (e.g., deliberative argument with its appeal to the desirable and opportune). There are also the stasis categories, customarily connected with invention, but as a modern system of argumentative "kinds" (definition, causation, etc.), they too come complete with formal standards for assessing the propriety of a particular appeal (see Fahnstock and Secor 1990). Then there is the Toulmin "method," which is clearly better suited to the analysis and judgment of arguments than it is to their invention. In sum, there is no shortage of frameworks for organizing and thus simplifying the work of judgment. The question for teachers is not which scheme to employ but, more directly, does knowledge of such categories contribute to better student practice? I tend to think a little order is a good thing, a guide to the big picture, like the map of a country with each state or province or category marked out and attended by its own flag or, in this case, by general comments on the basic nature and function of the particular category. The assumption here is that formal knowledge, knowledge about structures abstracted from context, can serve first to orient the student to new terrain and then to point in the direction of effective practice. On the other hand, we refer to "the exercise of judgment" because, in practice, coming to a decision is a form of action, in this case a dialogical action involving unpredictable contact with an independent "other" under unique circumstantial conditions. At issue, then, is the suitability of theoretical frameworks to a pedagogy of practical judgment where specific contingencies and dynamic events challenge the application of static, general structures.

Along with comprehensive organizational schemata, there are also categories of judgment distinguished by the standards of "exactitude" adopted in a particular field. At one end of the spectrum are sciences like botany or zoology with very precise standards for the verifiability of claims; at the other end are inquiries into practical human conduct in areas such as ethics and politics where the "threshold of certainty" for knowledge and judgment is lower because the data are variable, contingent, and only probable.[10] Of course, some cases in any field call for stricter approaches to judgment than others—and, admittedly, diverse approaches can exist within a single field. But these complexities aside, our students need, at the very least, to comprehend the field-specific standards of their chosen disciplines (even perhaps the standards of other fields as a counterpoint), a process to which rhetorical training can certainly contribute (cf. Graff 1992). Moreover, because all our students are public citizens with personal lives and impending professional careers, familiarity with and the capacity for practical judgment as outlined above seems to me central to their intellectual growth and, consequently, to any serious college curriculum. So, the pedagogical issue in this instance is not "what" one should study but rather "how" students can develop their capacity for practical judgment given the inextricable relation of judgment to concrete situations, specific details, and direct contact with particular persons. I will return to this topic shortly, but first there remains a final, critical feature of judgment theory to consider; namely specific formulae for decision-making and their pedagogical utility.

The options are abundant: there is judgment by deduction from general principles or maxims, by analogy with paradigm cases, by propriety, by preponderance, by appeal to authority, by formal validity, by consensus, by methods of all sorts and sizes. There are also systems that combine several methods and coordinate these with specific kinds or categories of judgment. Of course, such systems are replete with theoretical implications; but, from our present perspective, this topic returns us to our original question: to what extent does any systematic, theoretical approach to judgment actually serve pedagogical rather than analytical purposes? Or, more directly, does this or that or any system actually elucidate the methods by which reason *arrives at resolution*? Once again, I am inclined to be supportive: such systems can help students to comprehend the subject as a whole; and, on occasion, they can help teachers plan a course or a unit. But, on balance, the teacher's desire for a perfect system or structure—one that will simplify and elucidate the subject in order to inform and motivate students—seems to me a lingering Ramistic dream. Dreams aside, "the locus of judgment" in practical argument remains the specific case in all its detail, to which no road map will provide more than a very imprecise guide (Jonsen and Toulmin 330). To what extent, then, does a schematic outline serve to distort the picture, like a flat rendering of the globe? Or consider this: the formulae for judgment that most people adopt are typically matters of local custom, the patterns of one's community or discipline, learned on one's feet rather than at one's desk. Since we argue with others (who probably don't own our map), we may be better off investigating their practices and standards (which may well resist our categories anyway). Such speculations certainly don't overturn arguments on behalf of theory, schemata, and maps, which represent the grammar of argument. But when it comes to the actual conduct of judgment, grammar and theory are addressed to the understanding and only go so far as an index to action. Which is why, since Rome, the dominant method for the study of these matters has been to introduce students to general theories but to concentrate their attention primarily on argument in its dynamic aspect. In practical terms, this pedagogical concentration takes two forms. The first involves the observation and analysis of others in the "exercise" of their judgment; the second is contained in this manifesto: when seeking facility in judgment, invention, or anything else connected with rhetorical argumentation, follow the path marked "practice."

Just how teachers can promote a practice-based pedagogy of argument is the subject for the final section of this Epilogue. But we can glimpse some possibilities if, in closing this discussion of judgment, we consider our topic in specific relation to antilogic. To do so, I offer an antilogical definition of practical judgment with commentary on its phases, potential, and problems. Through these comments, I also try to advance my earlier argument that practical judgment is not simply "a knowledge affair" (thinking sequestered in its own space) but rather a reconstruction of the world in and through the dynamic offices of discourse. Here, then, is my antilogically-informed definition: practical judgment is a dialogical art in which interlocutors *test* arguments within a field of competing options, *calculate* the relative merits of opposing probabilities and, if they have not yet made a decision, *determine* their conclusions through a consensual commitment to argument itself. Dialogue is once again the "covering principle" with the agent/action ratio prominent throughout.

In phase one, during which arguments (actually opinions) are tested, we are clearly at the border between invention and judgment. In the former, scepticism tests every claim not only to identify and eliminate what is false or unconvincing but also to hone what is useful and urge new developments, all under the supportive sign of *epoche*. In the service of judgment, however, testing takes on a different function; i.e., it challenges each claim on offer for the purpose of corroborating those that are, in fact, viable. In *On Liberty*, Mill comments on the role of this kind of challenge in civic life by noting that only the complete freedom to contradict an opinion can "justify us in assuming its truth for purposes of action" (1996, 22). That is, no idea can achieve the level of "assurance" that warrants belief and commitment until it has been tested by rigorous critique (see Cherwitz and Hikins 1979). In fact, writes Mill, "the whole strength and value" of human judgment depends not just on critique but, more specifically, on "the steady habit of correcting and completing [one's] own opinion by collating it with those of others" (1996, 22-23). This habit of "correcting by collation" amounts essentially to argument by antilogic, the juxtaposition of a claim with other options. The purpose of this challenge is not (for Mill or for antilogic) to ascertain the truth, but rather to clarify whether an argument deserves to be taken seriously. And yet, while testing through critique can earn an argument a place at the table, only in uncomplicated circumstances will judgment be determined by an early challenge that reveals the clear supremacy of one claim over its rivals. More likely is a narrowing of the field to focus on those claims on various sides of the controversy that have met the "test." Of course, the dynamics of argument can develop in myriad ways and dialogical partners are free to determine a preferred solution at any point. For purposes of discussion, however, we will assume that in many arguments, judgment must be deferred because testing culminates in the need to adjudicate opposing positions still in play after their initial challenge. Under such circumstances, practical judgment turns to the calculation of strengths and weaknesses as the most appropriate way to further the progress of judgment.

Obviously, at this point in an argument, a solution has become elusive. The process of calculation (conducted through dialogue) is intended first of all to enumerate the balance of reasons for and against a claim and then to identify a preponderance in one direction or another. Then, if necessary (i.e., if a determination still has not been reached), calculation turns to a comparative assessment of the preponderance of support for one claim over another. This process is filled with nuance created by the unique circumstances of each situation, but I will try to be general and brief. Calculation remains a dialogical activity because opposing positions are still in play and because judgment, like inquiry, proceeds by question and answer (see Gadamer 1994, 365-67).[11] Cicero describes calculation as a process of adding and subtracting so as to see "by the sum, where our duty lies" (*De Officiis* 1.59; cf. *Nic. Ethics* 1140a). And indeed, in many critical matters, people can and do resort to the two-column ledger to calculate the comparative relation of strengths to weaknesses at play within and between claims. Nonetheless, the process of argumentative calculation does not require an exact tabulation of positive and negative points and can, instead, turn on the identification of a preponderant strength or weakness, that feature of the case that may prove decisive. Either way, determining the ratio of strengths to weaknesses requires a special form of holistic assessment; i.e., both sides of each argument should be reviewed by each interlocutor and added to the mix. Only in this way can the resulting calculation reflect the collaborative commitment that motivates dialogue and prepares the way for consensual assent. The process can get complicated; but the effort to itemize the analysis, while not without its drawbacks, can provide both a strategy for dealing with detail and a bulwark against potential confusion. In any case, as an interpretive category, the dyad of "strengths and weaknesses" avoids the ethical certainty connected with "good and bad" and the elitism of "better or worse."[12] On the other hand, "strengths and weaknesses" are themselves relative and need to be complemented with a more decisive standard to fulfill their role in calculation.

Those standards are provided by the ratio of utility and intrinsic merit (Cicero's *utilitas* and *honestus*), those values that are the special province of prudence. Admittedly, prudence and its evaluative criteria are also "ethically determined," but *utilitas/honestus* offer a standard tied to practical actions in particular circumstances, a framework of interpretive latitude over which opposing opinions can converge, rather than the this-or-nothing ultimatum of claims based on "the truth" or on the coercive binary of "better/worse." Indeed, the process of calculating strengths and weaknesses in a dialogical setting assumes that differences will arise and are admissible, that the feature of the claim under consideration is not strong or weak in an absolute sense, but that the preponderance of evidence or opinion will support or oppose the appropriateness of this point to the case at hand. And, as a result, this discrete calculation will add to/subtract from the weight of the case *in toto*. As you will no doubt imagine, there are abundant other intricacies connected to the pivotal process of calculating judgment. For now, however, I will have to content myself with addressing only one. Despite a denomination that seems to imply accounting-like objectivity, the practice of calculation is deeply humanistic—not only because it operates according to prudential criteria but less obviously in its pursuit of purposeful action that meets the specific demands of a particular human controversy. Clearly, initial testing was not able to identify the preponderance of evidence necessary for definition; nor will abstract, scientific knowledge be able to address the unique demands of the specific case. What remains are the resources and commitment of the dialogical partners; what they seek is not the best of all possible actions but, rather, an action appropriate for and acceptable to the people involved in this situation. What they actually deliberate over is an artifact, a human product that reconstructs the situation and the proposed response in and through discourse, an artifact they must judge both for the integrity of its discursive form and the propriety of the action it would urge. Given the complexity likely to swirl around the deliberative process at this moment, no wonder that judgment resorts to what may seem the rather pedestrian but also the more-or-less definite procedures of calculation for its aid in measuring the subtle weight of propriety.

And yet, as we all know, there is no guarantee that good offices will result in the desired effect. Thorough testing and careful calculations may still fall short of a final determination, a decision regarding an acceptable course of action. At which point, coming to a decision is as much about the relationship of those involved as it is about effective procedures. Whereas the focus of deliberation heretofore had been on the claims themselves, these claims have at this point received insufficient support to motivate a decision. The focus now shifts to the participants in dialogue and their capacity for agreement in spite of uncertainty. I will call this capacity "consensus," though *concordia discors* (harmony amidst discord) may be closer (see Quintilian 1.10.12). The determination of controversy at this point is, thus, a matter of consensus, the shared commitment of dialogical partners to act together in pursuit of resolution. Consensus thus conceived is not unanimity; it admits dissent and the provisionality of its decision. As an imperfect agreement, it assumes, on the one hand, that the process to date has been acceptable and the relations among those involved respectful. Indeed, if such has been and is the case, then (and this is the crux of the matter), dialogical collaboration is in itself a reasonable warrant for choice in the context of opinions in intractable disagreement. That is, opinion is fallible, always; but decisions must be made nonetheless. Consensus recognizes this fallibility and does not mistake the pragmatic need for a working agreement with a change in the epistemic status of the option that, at a particular moment, has preponderant support. A decision in Protagorean argument is a guarantee that—at this moment, for these people—this claim has a larger share of propriety than its competition, but it does not have it all. To enter into dialogical argument is to agree "up-front" that consensus is both a desirable goal and the court of last resort. To reject dialogical consensus (when not a retreat into dogmatism) is to refuse this agreement and the relationship that produced it. To refuse consensus is not simply to claim that you don't like the decision but also to claim that the relationship is now flawed and you cannot act together. To agree to consensus with reservations is to agree that the decision is the best available option for shared commitment, which is, at this point, the lynchpin of a purposeful response.

Either way, to determine argument by consensus places the people who measure at the center of the deliberative process and allows these agents to come to a decision which, while perhaps not perfect, is nonetheless "acceptable" to the participants. And yet, probable claims, tested in argument and accepted by the judgment of those involved, possess all the warrant we need or are likely to get. We act in the world on the basis of such warrants because no other standard is available to adjudicate endemic disagreements over human conduct. We can accept these standards and relish consensus when we achieve it, or we can maintain the nostalgic hope for a more perfect standard and rant against the loss of certainty and "the glib and oily" rhetoric that has replaced it. Having put our ideas up for challenge, having invited difficulty rather than avoided it, having looked and listened carefully and tried to comprehend what difference really looks like, having cooperated to prevent the process from being kidnapped by self-interest at critical moments, having made the choice to conduct our argument in goodwill, we have every right to think consensual judgment good enough to justify our actions.

Of course, all this is highly speculative and fully partisan, the Protagorean perspective on judgment in its best dress. Just how substantially my "phases" differ from traditional schemata is open to debate; just how these "speculations" would translate to pedagogy remains open to investigation. On a larger scale, it is also an open question just how all this philosophy and dialogism, all this Protagorean and Ciceronian history, all this invention and judgment theory would be best distilled and adapted to the contemporary classroom. Obviously, that won't happen in the closing pages of *Many Sides*. Nonetheless, the translation/adaptation of rhetoric's rich history and theory for pedagogical purposes constitutes a particularly important scholarly task: Protagoras, Plato, Aristotle, and Cicero all conduct their discussions of argument in direct relation to education. This relation has motivated *Many Sides* from the outset, and I hope *Many Sides* has, in turn, set the stage for more sustained efforts along these lines. For now, I must be content to make some suggestions about what kinds of classroom methods are likely to predominate in a neo-Protagorean pedagogy. And, as promised, my suggestions will honor the manifesto and reiterate the commitment to practice, practice, practice.

4. PROTAGOREAN ANTILOGIC IN THE CONTEMPORARY CLASSROOM

There are numerous issues related to classroom methodology that deserve attention. Some are intrinsic in the sense that they develop directly from the history and nature of antilogic itself; others connect antilogic with extrinsic issues prominent in contemporary rhetoric or pedagogy, in the modern college and university, or in the wider world from which argument ultimately derives its context(s). The list of "extrinsic" topics is both too long to itemize and beyond the scope of *Many Sides*, which has devoted its attention to addressing the history and developing the basic nature of its subject.[13] What remains in these closing pages is to offer some ideas on the "practicalities of antilogical instruction," ideas that complement my argument to date. In its basic form, this pedagogy draws on the abundance of Quintilian for three related practices that I have organized as an integrated sequence and refashioned to reflect the dialogical pragmatism of neo-Protagorean theory. This final section outlines an experimental pedagogy for readers to consider, critique, revise, or reject as a starting point for the antilogical classroom. The triad itself includes *critical interpretation, rhetorical reinvention*, and *case simulation*, all new names for practices of ancient pedigree.

As you may recall from the discussion of Roman pedagogy (Ch. 6, sec. 1), students began their language studies with a *grammaticus* who not only addressed topics of correct expression and felicitous style but who did so by examining exemplary models from the literary/historical canon. Learning to read critically with an eye to the strategic deployment of discursive resources was thus an early feature of language study. When Roman pupils matriculated to the study of rhetoric, this emphasis on detailed, critical reading remained the order of the day. Moreover, as Quintilian's students moved through the rhetorical curriculum, the texts for critical scrutiny became progressively more argumentative, beginning with speeches drawn from literary and historical sources and moving to more complex arguments such as Cicero's orations. In many schools today, close reading for style and strategy remains the prerogative of literary studies; while teachers of rhetoric interested in critical analysis are left with anthologies on popular topics of controversy, almost none of which contain literary and historical sources (cf. William Covino's *The Elements of Persuasion*, 1997). Whatever text a teacher chooses, the process of analysis itself, when guided by antilogical principles, can be divided into two phases, beginning with reconstruction. Recall from previous discussion that rhetors make choices about the development of their arguments based on what is most appropriate to the specific circumstances of a case. Critical readers who would investigate how authors approach invention and judgment must therefore begin by reconstructing the rhetorical situation that underlies authorial choice. When exemplary arguments are taken from history, literature, or other narratives (such as film), the model text comes already surrounded by circumstantial detail. Regardless of the nature of the model, the critical reader has "gotta know the territory" to effectively analyze the decorum on display in any particular text.

Once an argument has been firmly situated, students can begin to consider the model text *in utramque partem*; i.e., by contemplating a particular rhetorical choice from alternative perspectives. Individuals can experiment with this process in any number of ways—by contrasting a text's strengths and weaknesses, by developing arguments for and against the propriety of a particular choice, by juxtaposing one rhetorical feature with an imagined alternative, or, most importantly, by undertaking any of these approaches in dialogue with others according to the suggested protocols (see sec. 2). The pedagogical rationale for these methods should be clear enough: when students reproduce the situated detail and multiple, competing options that condition rhetorical choice, their analyses are most likely to press beyond a general appraisal of the completed product to an examination of the model as a working process, the embodiment of decorum in action. This examination can subsequently inform the student's own practice by offering both a goal to achieve and some idea of how to get there. Of course, teachers can lecture about general approaches to rhetorical strategy or offer precepts designed to facilitate effective choice. But the hermeneutical analysis of rhetorical models, when conducted well, can provide students with practice in the protocols of rhetorical production, a subject which lecture and precept can only address in the abstract. There are, of course, differences between the creation and analysis of arguments, since the former obviously requires the invention of strategic choices while the latter focuses on an artifact with its choices already intact. Nonetheless, in analyzing the effectiveness of an artifact, antilogical interpretation must grapple with the multivocality, oppositionality, and dynamism of argument immersed in specific circumstances. It follows that students of antilogic can begin to understand the process of argument by analyzing mature texts as well as by composing their own arguments.

According to Quintilian, it was the teacher's responsibility to first model then guide classroom practice in what I am calling critical interpretation. Typically, the teacher would select texts for analysis with a specific pedagogical agenda in mind (e.g., practice with stasis questions) and then provide commentary on "the methods employed . . . by individual [orators] under varying circumstances and conditions of time and place" (see *IO* 2.5.1-16; quote 2.5.15). Following the teacher's outline, students would prepare their own analysis of specific cases for delivery in front of the class; and, as a result, the class as a whole was exposed to a diversity of response against which individuals could compare their own efforts. Contemporary antilogical pedagogy would promote the invocation of multiplicity by fostering the dialogical interaction among students, either in discussions involving the class as a whole or in the more comfortable setting of small groups. Either way, the intended goal is to put a cornucopia of potential interpretations before the class for consideration, juxtaposition, and further discussion. This proliferation of views serves naturally as an incentive to invention; but it also illustrates the first tenets of neo-Protagorean theory. Edwin Black puts the case this way:

> It is inevitable that any expositor will approach a work from a certain point of view. His frame of reference may be subconscious and unsystematized, but it will assuredly be present, shaping the bias of his interpretation by influencing the direction of his attention, selectively sharpening some and dulling others of his sensibilities and molding the nuances of judgment in a thousand imperceptible ways. (1971, 171)

Variations in the "nuance of judgment" occupy more than a theoretical place in rhetorical pedagogy; i.e., since rhetoric is involved in the negotiation of meaning with real people whose perspectives are always more or less different from one another, this in-class exhibition of interpretive differences offers concrete illustration of the pluralism that conditions the argumentative process.[14] So, not only does careful reading alert students to an array of rhetorical strategies and accustom them to the operations of decorum, it can also reveal the many sides that attend most arguments, the relation of each claim to a particular perspective, and the inevitable contradictions that surround the search for social knowledge.

Having begun to study the conduct of argument through an analysis of models, students are ready for their own experimental compositions. Rather than ask his pupils to "jump right in," Quintilian employs a series of exercises (the *progynasmata*) that allow students to wade a little deeper into the contingencies of rhetoric with each new assignment and to do so "in the company" of model texts whose authors exhibit a certain kind of rhetorical excellence. These exercises fall under the classical heading of rhetorical imitation, an ancient pedagogy that, put briefly, asks students to reinvent the virtues of an exemplary model in the composition of an original text (see Ch. 5, sec. 4; Ch. 6, sec. 3). Quintilian includes multiple stages in his continuum of exercises; I will simplify the process to two: the imitation of a model argument set in a particular context and the imitation of a character responding to a particular controversy. In the first instance, students are initially provided with a model text and enough contextual detail about it (either in a case narrative or by the teacher) to identify the basic situational features of the case—time, place, personnel, a brief time-line, some sense of individual motives. These early models for imitation should also illustrate familiar forms of argumentative discourse—closing arguments in a legal case, the request for funds from a governing body, a half-time pep talk to a sports team, a speech for or against impending legislation, the review of a film. The composition process thus begins under the tutelage of a model text, chosen by the teacher for its employment of specific rhetorical strategies. The use of more than one model will, of course, expand the range of strategies for study, while the use of contrasting models can deepen the focus on rhetorical judgment.

In an antilogical classroom, students will undertake their analysis of the model text in dialogue with their classmates, dialogue that operates according to the protocols for invention and judgment outlined above. In particular, teachers will introduce their classes to the ethos of respectful opposition, to the willing suspension of judgment, and to the exercise of practical reason, all of which are intended to guide group discussion and allow dialogical exchange to prepare the way for the students' own compositions. As readers will readily imagine, introducing antilogical protocols to a new class requires careful preparation because the governing concepts of Protagorean argument run counter to conventional presumptions. In particular, the antilogical opposition to every claim can seem disconcerting or "against the grain" for some students who would prefer not to disagree with their peers. Nonetheless, for dialogue to" flourish," opposition is crucial (Billig 83). It is consequently the teacher's responsibility to introduce, explicate, even to model in her own actions the nature and function of respectful opposition as an ethically-just rhetorical strategy, a strategy intended to advance the collective effort at analysis without compromising the participants in discussion. Most students will quickly learn that the dialogical attitudes are both socially benign and rhetorically productive. More specifically, these protocols will lead to the collaborative discovery of understandings about the text in question that transcend the horizons of any single student. Equally important, the group discussion of the model text will reveal contours in the landscape of opinion to which students must respond in making effective rhetorical choices. As a result, then, of productive dialogical relations, students can begin to understand how different claims will function in actual practice and to contemplate their own potential strategies from a broader perspective than the "my-side bias." In sum, dialogical analysis prepares students for composition by pooling their rhetorical insights, by compounding their value through dialogue, and by consolidating this collective output as intellectual capital for future investment.

So, having undertaken the dialogical analysis of a model text, students should have both a working knowledge of the contextual issues that attend this rhetorical situation and a store of strategies ready for use in similar circumstances. Such knowledge is the base on which *imitatio* will build, but the lessons learned in group discussion must be modified in the act of composition. Quintilian would routinely provide one model for analysis, then another related or parallel problem for composition. In a contemporary classroom, the composition assignment that follows interpretation will vary the circumstances of the case in some modest way, through changes in time, place, persons, or motives. Given these changes, students need not abandon the precedent and start from scratch; they are still in the company of a helpful guide. But they will have to exercise practical judgment in translating the virtues of the model to the demands of a new setting. Typically, we think of imitation in the sense of duplicating, following, reproducing the original. But in modifying the rhetorical situation of the students' assignment, the teacher makes the model a point of departure rather than a subject for transcription. In fact, what is being imitated in these situational assignments is the ability of the student to follow the lead of the model in adjusting to circumstances, so that to imitate truly requires the student to reconsider the case and reinvent its strategies in ways that suit the new situation.

In a succession of such assignments, what is imitated is the "treasury house" of rhetorical resources, options to be added to the inventory of our students' rhetorical imaginations, where they can be called upon for use when needed (Quintilian 2.7.4). In the single assignment, what is practiced in imitation is the exercise of decorum as embodied in unique form by the model text, as identified by analysis, and finally as reproduced in what amounts to a musical "variation." That is, having analyzed the "theme" of the original, the student must transpose it to a different key or tempo, going beyond the artifact to the reinvention of its artistry (D. L. Clark 174). In so doing, students are learning what we cannot teach by precept because there is little we can say about decorum that will capture its actual function (Leff 1990, 202). Instead, rhetorical understanding follows from practice undertaken in the dialogical company of a well-selected model.

The second stage of rhetorical reinvention, the imitation of character, adopts the same basic pedagogy. This time, however, the process becomes more complex because the student must not only calculate what fits the general situation but also analyze the web of personal, social, rhetorical motivations to be evoked if the imitation of character is to succeed. Once again, the point of departure is a model text (e.g., Lady Macbeth to her husband, John F. Kennedy's inaugural speech, the closing arguments of *Inherit the Wind*, Hillary Clinton to the Beijing Conference on Women, Michael Jordan's retirement statement). Once again, the composition assignment will modify the situation (Kennedy's hypothetical second inaugural, Rosa Parks to the Beijing Conference). And once again, discussion would proceed according to dialogical protocols with an emphasis on *in utramque partem*; i.e., with alternative responses coming from the group's analysis of the ethos on display in the model. What is new here is the attempt by students to suspend their own inclinations, their perspective on this situation, and to take the measure of another knowing subject, another rhetor with a distinct set of priorities and plans for appropriate conduct in response to a particular controversy.

Ancient pedagogy assumed that learning originates in the imitation of actions by those whom we most admire. In the imitation of character, the student's attention is similarly trained on the holistic analysis of the rhetor in action. Of course, rhetorical ethos is one of many topics for scrutiny in the earlier exercises; but until the study of character is a focus of attention and progresses beyond speculation, understanding will remain incomplete. The imitation of character offers an opportunity to actually practice rhetoric from another perspective, to step outside one's own selfhood, identify with another subjectivity, and conduct an argument in accordance with someone else's standards. In brief, these exercises ask students to prepare for argumentative encounters by imagining the motives and assuming the commitments of another persona as though they were one's own. The intricate psycho-social dimensions of this activity are well beyond the scope of this Epilogue. Nonetheless, it is possible to say a little more about the implications of this exercise.

For these exercises to succeed, impersonation must be more than mere ventriloquism, with students employing the models as proxies for their own voices. Quintilian argues that for impersonation to be convincing, it must go beyond first impressions, superficialities of style, and general traits of character (10.2.1-28). That is, students must not only suspend their own attachments, they must commit themselves as much as possible to a new identity. The best analogy for this process is the drama, with the student as both playwright and actor attempting to convince an audience that it is someone else who walks the stage. And, in fact, when these imitations are actually performed before a classroom audience or in a discussion group, the dramatic component can make the learning experience more engaging for the student and more fun for the class as a whole.

Of course, there are many teachers whose goal is to cultivate the independence of their students and, in particular, their ability to think and speak for themselves. For these teachers, it may seem that when students take on the roles of others they are also subordinating their own ideas and values, even perhaps surrendering their self-reliance and selfhood (see Bender, 113-15).[15] However, even actors committed to their role never fully relinquish their own orientation, which they bring to bear during interpretation and which informs all performance, professional and amateur (cf. Gadamer 1994, 359-61). In effect, when we undertake an impersonation we are entering into a relationship with the model that is essentially bilateral; but in a dialogical relationship, it is the rhetor who reaches out to understand the other side, who seeks first of all to comprehend what she does not yet know. It is the promise of this understanding that is the principal motivation for the imitation of character.

Moreover, as students experiment with a diversity of roles and rhetorics, they work to enlarge the scope of their persona by looking beyond their own horizons and sampling the panorama of human character. Finally, what is sampled in imitation is not the subject's character in Romantic isolation but rather the unique way in which all subjects negotiate their relations with the wider community. That is, when our students attempt to imitate a "character responding to a particular controversy," they are also investigating the ways in which different people balance the claims of "private interest and public good" (Bender 115). It makes sense, then, for teachers to provide their students with a large cast of diverse characters for imitation. As rhetors in training, students learn about new strategies from each model and about diverse responses from a full cast of *dramatis personae* (see Quintilian 3.8.51). As members of various communities, they also learn how to think through problems from other positions and, hopefully, how to bridge the gap with their neighbors. In both roles, students practice their capacity for prudential action by auditioning for someone else's favorite role.

It is but a short step from the imitation of character to the third and final stage of the antilogical curriculum—the simulation of a complete dialogical argument. In this capstone exercise, students are once again presented with a hypothetical case in which they take on a dramatic role and act in accord with specific situational conditions. This time, however, the case situation is even more comprehensive, designed not only to create a specific context for a particular claim but also to recreate the dynamic conditions that surround argument and that motivate changes in claims as the terms of discussion evolve. In particular, a rhetorical case simulation will involve a full cast of characters whose interests are diverse, whose interactions plot an unpredictable course, and whose positions must all be considered in negotiating an agreement. Moreover, these capstone cases come to the student without benefit of a model case that establishes the goals and identifies strategies for emulation. Instead, students receive a portfolio of background data and research options to pursue and analyze for themselves. So, to come to grips with the intricacies of these simulated controversies, students must rely on their own capacities for interpretation, invention, and judgment. As in any actual controversy, however, students may draw on their most powerful rhetorical resource, their dialogical relations with others.

The parallel stage in Quintilian's curriculum was the move from the *progymnasmata* (preliminary exercises) to declamation, the mock judicial and deliberative debates that were the "crowning exercise" of Roman rhetorical study (5.12.17). Quintilian refers to the declamatory cases as "the most useful" of all exercises because they summarize all previous lessons, emphasize the need for actual practice, and approximate as closely as possible the kinds of argument that students would encounter outside the classroom (2.10.2). The topics of these exercises, like those that precede them, were drawn from Greek and Roman cultural history; so students were familiar with the characters, the context, and the conventional beliefs that conditioned the reception of a declamatory argument. And, once again, Roman students were expected to address each case *in utramque partem*, by preparing to argue both sides. In turn, to consider the case from opposing perspectives calls on specific manifestations of rhetorical decorum; i.e., the ability to imagine alternative personae, to choose rhetorical strategies in keeping with separate characters, and to suspend one's judgment while considering opposing probabilities. In brief, declamation was designed to solicit a holistic display of rhetorical expertise, which it could only accomplish if the challenges it posed were—as Quintilian repeatedly insisted—as close as possible "to the actual practice and experience of life" (12.2.7; cf. 5.12.17, 9. 2.10; 10.2).

The persistence of declamation as an instructional tool for over two millennia is tied directly to this verisimilitude, to the realistic exercise of what we now call problem-solving skills as opposed to any effort at realism in the case narrative (which in Rome were often fanciful, even exotic). In its modern form, however, case narratives themselves have taken a decisive turn toward the "real." First in law and medicine, then in business, and much later in composition studies (during the 1970s), case study problems began to reconstruct the declamatory tradition on a foundation of data and documents accumulated from the field study of actual problems (Mendelson, 1989; e.g., see Troyka and Nudelman; Couture and Rymer). This documentary material could be assembled into a unified, focused narrative or presented "raw," as a network of texts, profiles, images, and research options for further examination (Moore 91-92). In our own time, this contextual material and background information can be collected or referenced on a case-study website, where the potential range of available data gives new dimensions to the notion of realism.[16] In addition to verisimilitude, the contemporary case (like its Roman predecessor) must also address issues and involve situations with which students have some familiarity and engagement. Among the scenarios for case study I have found attractive for my own students are these: (a) members of a college admissions board (including student representatives) debate applications from minimally-qualified students deemed to merit special attention; (b) a family in Oregon, along with various advisors, discuss the possibility of euthanasia for a terminally ill parent; and (c) various "creative teams" develop, propose, and lobby the head of a production studio for a TV pilot directed at college-aged viewers. The topics employed, the ideological interests invoked, the complexity involved are all matters for every teacher to consider. These cases are never theoretically innocent, nor is the pedagogy neutral.[17] With all this in mind, here is one approach to the classroom practice of the simulated case study.

The following options are variable according to student need and course design. The whole process can be undertaken in a single class period or extended over several weeks; each particular step could be expanded or compressed; students can work with the guidance of the teacher or not, though it is assumed that they will operate in teams and that the teams will already be familiar with the dialogical protocols and the nature of practical judgment. I have divided the "lesson plan" into six steps, though at a certain point there is an opportunity to improvise. Step 1 begins with the examination of the case narrative by the work team (anywhere from two to six members), followed by research efforts prompted by the narrative, and then by the interpretation of the research data the team uncovers. Since the research component is new, many students will require guidance; and when research data is distributed within the team for analysis, the dialogical attitudes need to be recalled. The goals of this step are straightforward: to develop a preliminary appraisal of the topic, to generate *copia* for future use, and—not least important—to identify each student's a priori presumptions about the topic. In Step 2, with analysis in full swing, students can decide which roles they would like to play in the case drama. With larger groups, it can be helpful to have two or more people take up the same role so that they can discuss the character and take turns reviewing one another's performances. In some scenarios, it is possible for students to play parts on opposing sides of the case, or at least to take up alternative roles for the purpose of planning.

Planning itself starts in Step 3, when teams begin to think through both the group's basic strategy and the position of each character in relation to the argument. Dialogical protocols are yet again the guide to invention, though what is invented at this stage is an opening move for discussion rather than a conclusive proposition intended to determine the case. There may be dogmatic, intransigent characters in the drama (an absolutist on euthanasia, an admission board member who "goes by the book"), and students need to represent them accurately. But the first round of discussion among the parties is always a negotiation, and no participant should be allowed to kidnap the argument and insist on its premature resolution.

Step 4 is the first encounter among the work teams involved in the case (those allied either for and against euthanasia, the first meeting of the admissions board, the first meeting of the creative teams with the studio director). Like the presentation of arguments in the earlier imitations, these encounters are enacted before the class as a whole, with the students "performing" in character and with those who are not at present "on stage" as audience and reviewers. The narrative will set the terms for this encounter—its location, its goals, the conventions of exchange (e.g., who starts the discussion, how long the characters can speak, how long the discussion should last). But the participants themselves must end the meeting by making plans for further discussion (the who, when, where, and how of the next exchange).

Step 5 is outlined in the case narrative: teams regroup to review the events of the first meeting, critique one another's performances, and plan for the next encounter. In the process, opening arguments will undoubtedly be modified, as the team considers possibilities for further negotiations prompted by the responses of others. At this point, the process of argument becomes a matter of multilateral developments, with the relationship among the parties assuming a central role in the search for resolution. Given the parameters that the teacher has established for the exercise, teams can consider the option of repeating Steps 3-5 (planning, meeting, review and revision) based on the state of negotiations.

In all cases, the assignment narrative will set requirements for an eventual Step 6 in which each side/participant will make her best case, consensus will be sought, and decisions made according to criteria appropriate to the particular situation. In the case of the admissions board, decisions will depend upon consensus among a majority; for the TV pilot project, final presentations will occasion a comparative judgment by the studio director (played by a student or a group of students); and in the case of euthanasia, consensus is desirable, but the issue may have to go to an arbitrator, a state board which will listen to final appeals and explain its judgment in detail. Indeed, regardless of the case situation, the deciding authority is responsible for explicating the grounds for decision. These final "verdicts" are an obvious pretext for student writing, though written responses can come at any point in the process: e.g., as research reports following Step 1, as argumentative briefs prior to or following the planning of Step 3, as revised claims prior to a new round of negotiation. The form or the process of composition can also vary; argumentative essays may take shape as collaborative projects or as the arguments of individual students, as reports "in character" or as a comparison of opposing perspectives, as critiques of one's own performance or as an analysis of particular strategies. Perhaps the most interesting and instructive assignments are paired essays from each student: first a brief on the student's preferred position at the outset; second, a closing assessment based on the experience of the case as a whole. These compositions do more than demonstrate the students' grasp of the concrete case. They provide an opportunity to experiment with the distinctive features of antilogical practice; i.e., with the protocols for dialogical invention, with the decorum of practical judgment, even with Protagorean theory and its relation to argumentative practice. Nor are these experiments confined to written compositions, since the oral exchanges of each step offer opportunities to employ antilogical protocols and to inquire into the many sides in dispute (*multiplex ratio disputandi*).

As each case develops, and as new cases raise new challenges, students become increasingly familiar with the art of Protagorean argument in both its practical and academic attire. On the one hand, these simulated controversies provide a context for practice with a concrete set of rhetorical protocols: the ability to respect difference and cross boundaries, to initiate dialogue and weigh options, to negotiate conflict and exercise prudence, to adapt to changing conditions and judge according to well-reasoned standards. On the other hand, all this exercise goes on in the safe haven of a controlled environment, where practice is guided and action is connected to reflection, where performance is measured by personal growth rather than practical outcomes, where methods remain connected to principles and principles are themselves a subject for study, and where it is possible to experiment with new methods for the sake of discovery itself. The students' progress is thus informed by guiding principles, while the principles themselves gain substance in practice.

Of course, to theorize pedagogy in this way is the kind of thing that teachers and scholars do. For students, the simulated case amounts to a practicum in rhetorical decision making. After a little practice and with a little guidance, however, these rhetorical dramas can begin to address the experience of argument at a more fundamental level: i.e., these cases and this pedagogy provide students with an opportunity to rehearse a particular rhetorical attitude, a way of conducting themselves in controversy, an attitude that finds expression in particular methods but which is itself more comprehensive, more tangible, more meaningful than technique alone (cf. Burke 1969, 441). The attitude of antilogic may at first seem unfamiliar and unconventional; it is certainly a departure from what our students observe of argument in public discourse or what they are likely to encounter in most college courses. Nonetheless, this unconventional attitude is easy enough to comprehend in actual practice, where it is manifest, first and foremost, in the rhetor's willingness to reach out and make contact with the other side(s). Once familiar with antilogic through practice, a class can discuss as a group and students can judge for themselves the utility of antilogical methods and the value of the attitude that motivates them. Similarly, the simulated cases and the class discussion of them should provide teachers with some of the evidence they need to assess the potential of antilogical argument and the pedagogy that attends it.

In the process of critical assessment, teachers will naturally want to consider not only what this pedagogy tries to do but also what it leaves out. For example, all the talk of ethos and attitude limits the attention to concrete strategies, little consideration is given to the cultural conditions that influence the contemporary classroom, and the discussion of particular assignments is too brief to address its potential problems. These concerns and other critiques will all require discussion—an antilogical rejoinder and continued dialogue. In turn, dialogue will beget adjustments, corrections, additions and refinements in the plan on offer above. This Epilogue is only a step in the evolution of an antilogical pedagogy.

And yet, while I acknowledge that the pedagogy of the Epilogue is rudimentary, the rudiments count for something because the Protagorean tradition, the tradition of Cicero and Quintilian, of antilogic and *controversia,* is itself so substantial. In the course of this book, I have attempted to render this tradition by recounting its early history and controversial philosophy, by identifying its ethical and social values, by detailing its most enduring rhetorical features, and by tracing the development of its pedagogy. All these topics taken together attempt to reclaim that tradition, or more exactly, to proclaim the potential of Protagorean rhetoric to redirect and revitalize the practice and the pedagogy of argumentation.

Admittedly, my argument is only one among many expressions of the Protagorean spirit in modern form. Indeed, neo-Protagorean arguments now proliferate in almost every culture that cultivates pluralism, across the spectrum of disciplines, from physics and anthropology to architecture and cinema, from political theory to theology, in serious discourse and popular media, in all demographic nooks and crannies. As part of this modern chorus, *Many Sides* calls specific attention to the rhetorical side of the Protagorean tradition. In particular, to a form of argument especially suited to the multiplicity of the post/modern cosmo*polis*. In our communication intensive environment, where argument is as endemic as ever, antilogic stands poised to contribute more to our discursive relations than just its outrageous name.

NOTES

Introduction

[1] Compare this concept with Wayne Booth's notion of the "field of selves": "Even when thinking privately, 'I' can never escape the other selves which I have taken in to make 'myself,' and my thought will thus always be a dialogue" (1974, 126).

[2] Similarly, James Davison Hunter, in *Culture Wars* (1991), argues that given the polarized nature of contemporary moral discussion in the US "changing the environment of public discourse" from bellicose eristic to a rhetoric of "moderation and forbearance" is the first step toward a more rational consideration of important social issues (320).

[3] As the reader will soon see, the Protagorean term antilogic does not mean "against-logic," nor does it imply an argument that is irrational. Rather, antilogic will here mean the juxtaposition of opposing arguments, or anti-logoi. It is interesting to note in this instance that after Stephen Toulmin's *The Uses of Argument* appeared (1958), one of his colleagues at the University of Leeds was so annoyed by its move away from formal logic that he referred to the book as "Toulmin's anti-logic" (in Golden, et al, 375). Throughout *Many Sides*, however, when I refer to Protagorean antilogic, I refer strictly to argument based on the dialogical relation of two or more opposing positions.

[4] The majority of the Sophistic works and fragments are compiled in the *Die Fragmente der Vorsokratiker* of Diels-Kranz (edited in English by Rosamund Kent Sprague). These fragments are here referred to by their numbers in DK.

[5] It was only after my own book was well underway that Professor Sloane's *On the Contrary* (1997) was available. Nonetheless, his work and thought have been a major source of inspiration for *Many Sides*, which I would like to think extends the study of *controversia* both further back towards its origins and forward to the contemporary classroom.

Chapter 1

[1] Throughout this book, I have followed Friedrich Solmsen's practice and eliminated all diacritical marks from the English transcription of Greek words (1975).

[2] *Diels/Kranz* is the standard source of all Presocratic writing (cited in text as *DK*). The standard English translation of *DK* is edited by Rosamund Kent Sprague; see also Freeman. The Protagorean canon itself, however, exists only in a few, isolated fragments, all of which are subject to what Donovan calls "a prodigious number of interpretations" (34). The most thorough review of the fragments and their interpretations is to be found in Schiappa. For critical commentary on Protagoras, the reader can begin with Billig, de Romilly 1992, Donovan, Kerferd, Payne, Rankin, Schiller, Smith, Untersteiner, and Versenyi. Of this group, de Romilly provides the most substantive commentary on Protagoras' cultural context. For additional bibliographic materials dealing with Protagoras, see McComiskey 32-34 and Schiappa 218-34.

[3] Freeman reduces the number of Protagoras' books by claiming that many of the titles are actually sections of longer works (346-48) and Untersteiner collapses the entire corpus into two: *Antilogiae* and *Aletheia* [or Truth] (10-16).

[4] Protagoras' stature in the ancient world is indicated by the inclusion of his statue—along with those of Thales, Heraclitus, and Plato—in a monument at Memphis in Egypt that was uncovered in the mid-19C (see Kerferd 1981, 43-44). His present relevance is hinted at in Paul Feyerabend's dialogue in which Protagorean ideas are the point of departure for a discussion of the status of contemporary scientific and practical knowledge (1991, 1-45).

[5] As a measure of the statement's impact, consider the following: Plato devotes his dialogue *Theaetetus*, his most sustained inquiry into the nature of knowledge, to the human-measure fragment (see also *Cratylus* 385e-386a); and Aristotle, in his *Metaphysics*, refers to the issues raised by Protagoras in this fragment as "the hardest and most urgent of all problems" (999a, 24-9). More recently, Jacqueline de Romilly calls the human-measure fragment "the keystone to Sophists' thought" that "dictates" all the rest of their doctrine (1992, 98), while David Payne comments that the ideas implicit in the human-measure fragment, and in the "great speech" of the *Protagoras*, reveal Protagoras as "the first great Humanist" (192). Kerferd writes that the human-measure fragment "will take us directly to the heart of the whole fifth-century sophistic movement" (1981, 85-86); cf. Schiappa, 117-21.

[6] The human-measure fragment is the *locus classicus* of philosophical relativism. Consequently, the territory traversed in the rest of this chapter has been well-mapped by others (see Brett, Cornford, de Romilly, Feyerabend [1987 19-89], Kerferd, Maguire, Margolis, Untersteiner, Versenyi). To reiterate my own intentions: I would represent the bedrock concepts implicit in the human-measure doctrine from a distinctly rhetorical perspective and for an audience of teachers, students, and scholars of rhetoric, composition, and argumentation. I would urge my readers to be patient with these forays into philosophy; my motives are neither antiquarian nor purely theoretical. Rather, I would locate the philosophical principles that inform antilogic as an argumentative *praxis* and so establish a base for our ensuing rhetorical inquiry.

[7] For a comprehensive account of Presocratic philosophy, see Barnes 1972, 1-151; see also Kirk et al.; and Guthrie v. 3. On Parmenides, see Barnes 155-230; *Encyc. of Philosophy* v.6, 49; Kirk 263-85; *Oxford History* 137. In the *Phaedrus*, Plato reflects the Parmenidean

tradition when, in the palinode, he writes of the "colorless, formless, and intangible truly existing essence with which all true knowledge is concerned" (247c; Fowler trans.).

[8] "The Way of Truth" (ll. 344-46), in Kirk 270-72; *DK* 28 B6.

[9] Diogenes Laertius adds this second sentence to the fragment: "For many are the obstacles that impede knowledge, both the obscurity of the question and the shortness of human life" (9.51). See also *DK* 80 A2, A3, A12, A23.

[10] Among the issues raised by this statement is the (presumably) fictional, though long-believed account that this heretical notion lead to Protagoras' banishment from Athens (see Philostratus 1.10.1, *DK* 80 A2). In addition, recent historians have challenged the standard agnostic interpretation of the statement with an anthropological one; i.e., did Protagoras simply disclaim any ability to comprehend the ontology of the gods, or did he initiate a humanistic approach to the gods as social, anthropological phenomena (see Schiappa 143-48)?

[11] The *Theaetetus* does appear to authorize this expansion of "perception" to include additional cognitive activity beyond simple sensation (see the discussion of values and customs, 166d-172b). Cornford argues that Plato's objections to Protagorean epistemology "do not touch Protagoras, who did not limit knowledge to perception" and that Protagorean perception itself should be "stretched" to include a wider field of cognitive activity (65-68) (see also Jordan; Kerferd 1981, 85-87). For an alternative view, see Kaufman (17, n. 3).

[12] For additional commentary by Plato on matters connected with the human-measure fragment and the subject of knowledge, see the *Sophist* (which is a continuation of the dramatic conversation begun in the *Theaetetus*), as well as the *Laws* 4.716c; *Cratylus* 385 and 391b-c; and *Meno* 98a. On Plato's epistemology, see Guthrie v. 4-5; Cornford.

[13] I should note that despite the sarcasm of 161c, Socrates is consistently respectful of Protagoras, both in the *Protagoras* itself and in the *Theaetetus*, where he calls the great Sophist "the wisest of men" (160d).

[14] Early in the *Theaetetus*, Plato does concede that perception exists in relation "to someone" (160b), and he does allow Protagoras, in a famous apologia, to defend the viability of his perspectivism (166-168c). But once the discussion moves from sense perception to reflection, Plato conveniently replaces perception "for someone" with the knowledge of the "truth."

[15] Of course, perspectivism has a long and detailed philosophical history; e.g., in Montaigne's "Apology for Raymond Sebond," the "possible worlds" of Leibniz, the mentalism of Berkeley, and of course in the American pragmatic tradition of Emerson, Peirce, and Dewey; see Parker; see also Cherwitz and Hikins 1983, 250.

[16] It is interesting to note that Nietzsche lectured on classical rhetoric in 1872-73, the same year that he was preparing "On Truth and Lies." See Scott Consigny on Nietzsche's approach to the Sophists (1994).

[17] Discussions of Nietzsche's perspectivism and his theories of language can be found in the "Introduction" by Sander Gilman et al. to *Nietzsche on Rhetoric and Language*, in IJsseling (Ch. 13), and in De Man (Chs. 4-6). Nietzsche himself writes, "What is truth? a mobile army of metaphors, metonyms, anthropomorphisms, in short, a sum of human relations which were poetically and rhetorically heightened, transferred, and adorned, and after long use seem solid, canonical, and binding to a nation . . . . Everything that sets man off from the animal depends upon this capacity to dilute the concrete metaphors into a schema . . . [to build up] a new world of laws, privileges, subordinations, boundary

determinations, which now stands opposite the other, concrete world of primary impressions, as the more solid, more universal, more familiar, more human, and therefore as the regulatory and imperative world" (250). From the Nietzschean point of view, knowledge is inherently rhetorical (hence contingent and perspectival) because any assessment of the world, including the "concrete world of primary impressions," must be carried out with the aid of schemes initiated and organized by rhetoric.

[18] Burke provides this summary: "To live is to have a vocation, and to have a vocation is to have an ethics or scheme of values [orientation], and to have a scheme of values is to have a point of view, and to have a point of view is to have a prejudice or bias which will motivate and color our choice of means" (1984, 256).

[19] One typically Burkean method of reorientation is to be found in the dis-orientation of "perspectives by incongruity," a technique Burke claims to have found in Nietzsche (1984, 88; see *Permanence and Change*, Part II). I should also note that Burke's concept of the "terministic screen" is, to a large extent, a linguistic extension of the theory of orientation (see *Language as Symbolic Action* 44-62).

[20] Perspectivism is certainly not limited to humanistic studies. See Heisenberg's Principle of Uncertainty (in *Physics and Philosophy*) or Niels Bohr's theory of "complementarity," the latter of which is summarized by Richard Rhodes as follows: "'Physics concerns what we can *say* about nature.' And the best we can say, [Bohr] realized, is always partial and incomplete; only by entertaining multiple and mutually limiting *points of view*, building up a composite picture, can we approach the real richness of the world" (Rhodes 3, italics mine).

[21] Readers primarily interested in rhetorical issues may prefer to avoid the admittedly technical matters of this section and move directly to the discussion of relativism (sec. 3), a topic more directly linked to issues of argumentation.

[22] For Aristotle, the law of non-contradiction is the axiom on which apodeictic, or demonstrative (scientific) proof rests (see *Metaphysics* 1005b 5-34, esp. 23, and 1011b 13-14). For an interesting deconstruction of this principle, see De Man's Ch. 6 on Nietzsche and non-contradiction (119-131), which quotes Nietzsche as follows: the law of non-contradiction. "contains no criterion of truth, but an imperative concerning that which should count as true . . . . The conceptual ban on contradictions proceeds from the belief that we can form concepts, and that the concept not only designates the essence of the thing but comprehends it . . . . In fact, logic applies only to fictitious truths that we have created. Logic is the attempt to understand the actual world by means of a scheme of being posited by ourselves, more correctly: to make it easier to formalize and to compute" (120-21).

[23] I realize that in positing the tradition of Plato and Aristotle as the antithetical opposite of Protagoreanism I run the risk of replicating the appeal to binaries that the Protagorean program would disrupt. In this case, however, the juxtaposition of these alternative schemata is inscribed in philosophical history. In the next few pages, I will employ antithesis to identify the principal features of a controversy, the terms of which were established by Protagoras' opponents at the outset.

[24] Maguire notes that Plato's discussion of Protagoras and contradiction assumes that Protagoras is "making a statement about the nature of reality, or being, or 'what is'" that is incompatible with Protagoras' actual position (136). Similarly, Brett notes that the lack of contradiction does not "entail the impossibility of falsehood without the assumption that there are some true statements" (158, n. 34). Consider also this comment by Guthrie regarding Aristotle's response to opposing views (such as those of Protagoras): "although

he assumes the mantle of impartiality and claims to be only introducing clarity and order into his predecessors' accounts and bringing out their real intentions, in fact he is distorting their views to force them into his own different scheme" (6.96-97).

[25] Cherwitz and Hikins remark that "contradictory judgments are not really contradictions at all, since they are judgments about different aspects of the same object" (1984, 264).

[26] Shia Moser writes: "When, in Greece of the Fifth century, the force of customs was weakened and ethical dogmatism was attacked by the Sophists, a new foundation for moral action was sought and the philosophic ethics of Socrates came into being . . . . Not only ethics, but the whole philosophic system of Socrates and Plato was profoundly influenced by the desire to meet the challenge of ethical relativism" (3).

[27] In fairness to Hegel, he is more attentive to the Sophists as philosophers than most of his predecessors or contemporaries. See Schiappa and Kerferd (1981).

[28] Harvey Siegel's critique stands as representative of much contemporary response: " I will argue that, like Protagoras, the recent defenders [of relativism] fail to meet the challenge posed by the incoherence charge, and that the doctrine of epistemological relativism remains untenable because incoherent" (in *Relativism Refuted,* 1987, 3). For reviews of the contemporary debate over relativism, see Bernstein 1983, Feyerabend 1987, Krausz, and Margolis.

[29] From *A Treatise on Nature*, Bk. I; cited in Feyerabend (1987) 78, n. 47.

[30] See de Romilly, Havelock, Kerferd, Untersteiner, and Versenyi. For a bibliography of sources on Sophism and relativism, see Brett 139, n. 1 and Schiappa 19, n. 63.

[31] This ameliorative approach has an early precedent in the work of F. C. S. Schiller.

[32] Recall that in the standard translation of the human-measure fragment, "humanity" (*anthropos*) is equated with the individual. See sec. 1 above.

[33] On the topic of subjectivity, Richard Brett makes the following distinctions: the term relativism (which he notes is of 19C coinage) applies to perceptions, ideas, beliefs that are interpreted "relative to some framework." Often, this framework is "the one prevalent in the speaker's culture"; alternatively, however, the framework can be "specific to the individual speaker." Brett refers to the latter, more radical position as "Subjectivism" (141-42). Occasionally, this radical form is referred to as solipsism, or the complete isolation of the individual within the self-enclosed orbit of one's own perceptual orientation. See also Jonathan Harrison's entry on "Ethical Relativism" in the *Encyclopedia of Philosophy* (v. 3. 78-81). More on these topics in the next section.

[34] Naive or popular relativism is captured in such additional phrases as "when in Rome . . ." and "beauty is in the eye of the beholder." The distinguishing feature of naive relativism is, as Rorty puts it, the assumption that "every belief on a certain topic, or perhaps about any *topic*, is as good as every other" (1982, 1966). For a description of relativism which conforms to this popular definition, see N. L. Gifford's *When in Rome: An Introduction to Relativism and Knowledge*.

[35] Proclus attributes the following position to Gorgias: "[being] is not manifest if it does not involve opinion" (*DK* 82B 26). That is, *doxa*, or the opinion of the community, is the necessary vehicle for the appearance of knowledge about the real and true.

[36] For discussions of historical relativism, see Code 1982, Kaufman, Feyerabend 1987, Rorty 1979.

[37] Such a description, with its emphasis on cultural norms, naturally raises the issue of individual initiative. Modern relativism does indeed assume that the diverse worldviews revealed by historical (and anthropological) study are fully contingent upon their unique

place and moment in the process of cultural development. However, as *individuals* develop within these specific traditions, they are conditioned by but not confined to their culture's norms, customs, values, practices. Given a certain measure of freedom, individuals are capable of modifying, extending, "surprising" the standards at work in culture. And, to a significant extent, this effort at modification is the work of discourse.

[38] Of course, this picture of the knowing subject in relief against a backdrop of cultural norms and historical currents is much too static. For the contemporary relativist, the past is an active presence in the thinking process. That is, the ideas and procedures that characterize our response to present experience emerge from a background of prior relations and persistent memories, out of a history (even a pre-history) of contacts with personal, cultural, and historical forces reconstructed by the subject as the constitutive elements of one's personal orientation. Accordingly, the knowing subject is never a *tabula rasa*: when we think about something, we customarily begin *in medias res*, as though consciousness is a Burkean parlor and the "unending conversation" of human history has been going on in mind long before we are prepared to speak (1941, 110-111; cf. Bakhtin). So while rhetoricians rightly insist that the effective rhetor is always responsive to the kairotic demands of the present (see Thomas Kent's *Paralogic Rhetoric*), the relativist will argue that the past too is always present and makes its presence known in the way we orient ourselves in response to particular experiences.

[39] We can put the Protagorean position in a way more conducive to scepticism: if all thoughts are bound to their distinctive orientation, and if no individual can assume an omniscient/objective position beyond the limits of personal perception, then no particular knowledge claim possesses the requisite justification to allow it the status of indubitable, absolute truth (cf. *Encyc. of Philosophy* 7:449). This potential scepticism is the very point about the human-measure doctrine that so troubled Plato, who makes the distinction in the *Republic* between "doxophilists" (lovers of opinion) and philosophers (lovers of knowledge). The business of the latter is to attend to just those universal, immutable truths that lay beyond the reach of fallible opinion (479b-480a). Gorgias has been claimed as a sceptic on the basis of his three-part proposition that nothing exists; that even if it did exist, it could not be known; and even if knowable, it could not be communicated (*DK* 82 B3). I would agree with Sextus that the second proposition seems clearly sceptical, but the characterization of Protagoras as a sceptic—doubting our ability to know anything—seems to me dependent on the alien standards of the master tradition.

[40] For a more extensive discussion of scepticism in relation to Cicero, see Ch. 5, sec. 1. Cf. Epilogue, sec. 2.

[41] E.g., Karl Popper suggests that when knowledge is conceived of as following from disctinct perceptual orientations we become "prisoners caught in the framework of our theories; our expectations; our past experiences; our languages" (in Bernstein 1982, 44).

[42] To a significant extent, the Protagorean subject can be identified with the Aristotelian *phronemos*, the wise person capable of acting in behalf of goodness (*Nic. Ethics* 1140b).

[43] The linkage of Protagorean ideas and American pragmatism will be developed in Ch. 3. Cf. Paolo Freire: "World and man do not exist apart from each other, they exist in constant interaction" (36).

[44] Cf. Laszlo Versenyi's commentary on the human-measure doctrine: "there is no point in speaking in a grand manner about what things may or may not be in themselves; what we have to take into account and concentrate on is what they are for us, in the world we live in, in a world in which our relationship to things, our living in the world is decisive"

(1962, 182). Cf. also Polanyi's claim that "into every act of knowing there enters a passionate contribution of the person knowing what is being known" (viii).

[45] I am indebted throughout the following paragraph to the lexical distinctions made by philosopher Richard Bernstein in his discussion of incommensurability (1983, 82f).

[46] Feyerabend argues throughout *Against Method* that the concept of incommensurability is only vaguely understood (205-7, 211-13).

[47] Cf. Feyerabend 1975, 211-13, 262-63; see also Margolis' chapter on "Protagoras and Incommensurability," 1991, 87-118.

[48] According to Kuhn, contesting paradigms may initially seem as though "they practice their trade in different worlds," especially with regard to their definition of problems and their standards of evaluation (148-50). And yet, communication across such a "revolutionary divide," while it cannot be "forced by logic," is "partial" rather than absent (150-51). For practical rather than theoretical purposes, such "partiality" allows for some relationship between incommensurables, some point of contact on which comparison can be founded.

[49] In the *Nic. Ethics*, Aristotle writes that "it is the mark of an educated [person] to look for precision in each class of thing just so far as the nature of the subject admits. It is evidently equally foolish to accept probable reasoning from a mathematician and to demand from a rhetorician demonstrative proofs" (1094b 34-27). On the Aristotelian division of knowledge into kinds, see Consigny 1989; Guthrie 6. 130ff; and Gadamer 1994, 312-14. On the German disciplines, see Bernstein 1983, 30-37.

[50] Such propositions are routinely expressed in bivalent terms: X is or is not the case with regard to Y.

[51] A note of caution is in order here. The force of the "categorical" division of knowledge is that it tempers the notion that truth is inherently a function of human invention and so avoids the fear that all constraints placed by the world on human belief can be ignored (see Brinton 159). The weakness of the position is that distinctions in knowledge-types easily slide into hierarchy. As long as one holds (at least in part) the notion that in the category of the physical sciences meanings are dependent on self-substantiated facts or data, that the data are verifiable by systematic, formal analysis, and that this analysis will, in turn, guarantee the "truth" of an inductively-generated proposition, it is not hard to understand how one's faith in the determinate, verifiable nature of scientific knowledge would prompt the elevation of this "category" to paradigmatic status. Rorty writes that "there is nothing wrong with science, there is only something wrong with the attempt to divinize it": i.e., there is only something wrong with scientism, the effort to make the standards of technical knowledge the criteria for all knowing (1991, 34).

[52] Kerferd's translation, 1949, 23. Cf. Cornford 81-83.

Chapter 2

[1] Jaap Mansfeld refers to "the inspiring ambiguity of homo mensura" (39).

[2] Dialogical argument and its appropriate pedagogy are the specific topics of the Epilogue.

[3] In a 1912 review of Gomperz's *Sophistik und Rhetorik*, Schiller assigns explicit responsibility for these judgments regarding value to antilogic, "the rhetorical technique of arguing both sides" (112). I am indebted to Steven Mailloux's essay on "Sophistry and Rhetorical Pragmatism" for bringing this review to my attention (12). Mailloux's edited

collection of essays, *Rhetoric, Sophism, and Pragmatism*, offers helpful assessments of the affinity between Protagorean thought and contemporary philosophical trends, neo-pragmatism in particular.

[4] E.g., Jaap Mansfeld notes that "Protagoras' technique of argument and discussion [is] grounded in his epistemology" (47); and Untersteiner, if I read him correctly, refers to the mastery of "man as measure" over the potentially destructive force of antilogic (49) (cf. Solmsen 1975, 34). On the other hand, a number of critics give priority to the two-logoi fragment; e.g., Michael Billig, who presents one of the more interesting interpretations of antilogic, refers to the human-measure statement as "none too lucid" and identifies antilogic as the "central part" of Protagoras' worldview (5, 44-48); and Susan Jarratt writes that "the human-as-measure doctrine is an answer to the dissoi logoi" (1991, 50). David Payne effects an attractive coalescence of epistemology and rhetoric when he notes that for Protagoras "knowledge is an inherently rhetorical process because people necessarily value the items and objects of reality with categories and general ideas constituted in and through rhetoric" (3).

[5] The most complete textual analysis of the fragment is in Schiappa 89-102.

[6] For an interesting discussion of the role of key words in the study of classical rhetoric, see K. Welch 1990, 11-15.

[7] Cf. Havelock on "things that happen," 228. See also Versenyi on chremata, 1962.

[8] I will return to the consideration of stronger and weaker arguments in Chapter 3, during a longer discussion of Protagorean pragmatism.

[9] Protagorean teaching—insofar as we know it—should be qualified as democratic in theory, since, in practice, Protagoras himself seems to have taught primarily aristocratic males (cf. Jarrett 62). Alternatively, Protagoras does indicate that he will renounce his fee for "anyone" who comes to him for instruction and does not feel the benefit (*Protagoras* 328c). Schiappa rightly notes the irony in this account: "Though Protagoras' doctrines contributed to the democratization of Athens and can be read as justification for radical democracy, Ostwald is correct in concluding that 'only the wealthy could afford Protagoras'" (171). For the fullest and most dependable account of Protagorean political philosophy and its democratic orientation, see Cynthia Farrar's *The Origins of Democratic Thinking* (44-125).

[10] On sophistic education, see Jaeger 295-328; de Romilly 1992, 30-92; Marrou 24-60; James Jarrett 1-64; T. Poulakis 1997; and Schiappa 157-74.

[11] Nancy Struever points out that historical shifts in the relationship between rhetoric and philosophy are routinely accompanied by changes in, among other things, pedagogy (23).

[12] Marrou's charge should be balanced by Plato's expressed psychogogic purpose; i.e., to lead souls to the truth by means of words (*Phaedrus* 261a, 271c-d). For a very different assessment of Socratic education, see Martha Nussbaum's *Cultivating Humanity*.

[13] In *Knowledge and Human Interests*, Habermas refers to "emancipatory cognitive interests" that reveal "ideologically frozen relations of dependence that can in principle be transformed" (Wenzel 1988).

[14] The quote is from Thucydides' *Funeral Oration* (2.40-41): in Athenian democracy, "we think [discussion] an indispensable preliminary to any wise action."

[15] A synopsis of neo-Protagorean philosophy appears early in the Epilogue.

[16] Cf. Prodicus' distinction between discussion and debate in Protagoras 337b.

[17] Eric Havelock has claimed that the Socratic dialogues have, in fact, determined the outcome of the discussion before the dialectic even begins. So that in the *Protagoras*, the discussion is not assumed to be a conversation between equals but rather an interrogation driven by the questioner who knows in advance where he intends to go (Ch. 8).

[18] See Payne's notion that rhetoric is logically prior to dialectic (190), and Schiappa on antilogic as Protagoras' variant of dialectic (44).

[19] The following paragraphs on Athenian politics and culture draw from Cynthia Farrar (1-98); H. D. F. Kitto (109-35); and Oswyn Murray (240-46).

[20] For an interesting comment on the empiricism of Protagoras' contemporary Anaxagoras, see de Romilly 1992, 11.

[21] Mario Untersteiner would add the dramatist Aeschylus as a literary precursor who firmly understood the "double nature of reality" (22-23).

[22] The linkage between Heraclitus and Protagoras was begun by Plato in the *Theaetetus* (152e), but a good deal of 20C criticism has reaffirmed the relationship. See, in particular, Untersteiner (22-25) and Schiappa (90-98). My own reading of Heraclitus leans heavily on Charles Kahn, with additional references to Wheelwright and Barnes (1972; v. 2, 57-81). The classic discussion of "conceptualized" rhetoric is, of course, Kennedy (1980, 15-17).

[23] I am using primarily the translations of Heraclitus by Charles Kahn, though, on occasion, I employ the graceful renditions of Philip Wheelwright.

[24] Heraclitus' dominant symbol for this conjunction of opposites is fire, which provides for heating, cooking, and toolmaking, but also, in the funeral pyre and in conflagration, symbolizes death. In fire, then, the eternal process of extinguishing and rekindling finds a single, unified image (see *DK* 22 B30, 31a, 90; Kahn 145).

[25] De Romilly supports the priority of the human-measure doctrine when she writes that "Protagoras converted [antilogic] into . . . a method of argument in itself, for which the rest of his teaching paved the way" (1992, 76).

Chapter 3

[1] This linguistic application is clearly reflected in the many 20C translations of the two-logoi fragment ("there are two, opposing ideas present concerning everything"), translations that emphasize the rhetoricity of the concept by using language-related synonyms for logos and *pragma*. For example, W. K. C. Guthrie translates the fragment as "There are two opposite arguments on every subject" (III, 182), and George Kennedy as "Something can be said on both sides of every question" (1963, 31; see Schiappa for other examples, 90-91). Schiappa also champions the notion that the two-logoi fragment is "an extension of Heraclitus' thought into the realm of what we would now call linguistic theory" (98).

[2] Bloom 1982; Fish 1989; Mailloux 1995; Margolis 1991; Rorty 1979; C. West 1989.

[3] As noted at the end of the last chapter, the Milesian philosophers of the 6C BCE invoked oppositions like hot/cold as what Aristotle would term the "material cause" of natural processes. In a slightly later philosophy like that of Empedocles (early 5C BCE), the opposing forces of nature yield primacy to the conceptual principles of Love and Discord which function as the "efficient cause" of natural opposition (just as a potter is the "efficient" cause of change in the "material" of clay; see *Encyc. of Philosophy* 2.56).

[4] It is important to recall that the human-measure fragment is reputed to have been the first sentence of Protagoras' book *On Truth* and so operates as something of a manifesto.

[5] See Hardie 212-39; Kenny 160-89; and Dahl (passim). I found the detailed, but carefully reasoned discussion of *phronesis* by Sarah Broadie to be a particularly rewarding examination of this difficult subject; see 179-265. See also the last chapter of Garver, "Aristotle's Rhetoric and the History of Prudence" 232-48.

[6] In a detailed and fascinating application of this binary conception of *phronesis* to rhetoric, Garver posits both internal and external goals for rhetoric, the former answering to ethical standards of fulfillment; see 22-45.

[7] The conflict between a rhetoric of justification (or persuasion) and a rhetoric of inquiry has been a central feature of the rhetorical tradition. I continue to think of Whatley's *Elements of Rhetoric* as the former since it presupposes a Ramistic separation of logic (which establishes choice) from rhetoric (which is left only to defend one's choice). This conflict leads understandably (though I think mistakenly) to the condemnation of ancient rhetoric by writers like Knoblauch and Brannon (see 22-50).

[8] Throughout the following review of Peirce and Dewey, I am indebted to Bernstein's analysis and commentary. Unless otherwise noted, all references in this section are to *Praxis and Action* (1971). See also Bernstein's *John Dewey* (1966), esp. Chs. 5 and 6; and Robert Westbrook's *John Dewey and American Democracy* (1991).

[9] For an interesting discussion of the theory/*praxis* split in ancient and modern sources, see James Kinneavy's "Translating Theory into Practice in Teaching Composition: A Historical View and a Contemporary View" (1984). See also Nikolaus Lobkowicz's *Theory and Practice: History of a Concept from Aristotle to Marx.*

[10] Interestingly, both Peirce and Dewey tend to think of the relation between such dialectical opposites as mind and matter in Hegelian terms. Peirce wrote that if Hegel had looked on the first two stages of his dialectic (our logos and antilogos) with something other than "a smile of contempt," then "pragmatists might have looked upon him as the great vindicator of their truth" (in Bernstein 1971, 166). And Dewey writes that Hegel had demonstrated that "in a great variety of fields of experience, the supreme reality of this principle of living unity maintain[s] itself through the medium of differences and distinction" (in Bernstein 1971, 171). The emphasis of both comments is on the importance of contrast, of two-sidedness, of difference and distinction held in Heraclitean union.

[11] Peirce's comment bears comparison with the translation of the two-logoi fragment put forward by Mario Untersteiner, one of the first modern critics to break with the pejorative approach to sophism. Unstersteiner translates the fragment as follows: "in every experience there are two logoi in opposition to each other" (19).

[12] There is no exact translation of *aidos* into English, but de Romilly writes that it is best read as "all the kinds of consideration that may be shown for others when one is mindful of their due" (1992, 165, n.3).

[13] This last translation occupies a difficult-to-overlook position as the penultimate sentence in the selection from the *Rhetoric* in Bizzell and Herzberg's influential *Rhetorical Tradition*, a text that introduces a great many students to rhetorical history.

[14] Sprague and Freeman are the English companions to *Diels-Kranz*, my principal source for the Protagorean canon.

[15] The Loeb editors of Aristophanes add a footnote to this passage indicating its source in "the famous 'promise of Protagoras' . . . the Sophist of Abdera" (274-75, note b).

[16] If there is any doubt that Aristophanes himself, rather than simply the Loeb editors, has Protagoras in mind as his target (see n. 15 above), then the jokes about "measure," a term intimately associated with Protagoras through the human-measure statement, should resolve the uncertainty (see ll. 636-41).

[17] There is unquestionably a note of "Socratic" irony in almost all comments by Socrates on the life and character of Protagoras, but the basic tone of these references is one of qualified admiration which never questions the Sophist's moral stature. At the outset of the *Protagoras*, Socrates introduces the title character with admiration (309c-d); and in the *Theaetetus*, Socrates concludes his own imitation of a Protagorean "apologia" with the comment that "were he [Protagoras] alive to speak for himself [his speech] would be a much more impressive affair" (168c). And in the *Meno*, Socrates acknowledges that Protagoras was an esteemed teacher whose reputation was "consistently high" both throughout his life and after his death (91a-e).

[18] My own assessment of Protagorean ethics is indebted to Versenyi (see his *Socratic Humanism* 1963, 22-39; all references to his work in this section are to this text).

[19] In the *Theaetetus*, Protagoras does not appear in person but has his putative position represented by Socrates who, in turn, is recreated by Plato. There is much controversy over the accuracy of Plato/Socrates' account of Protagoras' views, though I believe the ideas presented are basically consistent with what we might expect from Protagoras, despite some excesses (see *Protagoras* 322d). On this topic, see F. C. S. Schiller's *Plato or Protagoras?* (1908) and Maguire (1973).

[20] Hippocrates is mentioned by Plato in the *Protagoras* and is probably a contemporary of the Sophist. For a discussion on the relation of Protagorean thought to Hippocratic ideas, see Chapter 2, section 1 above and Versenyi 33-35.

[21] For a modern program of ethics that is not based on universal, ahistorical principles, see the "distributive logic" of Iris Marion Young (1990) and the rhetorical application of such ethics in Christy Friend (1994). Compare also the discussion of "causuistry," or case ethics by Jonsen and Toulmin in *The Abuse of Causuistry* (1988) and Toulmin in *Cosmopolis* (1990).

[22] After a lecture of mine on antilogic at Providence University in Taichung, Taiwan, a wonderful student wrote this poem for me:

Unbelievably,
it starts with taking the opposite position,
defending your rival.
This exchange makes a tremendous . . . difference,
Or a miracle, in which there is no
Winner or loser in the process of negotiation.
This is not notorious argument.
It is a leap from hostility
To understanding your own ignorance,
A transference from the self to the other,
And a discovery that out of partiality comes wholeness.

[23] For an illuminating study of paradox (esp. in the Renaissance), see Rosalie Colie's *Paradoxica Epidemica*.

Chapter 4

[1] One recent writer maintains that there is a split in the two speakers' conceptions of the generic conventions involved, with Cleon speaking in a forensic and Diodotus in a deliberative mode. See Liu (1995, 19-20).

[2] We can see the debate here as the classic one for 5C Greek culture between nomos (the law) and phusis (nature), the latter in this case construed as human nature and its capacity for growth and development. This debate was directly related to Sophistic training and appears repeatedly throughout writing of the period; it is especially well represented in Euripides (see Kerferd 1981, 111-30; de Romilly 1992, 31-56, esp. 49).

[3] I am once again using the Robinson translation. See also Freeman, 417-23.

[4] There are, of course, precedents for the incorporation of alternative positions within a single speech, precedents in the form of proleptic comments or rhetorical questions (see Demonsthenes' "First Phillipic"). But typically, opposing positions or the voice of the "other" do not become much more than a shadow of the dominant argument. As for Plato's antilogical tour de force in the Phaedrus, though the first two speeches are entertaining on their own and clearly in opposition to one another, their ultimate rejection in the palinode hardly seems in doubt.

[5] In the next chapter, I will explore the origins of "suspended judgment" (epoche) in Hellenistic scepticism (see Schmitt 1972, Long 1974, Hankinson 1995).

[6] With regard to offering respect and granting a hearing to all sides, my position is not quite the same as that of Wayne Booth, who encourages the readers of Modern Dogma and the Rhetoric of Assent to accept what there is no good reason to doubt (1974, 40). Booth's position is in many ways similar to the Davidsonian "charity" of interpretation advocated by my friend Thomas Kent (1993). Protagorean theory would fully support the role of respect (aidos) in the process of communication but differs in believing that opposition is both an act of goodwill and an incentive to enhanced understanding.

[7] It is not necessary here to elaborate on the "conversational" aspect of antilogic because the conversational nature of rhetoric in general has been so thoroughly examined by contemporary scholars with an interest in the social context of discourse. See, in particular, Gregory Clark in Chapter 3 ("Writing in Conversation: The Conversational Model") of Dialogue, Dialectic, and Conversation 1990.

[8] Havelock spends a great deal of time in Chapter 8 of the Liberal Temper ("Human Opinion and Parliamentary Process") distinguishing between the Protagorean and the Socratic approach to argument (see 191-239), a critical effort to which I am indebted. Along the way, Havelock also notes that when Protagoras points out that he has often participated in public debates (335a), the Sophist implies that some other form of debate besides the Socratic method exists (211). That alternative method is antilogic (see discussion of dialectic in Ch. 2, sec. 3).

[9] It helps to keep in mind that "dialogism" is an "umbrella term" whose meanings are both broad and changeable (Schuster 605; see also Morson and Emerson 131-33).

[10] In terms of formal logic, contraries (like black and white) may or may not be predicated by the some subject; but, according to the "law of contradiction," they cannot be truly predicated by the same subject, at the same time, in the same respect (see Aristotle's Metaphysics 1005b 19-20). With contradictories, on the other hand, the truth of one position requires the falsity of the other, so that disproving position A (the number 24 is odd) is sufficient to prove the contradictory position B (24 is even). See Lloyd 1966, 87-

88. For discussions of the law of contradiction and its relation the "impossible to contradict" fragment (*ouk est antilegein*) and the human-measure doctrine of Protagoras, see Kerferd 1981, 88-89; Schiappa 134-40; and Chapter 1 above.

[11] An alternative translation for aidos is "shame," which some historians occasionally employ. But for obvious reasons aidos-as-respect is preferred here.

[12] An alternative translation of the middle sentence would be, "I think you should agree to discuss (*amphisbetein*) and not to fight (*erizein*)."

[13] Martin Buber argues that genuine communication involves a two-way transaction rather than one-way transmission. Moreover, he posits dialogue as the heart of the human condition, the means by which we acknowledge the authenticity of others (see Epilogue, sec. 2; Buber 1947).

[14] The motto of the Fourth International Conference on Argumentation (Amsterdam 1998) bears on this point: "to give in to a strong argument is a sign of strength."

[15] My own reading of Trimbur's provocative article is admittedly reductive in its focus on the kinship between the rhetoric of dissensus and the oppositional nature of antilogic. For a lively critique of Trimbur's concept of conversational interaction and discourse communities, see Kent 119-123.

[16] The foremost educational theorist of our own era seems to agree. John Dewey, in The Public and Its Problems, writes that "The essential need . . . is the improvement of the methods and conditions of debate, discussion, and persuasion. That is the problem of the public" (208). See Cornel West's The American Evasion of Philosophy (1989, 105).

Chapter 5

An earlier version of this chapter appeared in the *Journal of Education* 179:1 (1997): 15-47, as "Everything Must Be Argued: Rhetorical Theory and Pedagogical Practice in Cicero's *De Oratore*," copyright 1997 Boston University. The revised essay is included by permission of the publishers.

[1] Unless otherwise indicated, citations of Cicero are from the *De Oratore*. Other Ciceronian texts are noted in parenthesis. All translations are from the Loeb Classical Library editions, except for an occasional reference to Watson's *Brutus* (1970) which is specified in parenthesis.

[2] My reading of the *De Oratore* is informed by over a decade of conversations with graduate students at Iowa State. I wish I could thank them all, and I hope that the ideas advanced here do justice to their insight.

[3] Among Cicero's (mostly) completed works of this period are the *De Republica* (54-51 BCE, on political theory); *De Legibus* (51 BCE, on theories of the law); the *Academica* (45 BCE, on theories of knowledge); *De Finibus* (45 BCE, on morality); the *Tusculan Disputations* (45 BCE, on death, passion, and the virtues); *De Natura Deorum* (45 BCE, on the gods and their relation to nature); *De Fato* (44 BCE, on fate); and *De Officiis* (43 BCE, on moral conduct). The fact that these works deal with various, contradictory "theories of" their subject (e.g., *De Finibus* examines Epicurean, Sceptic, and Academic accounts of morality) not only indicates their orientation to comparative examination but also highlights the centrality of *controversia* to the philosophic enterprise.

[4] See the *Academica* (2.46.142); *Brutus* (7.30, 12.46, 85.192); *De Oratore* (3.32.128); *De Natura Deorum* (1.12; 1.23.63). In this last reference, Cicero calls Protagoras "the Greatest Sophist of the Age."

[5] See Chapter 1 for a complete list of critical works that treat Protagorean philosophy and rhetoric in detail. For quick reference, see Billig 1987; de Romilly 1992; Kerferd 1981; Schiappa 1992; Untersteiner 1950; and Versenyi 1963. To my knowledge, no one has traced a connection between Protagoras and Cicero in any detail.

[6] See the *Academica* and *De Officiis*. On Cicero's reading of philosophical history, see Buckley 1971; Hunt 1954; McKeon 1950; and Schmitt 1972.

[7] Cicero's adherence to Academic scepticism has been challenged, first by Rudolf Hirzel in the 19C, most recently by John Glucker and Peter Steinmetz, who argue independently that Cicero's early admiration for the Academy and Philo is replaced by an allegiance to Antiochus and the moral dogmatism of the old Academy (see Görler 1995). While this controversy need not detain us here, I find Görler's analysis persuasive; i.e., the various stands taken in the later philosophical works should be seen as a deliberately dramatic effort to render alternative points of view for the purpose of dialogue. I would also reiterate J. G. F. Powell's position that adherence to Academic scepticism does not entail the renunciation of strong commitments on Cicero's part, especially with regard to *virtus* or moral excellence, since Cicero never claims more than probable support for his positions (22).

[8] The charge of self-refutation leveled against a sceptic who maintains positive values will remind readers of a similar charge directed against *homo mensura* for implying a universal standard (see Ch. 1, sec. 2).

[9] On identification, see Burke in *The Rhetoric of Motives* 1951: xiv, 19-24, 51ff.

[10] For a discussion of the various stasis questions, see Bonner 1977, 296-300.

[11] On the spontaneous, paralogical adjustments of interlocutors to one another and to the situation, see Kent 1993.

[12] In technical terms, decorum effects an unprecedented alignment between the levels of style and the offices or duties of rhetoric (see *Orator* 69-72, 122-25; see also Douglas; Fantham 1984; Leff 1990).

[13] This distinction naturally recalls Fantham's dichotomy between intrinsic ornament and extrinsic embellishment, with the latter directed at the pleasure of the audience and the former organized according to the nature and form of the subject.

[14] With regard to this relevance, I need only point out here that if antilogic maintains, as it does, that all parties to an argument have a right to be heard, and if *controversia* depends, as I believe it does, on interaction among dialogical partners to reveal to all rhetors the limitations of their own initial logos or stand, then I hope it is clear that there is in the antilogical/controversial tradition a tolerance for, even a cultivation of dialogue between multiple voices that marks a significant step beyond the essentially monological patterns of formal and even informal argument, patterns which continue to dominate contemporary textbooks and, in turn, classroom conceptions of argumentation.

[15] From the *Ad Familiares* 1.9.23 (to Lentulus Spinther, Dec. 54 BCE; in *De Oratore* ix).

Chapter 6

[1] This 1998 conference was sponsored by POROI (The Project for the Rhetoric of Inquiry) at the University of Iowa to discuss Takis Poulakos' book *Speaking for the Polis: Isocrates' Rhetorical Education* (1997).

[2] All references to Quintilian's *Institutio Oratoria* are to the Butler translation in the Loeb Classical Library edition and will list specific passages in parenthesis. When necessary, the text is abbreviated as IO. As in Chapter 5, citations from Cicero are from the Loeb editions.

[3] See G. M. A. Grube's "Theodorus of Gadara." See also Grube 1965, 272-74.

[4] Cf. Quintilian's approach to his audience with Aristotle's notion that the rhetor should avoid "warping" the judgment of the jury (*Rhetoric* 1354a).

[5] In defense of Quintilian's stature as a critic, if not a theorist, I should note Saintsbury's comment that the *Institutio* is among "the fullest and most satisfactory applications of criticism" in the ancient world (290).

[6] Murphy refers to Roman pedagogy as a "collation" of Greek methods (1987, xxix).

[7] The popularity of declamation in the Greco-Roman world as a genre of public entertainment was in large part the result of its literary elements. In his study of Greek declamatory history, Russell notes that declamation in general, and suasoria in particular are inherently theatrical, featuring as they do an actor/rhetor performing the role of someone else, often under conditions of intense emotional distress (1983, 1-15). The emphasis of declamatory performance was often on the delivery of eloquent, even florid prose, heightened by routine appeals to sententiae (or ingenious language) and to the colors (expressions of pathos) rather than to persuasive value; i.e., by appeals to theatricality rather than argumentation (Bonner 1969, 71-83). Declamation, therefore, was a major contributor to the process that George Kennedy has called the letteraturizzazione of rhetoric, that shift in focus "from persuasion to narration, from civic to personal contexts, and from discourse to literature, including poetry" (1980, 5). It is because of this potential for "slippage" into fiction and theatricality that Quintilian is so insistent on the realism of declamation (2.10). However, insofar as the expansive, performative tendencies of declamation operate as an incentive to invention and a ludic complement to the constant demand of rhetorical instruction for common sense, they may well have served to keep their various audiences engaged, especially given imperial restrictions on ordinary discourse. And, in fact, the dual nature of declamation's history as both a literary and rhetorical practice may also have contributed to its place as the crown of the Roman curriculum, the skill that coordinates the Roman passion for eloquence with an equal commitment to public affairs.

Epilogue

[1] See, for example, the adaptation of the term in business motivational literature: e.g., in Linda Ellinor and Glenna Gerard's *Dialogue: Rediscover the Transforming Power of Conversation* (New York: John Wiley and Sons, 1998).

[2] The outline of dialogic theory that follows in the next few pages is a digest of ideas from a variety of modern philosophical traditions, including Buber's existential anthropology, Gadamer's philosophical hermeneutics, and Bakhtinian dialogism. See also A. Clark, J.

Poulakos 1971, and J. Stewart. Bibliographies of rhetorical studies in dialogue can be found in Stewart 1978 and Ward 1994.

[3] The same point is made by R. G. Collingwood: in order to find out what a person means by a certain statement, "you must also know what the question was to which the thing he has said or written was meant as an answer" (1939, 31).

[4] The general nature of the foregoing discussion gives rise to a great many questions: e.g., How is the object of inquiry related to or affected by dialogical relationships? Does dialogue require mutual commitment to the same protocols? How does dialogue accommodate disrespect, injustice, and inequities of power? Does dialogue function differently in forensic and deliberative contexts? What kind of dialogue is possible when we "converse" with texts rather than interlocutors? Can the dialogical principles I have outlined operate in scientific arguments in which universal concepts play a role? Given the scope of the Epilogue, I can only ask my readers to keep these potential complexities in mind as they contemplate the implications for argument of an emphasis on dialogical relations.

[5] See Michael Carter's forthcoming study of starting points for composition.

[6] Historically, the distinction made here between a purely negative scepticism and the generation of opposing counter-arguments has its roots in the difference between the negative dialectics of the Academician Arcesilaus and the positive approach of his follower Carneades during the Hellenic development of scepticism (see Ch. 5, sec. 1).

[7] I pass over the issue raised earlier that, under working conditions, it may be difficult to distinguish between invention and judgment. Nonetheless, the issue is interesting and deserves attention.

[8] Granted, the appeal to the moral "duties" in a homogeneous community like the ruling class of Cicero's Rome is likely to be less problematic than in most modern societies; but the *De Officiis* makes clear that, despite the importance of the virtues to judgment, their use is always attended with complications.

[9] For a careful analysis of the Ciceronian origins, the historical development, and the modern prospect for practical ethics or causuistic reasoning, see Jonsen and Toulmin (1988).

[10] I am referring to the study of practical ethics and working politics, not the academic versions of these two areas of study, which are generally quite abstract.

[11] I am, of course, assuming a great deal at this point, most notably that the dialogical relationship is still intact and characterized by respectful critique, and that, conversely, no one has hijacked the argument by dogmatic insistence or force. If granted the persistence of dialogue in good faith, the argumentative process should be no less dialogical during judgment than it is during invention. After all, testing requires counterpoint, while calculating preponderance will proceed by claims pro and con delivered by advocates and critics, not by someone speaking in proxy for a missing agent. It is hubris to assume we can speak for others, especially our opponents; it is despotism to refuse those involved the right to contribute on their own, especially those whose voices have not yet been heard. Finally, dialogue needs to persist because, in determining a decision, all those affected have a perspective that needs to be heard if judgment is indeed to be just.

[12] Readers will recall that the terms "better and worse" have been unfairly applied to Protagorean theory since Aristotle: as in "making the worse case the better." See Ch. 3, sec. 1.

[13] Put another way, I have chosen to confine myself to working out the "internal logic" of Protagorean pedagogy rather than consider my own proposals in relation to allied and

opposing trends in contemporary research. Much is lost in foregoing the latter approach, and antilogic itself argues that I have a responsibility to enter into dialogue with other contributors to argument pedagogy. Next time.

[14] Some groups of students will be more uniform in belief than others. But the teacher who employs the rhetorical/pedagogical strategies of *in utramque partem* and sceptical opposition should have more than an adequate share of diverse interpretations to work with.

[15] Throughout this paragraph and the next, I am indebted to Professor Bender's fine article on rhetorical imitation (1993).

[16] For theoretical discussion of the use of cases in rhetoric and composition classes, see Flower 1981 and Brockman 1984.

[17] In earlier chapters, I addressed the political and ethical biases of antilogical practice (see Ch. 2, sec. 1; Ch. 3 passim). In the present context, I will only rephrase what I have claimed before—that rhetorical knowledge follows practical experience, that method is a technical extension of theoretical commitments, and that when students practice specific methods they are experimenting with particular kinds of epistemic commitment, with particular conceptions of knowledge and value.

REFERENCES

Aeschylus. *Oresteia*. Translated by Richard Lattimore. New York: Washington Square Press, 1967.

American Heritage Dictionary of the English Language. 3rd ed. Boston: Houghton Mifflin, 1992.

Antiphon. "Tetralogy #3." In Sprague. 156-162.

Aristophanes. *The Clouds*. Translated by Benjamin Bickley Rogers. 1926. Cambridge: Harvard University Press, 1967.

Aristotle. *The Complete Works*. Edited by Jonathan Barnes. 2 vols. Princeton: Bollingen/Princeton University Press, 1984. (Titles abbreviated in text.)

Atwill, Janet M. *Rhetoric Reclaimed: Aristotle and the Liberal Arts Tradition*. Ithaca: Cornell University Press, 1998.

Augustine. *De Doctrina Christiana*. In Bizzell and Herzberg. 381-422.

Bacon, Francis. *The Advancement of Learning* and *Novum Organum*. In Bizzell and Herzberg. 625-33.

Bakhtin, M. M. *The Dialogical Imagination*. Edited by Michael Holquist. Translated by Caryl Emerson and M. Holquist. Austin: University of Texas Press, 1981.

_____. *The Problems of Dostoyevski's Poetics*. Edited and Translated by Caryl Emerson. Minneapolis: University of Minnesota Press, 1984.

_____. *Speech Genres and Other Late Essays*. Edited by Caryl Emerson and Michael Holquist. Translated by Vern W. McGee. Austin: University of Texas Press, 1986.

Barilli, Renato. *Rhetoric*. Translated by Guiliana Menozzi. Minneapolis: University of Minnesota Press, 1989.

Barnes, Jonathan. *Aristotle*. Oxford: Oxford University Press, 1982.

_____. *The Presocratic Philosophers*. 2 vols. London: Routledge and Kegan Paul, 1972.

Barrett, Harold. *The Sophists: Rhetoric, Democracy, and Plato's Idea of Sophistry*. Novato, CA: Chandler and Sharp, 1987.

Bator, Paul. "Aristotelian and Rogerian Rhetoric." *College Composition and Communication* 31 (1980): 427-32.

Beale, Walter. "Decorum." In *Encyclopedia of Rhetoric and Composition*. Edited by Theresa Enos. New York: Garland Press, 1996. 168-70.

Bender, Daniel. "Diversity Revisited, or Composition's Alien History." *Rhetoric Review* 12.1 (1993): 108-23.

Berkeley, George. *Principles of Human Knowledge and Three Dialogues*. Oxford: Oxford University Press, 1996.

Berlin, James. "Richard Whately and Current-Traditional Rhetoric." *College English* 42:1 (1980): 10-17.

Bernstein, Richard J. *Beyond Objectivism and Relativism: Science, Hermeneutics, and Praxis*. Philadelphia: University of Pennsylvania Press, 1985.

_____. *John Dewey*. New York: Washington Square Press, 1966.

_____. *Praxis and Action: Contemporary Philosophies of Human Activity*. Philadelphia: University of Pennsylvania Press, 1971.

Bialostosky, Don H. "Liberal Education, Writing, and the Dialogic Self." In Harkin and Shilb. 11-22.

Billig, Michael. *Arguing and Thinking: A Rhetorical Approach to Social Psychology*. Cambridge: Cambridge University Press, 1987.

Bineham, Jeffery L. "The Cartesian Anxiety in Epistemic Rhetoric: An Assessment of the Literature." *Philosophy and Rhetoric* 21.1 (1990): 43-62.

_____. "The Hermeneutic Medium." *Philosophy and Rhetoric* 28.1 (1995): 1-16.

Bitzer, Lloyd F. "Rhetoric and Public Knowledge." In *Rhetoric, Philosophy, and Literature: An Exploration*. Edited by Don Burks. West Lafayette: Purdue University Press, 1978. 67-93.

Bizzell, Patricia. "Beyond Foundationalism to Rhetorical Authority: Problems Defining 'Cultural Literacy.'" *College English* 52 (1990): 661-75.

_____, and Bruce Herzberg, eds. *The Rhetorical Tradition: Readings from the Classical Times to the Present*. 1st ed. Boston: St. Martin's Press, 1990.

Black, Edwin. "Plato's View of Rhetoric." In *Plato: True and Sophistic Rhetoric*. Edited by Keith V. Erickson. Amsterdam: Editions Rodopi, 1979. 171-91.

Blair, J. Anthony and Ralph H. Johnson. "Argumentation as Dialectical." *Argumentation* 1 (1987): 41-56.

Blake, William. *The Poetry and Prose*. Edited by David V. Erdman. New York: Doubleday, 1970.

Bloom, Harold. *Agon: Towards a Theory of Revisionism*. New York: Oxford University Press, 1982.

_____. *A Map of Misreading*. Oxford: Oxford University Press, 1975.

Boardman, John, Jasper Griffin, and Oswyn Murray, eds. *The Oxford History of Greece and the Hellenic World*. London: Oxford University Press, 1991.

Bolgar, R. R. *The Classical Heritage and its Beneficiaries*. Cambridge: Cambridge University Press, 1954.

Bonner, Stanley F. *Education in Ancient Rome: From the elder Cato to the younger Pliny*. London: Methuen and Company, 1977.

_____. *Roman Declamation in the Late Republic and Early Empire*. Liverpool: Liverpool University Press, 1969.

Booth, Wayne. *Modern Dogma and the Rhetoric of Assent*. Chicago: University of Chicago Press, 1974.

Borges, Jorge Luis. *Labyrinths: Selected Stories and Other Writings*. New York: New Directions, 1964.

Bouwsma, William J. "Socrates and the Confusion of the Humanities." In *The American Future and the Humane Tradition: The Role of the Humanities in Higher Education*. Edited by Robert E. Heideman. NP: Associated Faculty Press, 1982. 11-22.

Brent, Doug. "Rogerian Rhetoric: Ethical Growth Through Alternative Forms of Argumentation." In Emmel, Resch, and Tenney. 73-96.

_____. "Young, Becker and Pike's 'Rogerian' Rhetoric: A Twenty-Year Reassessment." *College English* 53:4 (April 1991): 452-466.

Brett, Richard. "Sophism and Relativism." *Phronesis* 34.2 (1989): 139-169.

Brinton, Allan. "William James and the Epistemic View of Rhetoric." *Quarterly Journal of Speech* 68 (1982): 158-69.

Broadie, Sarah. *Ethics with Aristotle*. New York: Oxford University Press, 1991.

Brockman, R. J. "What is a Case?" In *The Case Method in Technical Communication: Theory and Models*. Edited by R. J. Brockman. NP: The Association of Teachers of Technical Writing, 1984. 1-6.

Brockreide, Wayne E. and Douglas Ehninger. "Toulmin on Argument: An Interpretation and Application." *Quarterly Journal of Speech* 46:1 (1960): 44-53.

Brummet, Barry. "A Defense of Ethical Relativism as Rhetorically Grounded." *Western Journal of Speech Communication* 45 (1981): 286-98.

_____. "On Rhetorical Relativism." *Quarterly Journal of Speech* 68 (1982): 425-30.

_____. "Relativism and Rhetoric." In Cherwitz 1990. 79-103.

_____. "Some Implications of 'Process' or 'Intersubjectivity': Post-Modern Rhetoric." *Philosophy and Rhetoric* 9 (1976): 21-51.

Brunschwig, J. "Aristotle's Rhetoric as a 'Counterpart' to Dialectic." In *Essays on Aristotle's Rhetoric*. Edited by A. K. Rorty. Berkeley: University of California Press, 1996. 34-55.

Buber, Martin. *Between Man and Man*. Translated by Ronald Gregor Smith. London: Kegan Paul, 1947.

_____. *I-Thou*. 1923. Translated by Walter Kaufman. New York: Scribners, 1970.

Buckley, Michael. *Motion and Motion's God: Thematic Variations in Aristotle, Cicero, Newton, and Hegel*. Princeton: Princeton University Press, 1971.

Burke, Kenneth. *A Grammar of Motives*. 1945. Berkeley: University of California Press, 1969.

_____. *The Language of Symbolic Action: Essays on Life, Literature, and Method*. Berkeley: University of California Press, 1966.

_____. *Permanence and Change: An Anatomy of Purpose*. 1935. 3rd ed. Berkeley: University of California Press, 1984.

_____. *A Rhetoric of Motives*. 1950. Berkeley: University of California Press, 1962.

Burnett, Rebecca A. "Conflict in Collaborative Decision-Making." In *Professional Communication: The Social Perspective*. Edited by Nancy Roundy Blyler and Charlotte Thralls. Newbury Park: Sage Publications, 1993. 144-61.

Burnyeat, M. F. "Protagoras and Self-Refutation in Later Greek Philosophy." *Philosophical Review* 85 (1976): 44-69.

Carroll, Lewis. *Alice in Wonderland.* 1865. New York: W. W. Norton, 1971.

Carter, Michael. "*Stasis* and *Kairos*: Principles of Social Construction in Classical Rhetoric." *Rhetoric Review* 7:1 (1988): 97-112.

Cherwitz, Richard A., ed. *Rhetoric and Philosophy.* Hillsdale, NJ: Lawrence W. Erlbaum, 1990.

_____, and James W. Hikins. "John Stuart Mill's *On Liberty*: Implications for the Epistemology of the New Rhetoric." *Quarterly Journal of Speech* 65 (1979): 12-24.

_____. "Rhetorical Perspectivism." *Quarterly Journal of Speech* 69 (1983): 249-66.

Cicero. *Academica.* Translated by H. Rackham. Loeb Classical Library. Cambridge: Harvard University Press, 1933.

_____. *Brutus* and *Orator*. Translated by G. L. Hendrickson and H. M. Hubbell. 1939. Loeb Classical Library. Cambridge: Harvard University Press, 1971.

_____. *De Finibus Bonorum et Malorum.* Translated by H. Rackham. 1931. Loeb Classical Library. Cambridge: Harvard University Press, 1967.

_____. *De Inventione*. Translated by H. M. Hubbell. Loeb Classical Library. Cambridge: Harvard University Press, 1949.

_____. *Letters to Atticus*. 7 vols. Translated by D. R. Shakelton Bailey. Loeb Classical Library. Cambridge: Harvard University Press, 1965-70.

_____. *De Natura Deorum* and *Academica*. Translated by H. Rackham. 1933. Loeb Classical Library. Cambridge: Harvard University Press, 1967.

_____. *De Officiis*. Translated by Walter Miller. 1913. Loeb Classical Library. Cambridge: Harvard University Press, 1968.

_____. *De Oratore*. 2 vols. Translated by E. W. Sutton. 1942. Loeb Classical Library. Cambridge: Harvard University Press, 1988.

_____. *Topica*. Translated by H. M. Hubbell. Loeb Classical Library. Cambridge: Harvard University Press, 1949.

_____. *Tusculan Disputations*. Translated by J. E. King. 1927. Loeb Classical Library. Cambridge: Harvard University Press, 1966.

Clark, Allen. "Martin Buber, Dialogue, and the Philosophy of Rhetoric." In *Philosophers on Rhetoric*. Ed. Donald G. Douglas. Skokie, IL: National Textbook, 1973. 225-42.

Clark, Donald Lemen. *Rhetoric in Greco-Roman Education*. Morningside Heights, NY: Columbia University Press, 1957.

Clark, Gregory. *Dialogue, Dialectic, and Conversation: A Social Perspective on the Function of Writing*. Carbondale: So. Illinois University Press, 1990.

Clarke, M. L. *Rhetoric at Rome: A Historical Survey*. New York: Barnes and Noble, 1963.

Code, Lorraine. "The Importance of Historicism for a Theory of Knowledge." *International Philosophical Quarterly* 22:2 (1982): 157-74.

_____. *What Does She Know? Feminist Theory and the Construction of Knowledge*. Ithaca: Cornell University Press, 1991.

Cohen, Tom. "The 'Genealogies' of Pragmatism." In Mailloux. 94-108.

Cole, A. Thomas. *The Origins of Rhetoric in Ancient Greece*. Baltimore: Johns Hopkins University Press, 1991.

_____. "The Relativism of Protagoras." *Yale Classical Studies* 22 (1972): 19-45.

Colie, Rosalie L. *Paradoxica Epidemica: The Renaissance Tradition of Paradox*. Princeton: Princeton University Press, 1966.

Collingwood, R. G. *An Autobiography*. New York: Oxford University Press, 1939.

_____. *The Idea of History*. Oxford: Oxford University Press, 1946.

Colson, F. H. "Introduction." *M. Fabii Quintilianii: Institutionis Oratoriae, Liber I*. Cambridge: Cambridge University Press, 1924. i-xx.

Comprone, J. "Expressive Writing: Exercises in a New Progymnasmata." *Journal of Technical Writing* 4:1 (1985): 31-53.

Consigny, Scott. "Dialectical, Rhetorical, and Aristotelean Rhetoric." *Philosophy and Rhetoric* 21 (1988): 232-39.

_____. "Gorgias's Use of the Epideictic." *Philosophy and Rhetoric* 25:3 (1994): 281-97.

_____. "Nietzsche's Reading of the Sophists." *Rhetoric Review* 13.1 (1994): 5-26.

Copeland, Rita. *Rhetoric, Hermeneutics, and Translation in the Middle Ages*. Cambridge: Cambridge University Press, 1991.

Copi, Irving M. *Introduction to Logic*. 7th ed. New York: Macmillan, 1986.

Corbett, Edward P. J. "Classical and Modern Rhetoric: The Basic Issues." In *Discourse Studies in Honor of James L. Kinneavy*. Edited by Rosalind J. Gabin. Patomac, MD: Scripta Humanistica, 1995. 1-9.

_____. "The Theory and Practice of Imitation in Classical Rhetoric." *College Composition and Communication* 22 (1971): 243-250.

_____. James L. Golden, and Goodwin F. Berquist, eds. *Essays on the Rhetoric of the Western World*. Dubuque, IA: Kendall/Hunt, 1990.

Corder, Jim W. "Argument as Emergence, Rhetoric as Love." In *Professing the New Rhetorics: A Sourcebook*. Eds. Theresa Enos and Stuart C. Brown. Englewood Cliffs: Prentice Hall, 1994. 412-28.

Cornford, Francis. *Plato's Theory of Knowledge*. New York: Liberal Arts Press, 1957.

Couture, Barbara. "Against Relativism: Restoring Truth in Writing." *Journal of Advanced Composition* 13.1 (1993): 109-13.

——— and Jone Rymer Goldstein. *Cases for Technical and Professional Writing*. Boston: Little Brown, 1985.

Covino, William O. *The Art of Wondering: A Revisionist Return to the History of Rhetoric*. Portsmouth, NH: Boynton/Cook, Heinemann, 1988.

_____. *The Elements of Persuasion*. Boston: Allyn and Bacon, 1997.

Crosswhite, James. *The Rhetoric of Reason: Writing and the Attractions of Argument*. Madison: University of Wisconsin Press, 1996.

Crowley, Sharon and Debora Hawhee. *Ancient Rhetoric for Contemporary Students*. Boston: Allyn and Bacon, 1998.

Crusius, Timothy W. *A Teacher's Introduction to Philosophical Hermeneutics*. Urbana: National Council of Teachers of English, 1991.

Dahl, Norman O. *Practical Reason, Aristotle, and Weakness of the Will*. Minneapolis: University of Minnesota Press, 1984.

Dassenbrock, Reed Way. "Do We Write the Text We Read?" In *Falling into Theory: Conflicting Views on Reading Literature*. Edited by David H. Richter. Boston: Bedford Books, 1994. 238-248.

De Man, Paul. *Allegories of Reading: Figural Language in Rousseau, Nietzsche, Rilke, and Proust*. New Haven: Yale University Press, 1979.

Demosthenes. "The First Phillipic." In *A Synoptic History of Classical Rhetoric*. 2nd ed. Edited by James J. Murphy. Davis, CA: Hermagoras Press, 1994. 252-63.

Derrida, Jacques. "Differénce." In *The Margins of Philosophy*. Translated by Alan Bass. Chicago: University of Chicago Press, 1982. 3-27.

_____. "Signature Event Context." In Bizzell and Herzberg. 1168-84.

Descartes, René. *The Philosophical Works*. 2 vols. Translated by Elizabeth S. Haldane and G. R. T. Ross. New York: Dover, 1955.

Dewey, John. *Democracy and Education: An Introduction to the Philosophy of Education*. New York: Macmillan, 1916.

_____. *Experience and Nature*. La Salle, IL: The Open Court Publishing Co., 1925.

_____. "The Need for a Recovery of Philosophy." 1917. In *Creative Intelligence: Essays in the Pragmatic Attitude*. New York: Octagon Books, 1970. 3-69.

_____. *The Public and Its Problems*. 1927. Athens, OH: Swallow Press, 1954.

Diels, Hermann and Walther Kranz, eds. *Die Fragmente der Vorsokratiker*. 2 vols. 1956-59. Dublin and Zurich: Wiedman, 1972. (see Rosamund Kent Sprague for English translation)

Diogenes Laertius. *Lives of Eminent Philosophers*. v. 2. Translated by R. D. Hicks. Loeb Classical Library. Cambridge: Harvard University Press, 1965.

Donovan, Brian. "The Project of Protagoras." *Rhetoric Society Quarterly* 23.1 (1993): 35-47.

Downing, David. "The Political Consequences of Pragmatism; or, Cultural Pragmatics for a Cybernetic Revolution." In Mailloux. 180- 205.

Duff, John Wright. *The History of Rome in the Silver Age from Tiberius to Hadrian*. New York: Harcourt Brace and Company, 1929.

Eco, Umberto. *The Name of the Rose*. Translated by William Weaver. New York: Harcourt Brace, 1983.

Ede, Lisa. "Is Rogerian Rhetoric Really Rogerian?" *Rhetoric Review* 3 (1984): 40-48.

Edward, William A. *The Suasoriae of Seneca the Elder*. Cambridge: Cambridge University Press, 1928.

Ehninger, Douglas. "Argument as Method: Its Nature, Its Limitations, and Its Uses." *Speech Monographs* 37 (1979): 101-110.

_____. "Introduction." Richard Whately's *Elements of Rhetoric*. Carbondale: Southern Illinois University Press, 1963. ix-xxx.

Elbow, Peter. "The Uses of Binary Thinking." *Journal of Advanced Composition* 13:1 (1993): 51-78.

Ellinor, Linda and Glenna Gerard. *Dialogue: Rediscover the Transforming Power of Conversation*. New York: John Wiley and Sons, 1998.

Emerson, Ralph Waldo. *Selected Poetry and Prose*. 2nd ed. Edited by Reginald L. Cook. San Francisco: Rinehart Press, 1969.

Emmel, Barbara, Paula Resch, and Deborah Tenney, eds. *Argument Revisited: Argument Redefined: Renegotiating Meaning in the Composition Classroom*. Thousand Oaks: Sage Publications, 1996.

Emmeren, Frans H. van, Rob Grootendorst, Fancisca Snoek Henkemans. *Fundamentals of Argumentation Theory: A Handbook of Historical Backgrounds and Contemporary Developments*. Mahway, NJ: Lawrence Erlbaum, 1996.

Encyclopedia of Philosophy. 4 vols. Edited by Paul Edwards. New York: Macmillan, 1967.

Erasmus, Desiderius. "*Copia*: Foundations of the Abundant Style." In Bizzell and Herzberg. 499-556.

Euripides. *Orestes*. Translated by M. L. West. Wiltshire, England: Aris and Phillips Ltd., 1987.

Ewald, Helen Rothschild. "Waiting for Answerability: Bakhtin and Composition Studies." *College Composition and Communication* 44 (1993): 331-48.

Fahnstock, Jeanne and Marie Secor. *A Rhetoric of Argument*. 2nd ed. Boston: McGraw Hill, 1990.

_____. "Teaching Argument: A Theory of Types." *College Composition and Communication* 34:1 (1985): 20-30.

Fantham, Elaine. "*Orator* 69-74." In "Studies in Cicero's *Opera Rhetorica*." *Central States Speech Journal* 35 (1984):123-25.

_____. "*Varietas* and *Satietas*: *De Oratore* 3.96-103 and the limits of *Ornatus*." *Rhetorica* 4.3 (1988): 275-90.

_____, ed. *Women in the Classical World*. New York: Oxford University Press, 1994.

Farmer, Frank, ed. *Landmark Essays on Bakthin, Rhetoric, and Writing*. Mahwah, NJ: Hermagoras Press, 1998.

Farrar, Cynthia. *The Origins of Democratic Thinking: The Invention of Politics in Classical Athens*. Cambridge: Cambridge University Press, 1988.

Farrell, Thomas D. "Knowledge, Consensus, and Rhetorical Theory." In Golden, Berquist, and Coleman. 608-622.

_____. "Social Knowledge II." *Quarterly Journal of Speech* 62 (1978): 329-334.

Feyerabend, Paul. *Against Method*. 3rd ed. London: Verso, 1975.

_____. *Farewell to Reason*. London: Verso, 1987.

_____. *Science in a Free Society*. London: New Left Books, 1978.

_____. *Three Dialogues on Knowledge*. London: Basil Blackwell, 1991.

Fish, Stanley. *Doing What Comes Naturally: Change, Rhetoric, and the Practice of Theory in Literature and Legal Studies*. Durham: Duke University Press, 1989.

_____. "Is There a Text in This Class?" *Falling into Theory: Conflicting Views on Reading Literature*. Edited by David H. Richter. Boston: Bedford Books, 1994. 227-37.

Flower, Linda. "Communication Strategy in Professional Writing: Teaching a Rhetorical Case." In *Courses, Components and Exercises in Technical Communication*. Edited by D. W. Stevenson. Urbana: National Council of Teachers of English, 1981. 34-46.

Freeman, Kathleen, ed. *The Pre-Socratic Philosophers: A Companion to Diels, Fragmente der Vorsokratiker*. Cambridge: Harvard University Press, 1946.

Freire, Paolo. *Pedagogy of the Oppressed*. New York: Seabury, 1970.

Frey, Olivia. "Beyond Literary Darwinism: Women's Voices and Critical Discourse." *College English* 52 (1990): 507-26.

Friedman, Maurice S. *Martin Buber: A Life in Dialogue*. Chicago: University of Chicago Press, 1955.

Friend, Christy. "Ethics in the Writing Classroom: A Nondistributive Approach." *College English* 56:5 (1994): 548-67.

Frye, Northrop. *Anatomy of Criticism*. Princeton: Princeton University Press, 1957.

Fulkerson, Richard. *Teaching the Argument in Writing*. Urbana: National Council of Teachers of English, 1996.

_____. "Technical Logic, Comp-Logic, and the Teaching of Writing." *College Composition and Communication* 39:4 (1988): 436-52.

_____. "Transcending Our Conception of Argument in Light of Feminist Critiques." *Argument and Advocacy* 32:4 (1996): 199-217.

Gadamer, Hans-Georg. *Hegel's Dialectic: Five Hermeneutical Studies*. 1971. Translated by P. Christopher Smith. New Haven: Yale University Press, 1976.

____. "The Problem of Historical Consciousness." In *Interpretive Social Science: A Reader*. Edited by Paul Rabinow and William M. Sullivan. Berkeley: University of California Press, 1979.

_____. *Truth and Method*. 1960. 2nd Rev. Edition. Translated by Joel Weinsheimer and Donald G. Marshall. New York: Continuum Publishing Co., 1994.

Gage, John T. "An Adequate Epistemology for Composition: Classical and Modern Perspectives." In *Essays on Classical Rhetoric and Modern Discourse*. Edited by Robert J. Connors, Lisa S. Ede, and Andrea A. Lunsford. Carbondale: So. Illinois University Press, 1984. 152-69.

Gaines, Robert, ed. "Studies in Cicero's *Opera Rhetorica*." Special issue of *Central States Speech Journal* 35 (1984): 120-31.

Garver, Eugene. *Aristotle's Rhetoric: An Art of Character*. Chicago: University of Chicago Press, 1994.

Geertz, Clifford. "Anti Anti-Relativism." In *Relativism: Interpretation and Confrontation*. Ed. Michael Krausz. Notre Dame: University of Notre Dame Press, 1989. 12-34.

Gilford, N. L. *When in Rome: An Introduction to Relativism and Knowledge*. Albany: State University of New York Press, 1983.

Gilman, Sander L. "Introduction." *Fredrich Nietzsche On Rhetoric and Language*. Edited and Translated by Sander L. Gilman, Carole Blaire, and David J. Parent. New York: Oxford University Press, 1989. viii-xxvii.

Glucker, John. "*Probabile, Veri Simile*, and Related Terms." In Powell, 115-43.

Golden, James. L., Goodwin F. Berquist, and William E. Coleman, eds. *The Rhetoric of Western Thought*. 4th ed. Dubuque, IA: Kendall Hunt, 1978.

Gomperz, Theodor. *Greek Thinkers: A History of Ancient Philosophy*. Vol. 1. Translated by Laurie Magnus. London: John Murray, 1901.

Görler, Waldeman. "Silencing the Troublemaker: *De Legibus* 1.39 and the Continuity of Cicero's Scepticism." In Powell. 85-113.

Govier, Trudy. "Are There Two Sides to Every Question?" In *Selected Issues in Logic and Communication*. Edited by Trudy Govier. Belmont, CA: Wadsworth, 1988. 43-54.

Graff, Gerald. *Beyond the Culture Wars: How Teaching the Conflicts Can Revitalize American Education*. New York: Norton, 1992.

_____. *Professing Literature: An Institutional History*. Chicago: University of Chicago Press, 1987.

Greer, W.J. "Quintilian and the Declamation." *The Classical Weekly* 19:4 (1925): 27-31.

Griswold, Charles I., Jr. *Self-Knowledge in Plato's "Phaedrus."* New Haven: Yale University Press, 1986.

Grote, George. *Aristotle*. 2 vols. London: John Murray, 1872.

Grube, G.M.A. *The Greek and Roman Critics*. Toronto: University of Toronto Press, 1965.

_____. "Theodorus of Gadara." *The American Journal of Philology* 80 (1959): 337-65.

Gross, David. "Infinite Indignation: Teaching, Dialectical Vision, and Blake's 'Marriage of Heaven and Hell.'" *College English* 48:2 (1986): 175-186.

Guthrie, W. K. C. *A History of Greek Philosophy*. 8 vols. Cambridge: Cambridge University Press, 1969-81.

Hagaman, L. "The Modern Use of the Progymnasmata in Teaching Rhetorical Invention." *Rhetoric Review* 5:1 (1986): 22-29.

Hairston, Maxine. "Carl Rogers' Alternative to Traditional Rhetoric." *College Composition and Communication* 27 (1976): 373-77.

Hamilton, Edith and Huntington Cairns, eds. *Plato: Collected Dialogues*. New York: Bollingen Foundation, 1963.

Hankinson, R. J. *The Sceptics*. London: Routledge, 1995.

Hardie, W. F. R. *Aristotle's Ethical Theory*. Oxford: Clarendon Press, 1968.

Harkin, Patricia and John Schilb, eds. *Contending with Words: Composition and Rhetoric in a Postmodern Age*. New York: Modern Language Association, 1991.

Harré, Rom and Michael Krausz. *Varieties of Relativism*. Oxford: Blackwell, 1996.

Havelock, Eric A. *The Liberal Temper in Greek Politics*. New Haven: Yale University Press, 1957.

Hegel, G. W. F. *Lectures on the Philosophy of History*. Translated by J. Sibree. London: Henry G. Bohn, 1961.

Heidegger, Martin. *Being and Time*. 1927. Translated by John Macquarrie and Edward Robinson. New York: Harper and Row, 1962.

Heisenberg, Werner. *Physics and Philosophy: The Revolution in Modern Science*. 1938. New York: Harper and Row, 1962.

Herodotus. *The Histories*. Translated by Walter Brisco. New York: W. W. Norton and Co., 1992.

Hesiod. *Theogony and Works and Days*. Translated by M. L. West. New York: Oxford University Press, 1988.

Hikins, James W. "Realism and Its Implications for Rhetorical Theory." In Cherwitz 1990. 21-77.

Hock, R. and E. N. O'Neill. *The Chreia in Ancient Rhetoric*. Athens, Georgia: Scholars Press, 1986.

Homer. *The Iliad*. Translated by Richard Lattimore. 1951. Chicago: University of Chicago Press, 1976.

_____. *The Odyssey*. Translated by Richard Lattimore. 1965. New York: Harper and Row, 1977.

Howells, Wilber Samuel. *Logic and Rhetoric in England, 1500 - 1700*. Princeton: Princeton University Press, 1956.

Hunt, Everett Lee. "On the Sophists." In *The Province of Rhetoric*. Edited by Joseph Schwartz and John Rycenga. New York: Ronald Press, 1965. 69-84.

_____. "Plato and Aristotle on Rhetoric and Rhetoricians." In Corbett, Golden, and Berquist. 129-161.

Hunt, H. A. *The Humanism of Cicero*. Melbourne: Melbourne University Press, 1954.

Hunter, James Davison. *Culture Wars: The Struvvle to Define America.* New York: Basic Books, 1991.

IJsseling, Samuel. *Rhetoric and Philosophy in Conflict: An Historic Survey.* Translated by Paul Dunphy. The Hague: M. Nijhoff, 1976.

Isocrates. 3 vols. Translated by George Norlin. 1927. Loeb Classical Library. Cambridge: Harvard University Press, 1992.

Jaeger, Werner. *Paideia: The Ideals of Greek Culture.* 3 vols. Translated by Gilbert Highet. 2nd ed. New York: Oxford University Press, 1939-44.

Jarratt, Susan C. "Aspasia: Rhetoric, Gender, and Colonial Ideology." In *Reclaiming Rhetorica: Women in the Rhetorical Tradition.* Edited by Andrea A. Lunsford. Pittsburgh: Pittsburgh University Press, 1995. 9-24.

_____. "Feminism and Composition: The Case for Conflict." In Harkin and Schilb. 105-123.

_____. *Rereading the Sophists: Classical Rhetoric Refigured.* Carbondale: So. Illinois University Press, 1991.

Jarrett, James L. *The Educational Theory of the Sophists.* New York: Teachers College Press, 1969.

Johannesen, Richard L. "The Emerging Concept of Communication as Dialogue." *Quarterly Journal of Speech* 57 (1971): 373-82.

Johnstone, Henry W., Jr. *Validity and Rhetoric in Philosophical Argument: An Outlook in Transition.* University Park: Dialogue Press, 1978.

Jonsen, Albert R. and Stephen Toulmin. *The Abuse of Causuistry: A History of Moral Reasoning.* Berkeley: University of California Press, 1988.

Jordan, James N. "Protagoras and Relativism: Criticisms Bad and Good." *Southwestern Journal of Philosophy* 2.1 (1971): 7-30.

Jost, Walter and Michael J. Hyde, eds. *Rhetoric and Hermeneutics in our Time.* New Haven: Yale University Press, 1997.

Jowett, Benjamin, trans. *Plato's Protagoras.* Indianapolis: Bobbs-Merill, 1956.

Kahn, Charles H. *The Art and Thought of Heraclitus: An Edition of the Fragments with Translation and Commentary.* Cambridge: Cambridge University Press, 1979.

Kahn, Victoria. "Humanism and the Resistance to Theory." In Jost and Hyde. 149-170.

_____. *Rhetoric, Prudence, and Scepticism in the Renaissance.* Ithaca: Cornell University Press, 1985.

Kant, Immanuel. *The Critique of Practical Reason.* Translated by Mary Gregor. Cambridge: Cambridge University Press, 1997.

Kastely, James L. "Violence and Rhetoric in Euripides's *Hecuba* ." *PMLA* 108:5 (1993): 1036-49.

Kennedy, George A. trans. *Aristotle On Rhetoric: A Theory of Civic Discourse.* New York: Oxford University Press, 1991.

_____. *The Art of Persuasion in Greece.* Princeton: Princeton University Press, 1963.

_____. *The Art of Rhetoric in the Roman World: 300 B.C. - 300 A.D.* Princeton: Princeton University Press, 1972.

_____. *Classical Rhetoric in Its Christian and Secular Tradition from Ancient to Modern Times*. Chapel Hill: University of North Carolina Press, 1980.

_____. "A Hoot in the Dark: The Evolution of General Rhetoric." *Philosophy and Rhetoric* 25.1 (1992): 1-21.

_____. *Quintilian*. New York: Twayne Publishers, 1969.

Kenny, Anthony. *The Aristotelian Ethics: A Study of the Relationship between the Eudemian and Nicomachean Ethics of Aristotle*. Oxford: Clarendon Press, 1978.

Kent, Thomas. *Paralogic Rhetoric*. Lewisberg: Bucknell University Press, 1993.

Kerferd. G. B. "Plato's Account of the Relativism of Protagoras." *Durham University Journal* (1949): 20-26.

_____. *The Sophistic Movement*. Cambridge: Cambridge University Press, 1981.

Kimball, Bruce A. *Orators and Philosophers: A History of the Idea of Liberal Education*. Expanded Edition. New York: College Entrance Exam Board, 1995.

Kinneavy, James L. "The Process of Writing: A Philosophical Base in Hermeneutics." *Journal of Advanced Composition* 7 (1987): 1-9.

_____. "Translating Theory into Practice in Teaching Composition: A Historical View and a Contemporary View." In *Essays on Classical Rhetoric and Modern Discourse*. Edited by Robert J. Connors, Lisa S. Ede, and Andrea A. Lunsford. Carbondale: So. Illinois University Press, 1984. 69-81.

Kirk, G. S., and J. E. Raven. *The Presocratic Philosophers: A Critical History with a Selection of Texts*. Cambridge: Cambridge University Press, 1957.

Kitto, H. D. F. *The Greeks*. New York: Penguin Books, 1951.

Knoblauch, C. H. and Lil Brannon. *Rhetorical Traditions and the Teaching of Writing*. Upper Montclair, NJ: Boynton, 1984.

Krausz, Michael, ed. *Relativism: Interpretation and Confrontation*. Notre Dame: University of Notre Dame Press, 1987.

Krutch, Joseph Wood. *The Modern Temper*. 1929. New York: Harcourt, Brace, and World, 1956.

Kuhn, Deanna. *The Skills of Argument*. Cambridge: Cambridge University Press, 1991.

Kuhn, Thomas S. *The Structure of Scientific Revolutions*. 1962. 2nd ed. Chicago: University of Chicago Press, 1970.

Lakoff, George and Mark Johnson. *Metaphors We Live By*. Chicago: University of Chicago Press, 1980.

Lamb, Catherine E. "Beyond Argument in Feminist Composition." *College Composition and Communication* 42 (1991): 11-24.

_____. "Feminist Reponses to Argument." In *Perspectives on Written Argument*. Edited by Deborah P. Berrill. Cresskill, NJ: Hampton Press, 1996. 257-69.

Lassner, Phyllis. "Feminist Responses to Rogerian Argument." *Rhetoric Review* 8 (1990): 220-32.

Leff, Michael. "Burke's Ciceronianism." In *The Legacy of Kenneth Burke*. Edited by Herbert W. Simons and Trevor Melia. Madison: University of Wisconsin Press, 1989. 115-21.

_____. "Decorum and Rhetorical Interpretation: The Latin Humanistic Tradition and Contemporary Critical Theory. " *Vichiana* 3 (1990): 109-26.

_____. "Genre and Paradigm in the Second Book of *De Oratore*." *Southern Speech Communication Journal* 51 (1986): 308-25.

_____. "The Habitation of Rhetoric." *Argument and Critical Practices: Proceedings of the 5th SCA/AFA Conference on Argumentation*. Edited by Joseph W. Wenzel. Annandale, VA: SCA/AFA, 1987. 1-8.

_____. "Hermeneutical Rhetoric." In Jost and Hyde. 196-214.

Levi, Adolfo. "The Ethical and Social Thought of Protagoras." *Mind* 49 (1940): 284-302.

Liddell and Scott's Greek English Lexicon. 1871. Abridged edition. Oxford: Clarendon Press, 1974.

Liu, Yameng. "The Generic Triad in Classical Rhetoric." *College English* 57 (1995): 9-26.

Lloyd, G. E. R. *Polarity and Analogy: Two Types of Argumentation in Early Greek Thought*. Cambridge: Cambridge University Press, 1966.

Lobkowicz, Nickalaus. *Theory and Practice: History of a Concept from Aristotle to Marx*. London: University of Notre Dame Press, 1967.

Long, A. A. "Cicero's Plato and Aristotle." In Powell. 37-62.

_____. *Hellenistic Philosophy: Stoics, Epicureans, Sceptics*. London: Duckworth, 1974.

Loux, Michael J. *Universals and Particulars: Readings in Ontology*. Garden City, NY: Anchor Books, 1970.

Lukasiewicz, Jan. *Aristotle's Syllogism from the Standpoint of Modern Logic*. 1957. 2nd ed. Oxford: Oxford University Press, 1967.

Lunsford, Andrea A. "Aristotelian vs. Rogerian Argument: A Reassessment." *College Composition and Communication* 30 (1978): 146-51.

Lynch, Dennis, Diana George, and Marilyn Cooper. "Moments of Argument: Agonistic Inquiry and Confrontational Cooperation." In *On Writing Research: The Braddock Essays 1975-1998*. Edited by Lisa Ede. Boston: Bedford/St. Martins, 1999. 390-412.

Lyotard, Jean-Francois. *The Postmodern Condition: A Report on Knowledge*. Translated by Geof Bennington and Brian Massumi. Minneapolis: University of Minnesota Press, 1984.

Macintyre, Alasdair. *After Virtue: A Study in Moral Theory*. Notre Dame: University of Notre Dame Press, 1981.

Maguire, Joseph P. "Protagoras--or Plato?" *Phronesis* 17.2 (1973): 115-37.

Mailloux, Steven. "Introduction: Sophistry and Rhetorical Pragmatism." In Mailloux 1995. 1-32.

_____, ed. *Rhetoric, Sophistry, Pragmatism*. Cambridge: Cambridge University Press, 1995.

Margolis, Joseph. *The Truth about Relativism*. Cambridge: Basil Blackwood, Ltd., 1991.

Marrou, H. I. *A History of Education in Antiquity*. Translated by G. Lamb. London: Sheed and Ward, 1956.

Masefeld, Jaap. "Protagoras on Epistemological Obstacles and Persons." In *The Sophists and Their Legacy*. Ed. G. B. Kerferd. Wiesbadin: Franz Steiner, 1981. 38-53.

May, James. *Trials of Eloquence: The Eloquence of Ciceronian Ethos*. Chapel Hill: University of North Carolina Press, 1986.

McComiskey, Bruce. "Sophistic Rhetoric and Philosophy: A Selective Bibliography of Scholarship in English Since 1900." *Rhetoric Society Quarterly* 24:3-4 (1994): 25-38.

McKeon, Richard. "Introduction." In *Marcus Tullus Cicero: 'Brutus,' 'On the Nature of the Gods,' 'On Divination,' 'On Duties.'* Translated by Hubert M. Poteat. Chicago: University of Chicago Press, 1950.

_____. "The Methods of Rhetoric and Philosophy: Invention and Judgment." In *Richard McKeon on Rhetoric: Essays in Invention and Discovery*. Edited by Mark Backman. Woodbridge, CT: Oxbow Press, 1987. 56-65.

Meiland, Jack W. "Concepts of Relative Truth." *Monist* 60 (1977): 564-74.

_____. "On the Paradox of Cognitive Relativism." *Metaphilosophy* 11:2 (1980): 115-126.

Mendelson, Michael. "Declamation, Context, and Controversiality." *Rhetoric Review* 13.1 (1994): 92-107.

_____. "The Rhetorical Case: Its Roman Precedent and the Contemporary Debate." *The Journal of Technical Writing and Composition* 19:3 (1989): 203-26.

______ and Neil Lindeman. "The Absence of Dialogue: An Assessment of Argument Theory in Argument Composition Texts." In *Argument at Century's End: Reflecting on the Past and Envisioning the Future*. Edited by Thomas A. Hollihan. NCA/AFA: University of Utah, 1999. 316-24.

Mill, John Stuart. *Autobiography and Other Writings*. Edited by Jack Stillinger. Boston: Houghton Mifflin, 1969.

_____. *On Liberty and The Subjection of Woman*. 1859/1869. Hertfordshire: Wordsworth, 1996.

The Monist. General Topic: "Is Relativism Defensible?" 67.3 (1984).

Montaigne, Michel de. *The Complete Essays*. Translated by Donald M. Frame. Stanford: Stanford University Press, 1965.

Mooney, Michael. *Vico in the Tradition of Rhetoric*. Davis, CA: Hermagoras Press, 1994.

Moore, P. "The Raw and the Cooked: Using More Realistic Cases in Technical Writing Courses." *Journal of Business and Technical Communication* 1:2 (1987): 91-97.

Morson, Gary and Caryl Emerson. *Mikhail Bakhtin: Creation of a Prosaics*. Stanford: Stanford University Press, 1990.

Moser, Shia. *Absolutism and Relativism in Ethics*. Springfield, IL: Charles C. Thomas, 1968.

Moss, Jean Dietz. "Dialectics and Rhetoric in Renaissance Pedagogy." In *Learning from the Histories of Rhetoric: Essays in Honor of Winifred Bryan Horner*. Edited by Theresa Enos. Carbondale: So. Illinois University Press, 1993. 114-132.

Murphy, James J. "Introduction." *Quintilian on the Art of Speaking and Writing*. Edited by James J. Murphy. Carbondale: So. Illinois University Press, 1987. ix-xlxiii.

_____. "Roman Writing Instruction as Described by Quintilian." In *A Short History of Writing Instruction*. Edited by James. J. Murphy. Davis: Hermagoras Press, 1990. 19-76.

Murray, Oswyn. "Life and Society in Classical Greece." In *The Oxford History of Greece and The Hellenistic World*. Edited by John Boardman, Jasper Griffin, and Oswyn Murray. Oxford: Oxford University Press, 1991. 240-76.

Nadeau, Ray. "The Progymnasmata of Apthonius." *Speech Monographs* 19:3 (1952): 264-85.

Natanson, Maurice. "Claims of Immediacy." In *Philosophy, Rhetoric, and Argumentation*. Edited by M. Natanson and Henry Johnstone Jr. University Park: Penn State University Press, 1965.

Nehamas, Alexander. "Eristic Antilogic, Sophistic, Dialectic: Plato's Demarcation of Philosophy from Sophistry." *History of Philosophy Quarterly* 7:1 (1990): 3-16.

Nietzsche, Frederich. *On Rhetoric and Language*. Edited and Translated by Sander L. Gilman, Carole Blaire, and David J. Parent. New York: Oxford University Press, 1989.

Nill, Michael. *Morality and Self-Interest in Protagoras, Antiphon, and Democritus*. Leiden: E. J. Brill, 1985.

Nussbaum, Martha. *Cultivating Humanity: A Classical Defense of Reform in Liberal Education*. Cambridge: Harvard University Press, 1997.

_____. "Sophistry about Conventions." In *Love's Knowledge: Essays on Philosophy and Literature*. New York: Oxford University Press, 1990.

Nye, Andrea. *Words of Power: A Feminist Reading of the History of Logic*. New York: Routledge, 1990.

Oakley, Frances. *Community of Learning: The American College and the Liberal Arts Tradition*. New York: Oxford University Press, 1992.

O'Banion, John. *Reorienting Rhetoric: The Dialectic of List and Story*. University Park: Pennsylvania State University Press, 1992.

O'Keefe, Daniel. "Two Concepts of Argument." *Journal of the American Forensic Association* 13:3 (1977): 121-28.

Okrent, Mark B. "Relativism, Context, and Truth." *The Monist* 67.3 (1984): 341-58.

Ong, Walter J. *Orality and Literacy: The Technologizing of the Word*. New York: Methuen, 1982.

Owen, G. E. L. *Aristotle on Dialectic: The Topics*. Oxford: Clarendon Press, 1968.

Oxford English Dictionary: Compact Edition. New York: Oxford University Press, 1971.

Palmer, Richard E. *Hermeneutics: Interpretation Theory in Schleiermacher, Dilthey, Heidegger, and Gadamer*. Evanston: Northwestern University Press, 1969.

Parker, Francis H. *The Story of Western Philosophy*. Bloomington: University of Indiana Press, 1967.

Payne, David. "Rhetoric, Reality, and Knowledge: A Re-Examination of Protagoras' Concept of Rhetoric." *Rhetoric Society Quarterly* 16 (1986): 187-197.

Peaden, Catherine. "Understanding Differently: Rereading Locke's *Essay Concerning Human Understanding*." *Rhetoric Society Quarterly* 22:1 (1992): 75-90.

Perelman, Chaim and L. Obrechts-Tyteca. 1958. *The New Rhetoric: A Treatise on Argumentation*. Translated by John Wilkinson and Purcell Weaver. Notre Dame: University of Notre Dame Press, 1969.

Petronius Arbiter. *Satyricon*. Translated by W. C. Firebaugh. New York: Liveright Publishing Company, 1927.

Philostratus. *Lives of the Sophists*. Translated by W. C. Wright. Loeb Classical Library. Cambridge: Harvard University Press, 1968.

Plato. *Collected Dialogues*. Edited by Edith Hamilton and Huntington Cairns. New York: Bollingen Foundation, 1963.

Polanyi, Michael. *Personal Knowledge: Towards a Post-Critical Philosophy*. Chicago: University of Chicago Press, 1962.

Popper, Karl. *The Open Society and its Enemies*. 2 vols. 5th ed. New York: Torchbooks, 1966.

Poulakos, John. "The Components of Dialogue." *Western Speech* 38 (1974): 199-212.

Poulakos, Takis. *Speaking for the Polis: Isocrates' Rhetorical Education*. Columbia: University of South Carolina Press, 1997.

Powell, J. G. F. "Introduction." *Cicero the Philosopher: Twelve Essays*. Edited by J. G. F. Powell. Oxford: Clarendon Press, 1995. 1-35.

Putnam, Hillary. *Reason, Truth, and History.* Cambridge: Cambridge University Press, 1981.

Quintilian. *Institutio Oratoria.* 4 vols. Translated by H. E. Butler. 1920. Loeb Classical Library. Cambridge: Harvard University Press, 1989-92.

_____. *Major Declamations.* Translated by L.W. Sussman. Frankfurt: Verlag/Peter Lang, 1987.

Railsback, Celeste Condit. "Beyond Rhetorical Relativism: A Structural-Material Model of Truth and Objective Reality." *Quarterly Journal of Speech* 69 (1983): 351-63.

Rankin, H. D. *Sophists, Socratics and Cynics.* London: Croom Helm, 1983.

Ramage, John D. and John C. Bean. *Writing Arguments: A Rhetoric with Readings.* 4th ed. Boston: Allyn and Bacon, 1995.

Relativism Refuted: A Critique of Contemporary Epistemological Relativism. Edited by Harvey Siegel. Dordrecht/Boston: D. Reidel Publishing Company, 1987.

Rhodes, Richard. Review of *Niels Bohr's Times: In Physics, Philosophy, and Polity*, by Abraham Pais. *New York Times Book Review.* Jan. 26, 1992. 3.

Richards, I. A. *The Philosophy of Rhetoric.* New York: Oxford University Press, 1936.

Robinson, T. M. *Contrasting Arguments: An Edition of the Dissoi Logoi.* New York: Arno Press, 1979.

Rogers, Benjamin Bickley. "Introduction" to *Aristophanes.* 1926. Cambridge: Harvard University Press, 1967. 262-264.

Rogers, Carl. *On Becoming a Person.* Boston: Houghton, 1961.

Romilly, Jacqueline de. *The Great Sophists in Periclean Athens.* Translated by Janet Lloyd. Oxford: Clarendon Press, 1992.

_____. *Thucydides and Athenian Imperialism.* Translated by Philip Thody. 1963. Salem, NH: Ayer Co. Publishers, 1988.

Roochnik, David. *The Tragedy of Reason: Toward a Platonic Conception of Logos.* New York: Routledge, 1990.

Rorty, Richard. *Consequences of Pragmatism: Essays 1972-1980.* Minneapolis: University of Minnesota Press, 1982.

_____. "The Historiography of Philosophy: Four Genres." In *Philosophy in History: Essays on the Historiography of Philosophy.* Edited by Richard Rorty, J. B. Schneewind, and Quentin Skinner. Cambridge: Cambridge University Press, 1984. 49-75.

_____, ed. *The Linguistic Turn: Recent Essays in Philosophical Method.* Chicago: University of Chicago Press, 1967.

_____. *Philosophy and the Mirror of Nature.* Princeton: Princeton University Press, 1979.

_____. "Science as Solidarity." In *The Rhetoric of the Human Sciences.* Edited by John Nelson, Allan Megill, and Donald N. McClosky. Madison: University of Wisconsin Press, 1987. 38-52.

_____. "Solidarity or Objectivity?" In Rorty. *Objectivity, Relativism, and Truth.* New York: Cambridge University Press, 1991. 21-34.

Rosen, Stanley. "Herodotus Reconsidered." In *Herodotus: 'The Histories.'* Translated by Walter Brisco. New York: W. W. Norton and Company, 1992. 332-56.

Rottenberg, Annette. *Elements of Argument: A Text and a Reader.* 5th ed. New York: Bedford Books, 1996.

Russell, D. A. *Criticism in Antiquity.* London: Duckworth, 1981.

_____. *Greek Declamation.* Cambridge: Cambridge University Press, 1983.

Ryle, Gilbert. "Dialectic in the Academy." In Owen. 69-79.

Saintsbury, George. *A History of Criticism and Literary Taste in Europe from the Earliest Times to the Present.* vol. 2. Oxford: Basil Blackwell and Sons, 1900.

Saussure, Ferdinand de. *Course in General Linguistics.* Translated by Wade Baskin. Edited by Charles Bally and Albert Sechehaye. New York: Philosophical Library, 1959.

Schaeffer, John D. *Sensus Communis: Vico, Rhetoric, and the Limits of Relativism.* Durham, NC: Duke University Press, 1990.

Schiappa, Edward. *Protagoras and Logos: A Study in Greek Philosophy and Rhetoric.* Columbia: University of South Carolina Press, 1992.

Schiller, F. C. S. "Plato and his Predecessors." *Quarterly Review* 204 (1906): 62-68.

_____. *Plato or Protagoras?* London: Basil Blackwell, 1908.

_____. *Studies in Humanism.* 2nd ed. 1912. Westport, Connecticut: Greenwood Press, 1970.

Schmitt, Charles B. *Cicero Scepticus: A Study of the Influence of the Academica in the Renaissance.* The Hague: Martinus Nijhoff, 1972.

Schuster, Charles I. "Mikhail Bakhtin as Rhetorical Theorist." *College English* 47:6 (1985): 594-607.

Scott, Robert L. "On Viewing Rhetoric as Epistemic." *Central States Speech Journal* 18 (1967): 9-17.

_____. "On Viewing Rhetoric as Epistemic: Ten Years Later." *Central States Speech Journal* 27 (Winter 1976): 267-78.

_____. "Some Implications of Existentialism for Rhetoric." *Central States Speech Journal* 15:4 (November 1964): 267-77.

Secor, Marie J. "Recent Research in Argumentation Theory." *The Technical Writing Teacher* 14:3 (Fall 1987): 337-54.

Seneca the Elder. *Controversiae and Suasoriae.* Translated by M. Winterbottom. Loeb Classical Library. Cambridge: Harvard University Press, 1974.

Sesonsky, Alexander. "To Make the Weaker Argument Defeat the Stronger." In *Plato: True and False Rhetoric.* Edited by Keith V. Erickson. Amsterdam: Rodolphi, 1979. 71-90.

Sextus Empiricus. *Outlines of Scepticism.* Translated by Julia Annas and Jonathan Barnes. Cambridge: Cambridge University Press, 1994.

Siegel, Harvey. *Relativism Refuted: A Critique of Contemporary Epistemological Relativism.* Dordrecht: D. Riedel/Kluwer, 1987.

Sloane, Thomas O. *Donne, Milton, and the End of the Humanist Tradition.* Berkeley: University of California Press, 1985.

_____. *On the Contrary: The Protocols of Traditional Rhetoric.* Washington D. C.: Catholic University Press, 1997.

_____. "Reinventing *inventio.*" *College English* 51:5 (September 1989): 461-73.

Smith, Bromley. "The Father of Debate: Protagoras of Abdera." *Quarterly Journal of Speech Education* 4 (1918): 196-215.

Solmsen, Friedrich. "Dialectic without Forms." In Owen. 49-68.

_____. *Intellectual Experiments of the Greek Enlightenment.* Princeton: Princeton University Press, 1975.

Sprague, Rosamund Kent, ed. *The Older Sophists.* Columbia: University of South Carolina Press, 1972.

Spelman, Elizabeth V. *Inessential Woman: Problems of Exclusion in Feminist Thought.* Boston: Beacon, 1988.

Starr, C. G. *Civilization and the Caesars: The Intellectual Revolution of the Roman Empire.* New York: Norton and Company, 1965.

Stewart, John. "Foundations of Dialogic Communication." *Quarterly Journal of Speech* 64 (1978): 183-201.

Struever, Nancy S. *The Language of History in the Renaissance: Rhetorical and Historical Consciousness in Florentine Humanism.* Princeton: Princeton University Press, 1970.

Sullivan, Dale. "Attitudes toward Imitation: Classical Culture and the Modern Temper." *Rhetoric Review* 8:1 (1989): 5-21.

Sussman, L. W. "Introduction." Quintilian's *Major Declamations.* Frankfurt: Verlag/Peter Lang, 1987.

Swearingen, C. Jan. *Rhetoric and Irony: Western Literacy and Western Lies.* New York: Oxford University Press, 1991.

Tacitus. *Dialogue on Orators.* Translated by Herbert W. Benario. Norman: University of Oklahoma Press, 1991.

Tannen, Deborah. *Argument Culture: Moving from Debate to Dialogue.* New York: Random House, 1998.

Teich, Nathaniel. "Rogerian Problem-Solving and the Rhetoric of Argumentation." *Journal of Advanced Composition* 7 (1987): 52-61.

Thucydides. *History of the Peloponnesian War.* Translated by Richard Crawley. 1910. London: J. M. Dent, 1993.

Tolstoy, Leo. *War and Peace.* Translated by Louise and Aylmer Maude. Oxford: Oxford University Press, 1991.

Tompkins, Jane. "Fighting Words: Unlearning to Write the Critical Essay." *Georgia Review* 42 (1988): 585-90.

Toulmin, Stephen. *Cosmopolis: The Hidden Agenda of Modernity.* Chicago: University of Chicago Press, 1990.

_____. *The Uses of Argument.* Cambridge: Cambridge University Press, 1969.

Trimbur, John. "Consensus and Difference in Collaborative Learning." *College English* 51:6 (1989): 602-616.

Troyka, Lynn Quitman and Jerrold Nudelman. *Taking Action: Writing, Reading, Speaking, an Listening through Simulation Games.* Englewood Cliffs: Prentice-Hall, 1975.

Untersteiner, Mario. *The Sophists.* Translated by Kathleen Freeman. Oxford: Basil Blackwell, 1954.

Versenyi, Laszlo. "Protagoras' Man-Measure Fragment." *American Journal of Philology* 83 (1962): 178-84.

_____. *Socratic Humanism.* New Haven: Yale University Press, 1963.

Vitanza, Victor J. "Notes Towards the Historiographies of Rhetorics; or the Rhetorics of the Histories of Rhetorics: Traditional, Revisionary, and Sub/Versive." *PRE/TEXT* 8 (1987): 63-125.

Vlastos, Gregory. *Plato's "Protagoras."* Indianapolis: Bobs-Merrill, 1956.

_____. "The Socratic Elenchus." *Oxford Studies in Ancient Philosophy* 1 (1983): 27-74.

Ward, Irene. *Literacy, Ideology, and Dialogue: Towards a Dialogic Pedagogy*. Albany: State University of New York Press, 1994.

Watson, J. S. trans. *Cicero on Orators and Oratory*. Carbondale: So. Illinois University Press, 1970.

Welch, Kathleen E. *The Contemporary Reception of Classical Rhetoric: Appropriations of Ancient Discourse*. Hillsdale, NJ: Lawrence Erlbaum, 1990.

_____. "Dialectic/Rhetoric/Writing." In *Learning from the Histories of Rhetoric*. Edited by Theresa Enos. Carbondale: So. Illinois University Press, 1993. 133-143.

Wenzel, Joseph. "Jürgen Habermas and the Dialectical Perspective on Argumentation." *The Journal of the American Forensic Association* 16 (1979): 83-94.

_____. "The Rhetorical View of Argumentation: Exploring a Paradigm." *Argumentation* 1 (1987): 73-88.

West, Cornel. *The American Evasion of Philosophy: A Genealogy of Pragmatism*. Madison: University of Wisconsin Press, 1989.

West, Martin. "Early Greek Philosophy." In *The Oxford History of Greece and the Hellenic World*. Edited by John Boardman, Jasper Griffin, and Oswyn Murray. Oxford: Oxford University Press, 1986. 126-141.

Westbrook, Robert B. *John Dewey and American Democracy*. Ithaca: Cornell University Press, 1991.

Whately, Richard. *Elements of Rhetoric*. 1828. In *The Rhetoric of Blair, Campbell, and Whately*. Edited by James L. Golden and Edward P. J. Corbett. Carbondale: So. Illinois University Press, 1990.

Wheelwright, Philip. *Heraclitus*. Princeton: Princeton University Press, 1959.

Wilkins, Augustus S. "Introduction and Notes." *M. Tulli Ciceronis De Oratore*. New York: Arno Press, 1979

Wood, Neal. *Cicero's Social and Political Thought*. Berkeley: University of California Press, 1988.

Woods, Marjorie Curry. "Teaching the Tropes in the Middle Ages: The Theory of Metaphoric Transference in the Commentaries on the *Poetria nova*." In *Rhetoric and Pedagogy: Its History, Philosophy, and Practice*. Edited by Winifred Bryan Horner and Michael Leff. Mahwah, NJ: Lawrence Erlbaum, 1995. 73-82.

Wright. M. R. "Introduction." In *Cicero: On Stoic Good and Evil*. Translated by M. R. Wright. Wiltshire, England: Aris and Phillips Ltd, 1991.

Young, Iris Marion. *Justice and Politics of Difference*. Princeton: Princeton University Press, 1990.

Young, Richard E., Alton L. Becker, and Kenneth L. Pike. *Rhetoric: Discovery and Change*. New York: Harcourt, 1970.

INDEX

Breinigsville, PA USA
21 October 2009

226219BV00005B/31/A